KT-491-126

WITHDRAWN

Practical Recording Techniques, Second Edition

THE LIBRARY
GUILDFORD COLLEGE
of Further and Higher Education

Practical Recording Techniques

Second Edition

Bruce Bartlett
Jenny Bartlett

Focal Press
Boston   Oxford   Johannesburg   Melbourne   New Delhi   Singapore

621.3893
BAR
96147

Focal Press is an imprint of Butterworth–Heinemann.

Copyright © 1998 by Butterworth–Heinemann

 A member of the Reed Elsevier group

All rights reserved.

No part of this publication may be reproduced, stored in a retrieval system, or transmitted in any form or by any means, electronic, mechanical, photocopying, recording, or otherwise, without the prior written permission of the publisher.

 Recognizing the importance of preserving what has been written, Butterworth–Heinemann prints its books on acid-free paper whenever possible.

 Butterworth–Heinemann supports the efforts of American Forests and the Global ReLeaf program in its campaign for the betterment of trees, forests, and our environment.

Library of Congress Cataloging-in-Publication Data
Bartlett, Bruce.
  Practical recording techniques / Bruce Bartlett, Jenny Bartlett. – 2nd ed.
    p. cm.
  Includes index.
  ISBN 0-240-80306-X (pbk. : alk. paper)
  1. Magnetic recorders and recording—Handbooks, manuals, etc.
I. Bartlett, Jenny. II. Title.
TK7881.6.B367 1998
621.389'3—dc21   97-41861
                                                            CIP

British Library Cataloguing-in-Publication Data
A catalogue record for this book is available from the British Library.

The publisher offers special discounts on bulk orders of this book.
For information, please contact:
Manager of Special Sales
Butterworth–Heinemann
225 Wildwood Avenue
Woburn, MA 01801-2041
Tel: 781-904-2500
Fax: 781-904-2620

For information on all Focal Press publications available, contact our World Wide Web home page at: http://www.bh.com/focalpress

10 9 8 7 6 5 4 3 2

Printed in the United States of America

To family, friends, and music.

TABLE OF CONTENTS

3   Sound, Signals, and Studio Acoustics   *39*

4 Monitoring   *57*

5 Hum Prevention   *69*

6 Microphones   *89*

7   Microphone Technique Basics   *107*

9 Analog Tape Recording *157*

10 Digital Recording *181*

11 Effects And Signal Processors   *203*

14 Session Procedures   *265*

15 Recording the Spoken Word   *281*

16 The MIDI Studio: Equipment and Recording Procedures   *291*

18 On-Location Recording of Classical Music *337*

19 Judging Sound Quality *347*

C Further Education   *395*

Glossary   *401*

Index   *439*

PREFACE

Recording is a highly skilled craft combining art and science. It requires technical knowledge as well as musical understanding and critical listening ability. By learning these skills, you can capture a musical performance and reproduce it with quality sound for the enjoyment and inspiration of others.

Your recordings will become carefully tailored creations of which you can be proud. They will be a legacy that can bring pleasure to many people for years to come.

This book is intended as a hands-on, practical guide for beginning and intermediate recording engineers, producers, musicians—anyone who wants to make better recordings by understanding recording equipment and techniques. I hope to prepare the reader for work in a home studio, a small professional studio, or an on-location recording session.

Practical Recording Techniques offers up-to-date information on the latest recording technology, such as digital tape recording, hard-disk recording, keyboard and digital workstations, SMPTE and MIDI. But it also guides the beginner through the basics, showing how to make quality recordings with the new breed of inexpensive home-studio equipment.

The book first overviews the recording-and-reproduction chain to instill a system concept. Next, advice is given on equipping a home studio, from low-budget to advanced.

The basics of sound and signals are explained so that you'll know what you're controlling when you adjust the controls on a piece of recording equipment. Studio setup is covered next, including suggestions for improving your studio acoustics, choosing monitor speakers, and preventing hum.

Each piece of recording equipment is explained in detail, as well as the control-room techniques you'll use during actual sessions. Two chapters are devoted to the recent technology of digital recording and MIDI sequencing. Two sections on remote recording cover techniques for both popular and classical music.

A special chapter explains how to judge recordings and improve them. The engineer must know not only how to use the equipment, but also how to tell good sound from bad.

The last chapter answers the question, "Why do we record?" Finally, three appendices explain the decibel, introduce SMPTE time code, and suggest further education.

Based on my work as a professional recording engineer, the book is full of tips and shortcuts for making great-sounding tapes, whether in a professional studio, on location, or at home.

You'll find many topics not covered in similar texts:

How to choose and operate cassette recorder-mixers

Hum prevention

The latest monitoring methods

Microphone selection guide

Miniature microphones

Tonal effects of microphone placement

Glossary of sound-quality descriptions

Documenting the recording session

Audio-for-video techniques

Recording the spoken word

On-location recording

Troubleshooting bad sound; guidelines for good sound

ACKNOWLEDGMENTS

Thank you to Nick Batzdorf of *Recording* magazine for giving me permission to draw from my "Take One" series. Thanks to Pat Brown of Syn-Aud-Con for checking the chapter on hum prevention.

For my education, thank you to The College of Wooster, Crown International, Shure Brothers Inc., Astatic Corporation, and all the studios I've worked for.

Thank you to Marie Lee, Maura Kelly, and Tammy Harvey at Focal Press for working with me on this new edition.

My deepest thanks to Jenny Bartlett for her many helpful suggestions as a layperson consultant and editor. She made sure the book could be understood by beginners.

A note of appreciation goes to the Pat Metheny Group and Samuel "Adagio for Strings" Barber, among many others, whose music inspired the chapter, "Music: Why You Record."

Finally, to the musicians I've recorded and played with, a special thanks for teaching me indirectly about recording.

1

THE RECORDING CHAIN

Making a quality recording gives you a real sense of pride and achievement. With your skills, you can help artists realize their visions in sound. You can capture a musical performance on tape with exciting realism, or enhance the music with creative effects.

Thanks to the shift from analog to digital technology, the excitement and satisfaction of recording is accessible to more people than ever before. It used to take a whole roomful—or truckful—of expensive equipment to produce a good recording. But the new generation of smaller, cheaper gear means you may be able to tuck your studio into a corner of your bedroom or the back seat of your Toyota. As a result, many more people are involved in the process of recording—as musicians recording their own albums, or as engineers offering services to others.

Making a good recording, however, involves more than plugging a microphone into a tape deck. Let's face it: modern recording equipment and techniques are sophisticated. Before you can achieve a quality recording, you must understand the equipment, learn the techniques, and know the jargon.

This book separates the multitude of equipment and procedures into easily understandable parts. It lists the equipment you need, tells what it does, and suggests how to use it effectively. After studying this book and practicing with actual recording equipment, you'll be making great sounding tapes that you can be proud of.

Types of Recording

Currently there are five main ways to record music:

1.   **Live stereo recording:** Record with a stereo microphone or two microphones into a tape recorder.

2.   **Live mixed recording:** Record with several mics into a mixer, which is connected to a tape recorder.

3.   **Multitrack tape recording:** Record with several mics into a mixer, which is connected to a multitrack tape recorder. Each track or path on tape contains the sound of a different instrument. After the recording is done, you mix or combine the tracks to stereo.

4.   **Random-access recording:** You record the music onto a computer hard disk or MiniDisc. You can access any part of the recording instantly, and edit the audio program on a screen or video monitor.

5.   **MIDI sequencing:** A musician performs on a MIDI controller, such as a keyboard or drum pads. The controller puts out a MIDI signal, a series of numbers which indicates which keys were pressed and when they were pressed. The MIDI signal is recorded into computer memory by a sequencer. When you play back the sequence, it plays the tone generators in a synthesizer, or plays samples: digital recordings of musical notes.

A sequencer can be synched to a multitrack recorder (tape, hard disk, or MiniDisc). You record MIDI instruments with the sequencer, and record vocals and acoustic instruments with the multitrack recorder. Let's look at each type of recording in more detail.

Live Stereo Recording

You can use this method to record an orchestra, symphonic band, pipe organ, small ensemble, quartet, or soloist. The microphones pick up the overall sound of the instruments and the concert-hall acoustics. You might use this minimalist technique to record a folk group or acoustic jazz group in a good-sounding room.

Figure 1.1 shows the stages of this method—the links in the recording chain. Let's look at each stage from left to right (beginning to end).

1.   The musical instruments make sound waves.

2.   The sound waves travel through the air and bounce or reflect off the walls, ceiling, and floor of the concert hall. These reflections add a pleasing sense of spaciousness.

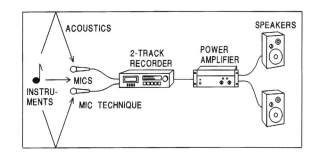

Figure 1.1
The recording chain for live stereo recording.

3.   The sound waves from the instruments and the room reach the microphones, which convert the sound into electrical signals.

4.   The sound quality is affected by mic technique—microphone choice and placement.

5.   The signals from the microphones go to a tape recorder. It may be a cassette deck, open-reel deck, or DAT (digital audio tape) recorder. The signal changes to a magnetic pattern stored on magnetic tape. During playback, the magnetic patterns on tape are converted back into a signal.

As the tape moves during recording, magnetic signals are stored on tape along a track—a path on tape containing a recorded signal. One or more tracks can be recorded side-by-side on a single tape. For example, a 2-track tape recorder can record two tracks on tape, such as the two different audio signals required for stereo recording.

6.   To hear the signal you're recording, you need a monitor system: a stereo power amplifier and loudspeakers or headphones. You use the monitors to judge how well your mic technique is working. The speakers or headphones convert the signal back into sound. This sound resembles that of the original instruments. Also, the acoustics of the listening room affect the sound reaching the listener.

Recording with Several Mics and a Mixer

Now let's look at a more complex way to record (Figure 1.2).

1.   You use several microphones. Each one is placed close to each instrument or singer. As a result, each mic picks up very little room acoustics. This gives a close, clear sound that's desirable in recorded pop music or narration. For more clarity, you might add some sound-absorbent material on the floor, walls, and ceiling.

2.   All the mics plug into a mixer, which blends all the microphone signals into one signal, stereo or mono. The mixer also has a volume control

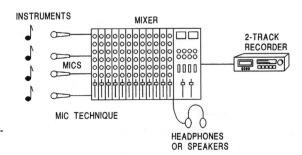

Figure 1.2
The recording chain for several mics and a mixer.

for each microphone. While listening to the mixer's signal, you adjust the volume of each instrument to make a pleasing loudness balance.

For example, if the guitar is too quiet relative to the voice, simply turn up the volume control for the guitar microphone until it blends well with the voice. That's a lot easier than grouping the musicians around a single microphone and moving the musicians until you hear a good balance.

Some mixers let you control other aspects of sound besides volume. You can control tone quality (bass and treble), stereo position (left, right, or center), and special effects (such as artificial reverberation, which sounds like room acoustics). You monitor the mixer's signal with headphones or speakers.

3.   When the mix sounds okay, you record the mixer's output signal with a tape recorder.

Multitrack Tape Recording

One problem with the previous setup is that you have to mix while the musicians are playing. If you make a mistake while mixing—say, one instrument is too quiet—the musicians have to play the song again until you get the balance right.

The solution is to use a multitrack recorder, which records 4 to 48 tracks side-by-side on a single tape. It's as if several 2-track recorders were locked together. You record the signal of each microphone on its own track, then mix these recorded signals after the performance is done. You can either record a different instrument on each track or record different groups of instruments on each track.

Here are the stages in this method (Figure 1.3):

1.   Place microphones near the instruments.

2.   Plug the mics into a mixing console: a big, sophisticated mixer. During multitrack recording, the mixing console amplifies the weak microphone

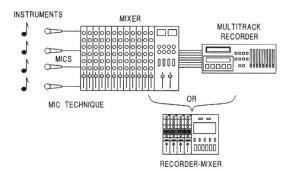

Figure 1.3
The recording chain for multitrack tape recording.

signals up to the level needed by the tape recorder. The console is also used to send each microphone signal to the desired tape track.

3. Record the amplified mic signals on the multitrack tape recorder. A recorder-mixer or portable studio combines the mixer and multitrack recorder in a single chassis. It's a convenient system for home studios. You can record more instruments later on unused tracks—a process called over-dubbing. Wearing headphones, you listen to the recorded tracks to keep your place in the song, and play along with them. You record your performance on an unused track.

4. After the recording is done, play all the tape tracks through the mixing console to mix them with a pleasing balance (Figure 1.4).

5. Play back the multitrack tape of the song several times, adjusting the track volumes and tone controls until the mix is just the way you want it. You can add special effects to enhance the sound quality. Some examples are echo, reverberation, and compression (explained in Chapter 2). Effects are made by signal processors which connect to your mixer.

6. Record your final mix on a 2-track stereo tape recorder (cassette, open-reel, or DAT).

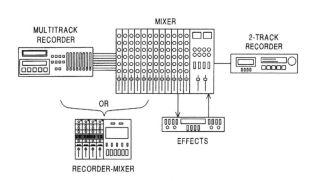

Figure 1.4
The recording chain for a multitrack mixdown.

7.   After all the songs are mixed to tape, you may want to edit the tape—remove noises and count-offs between songs, and put the songs in a suitable order or sequence. You edit an analog tape recording with a razor blade and splicing tape. You edit a digital recording with your personal computer and editing software (Figure 1.5). This system is called a DAW (Digital Audio Workstation).

8.   The edited tape is the final product, ready to be duplicated on CD or cassette.

To illustrate all the equipment mentioned, Figure 1.6 shows a small home studio setup using a recorder-mixer. Figure 1.7 shows a typical layout

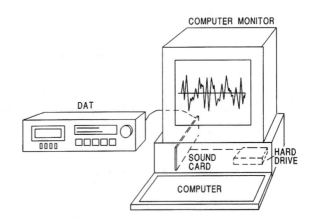

Figure 1.5
Digital Audio Workstation.

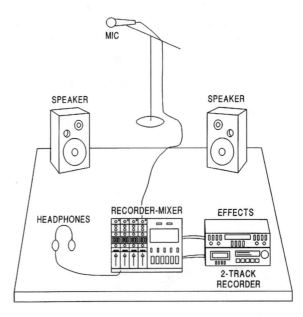

Figure 1.6
A small home studio.

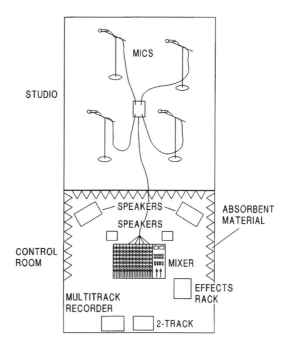

Figure 1.7
Typical layout of a larger
recording studio.

of a larger recording studio. Often, the tape recorders are operated by remote controls near the mixing console.

Random Access Recording

With this method, you record audio onto a computer hard disk or Mini-Disc. It's called "random access" because you can instantly play, or access, any part of the recorded program. With a tape recording, you must fast-forward or rewind to the part you want to hear.

One random-access setup uses a personal computer with a sound card and editing software (Figure 1.8). The sound card converts audio into a signal which is recorded on the computer's hard disk as a magnetic pattern.

Here are the steps in random-access recording:

1. Plug a mic or electric instrument into a mixer to amplify the signal. Plug the mixer output into the sound card. If the sound card has a digital input, first plug the mixer output into an A/D converter. This device changes the analog signal into a digital one. You might use the A/D converter in a DAT machine for this function.

2. Use your recording software to record the audio onto your hard drive.

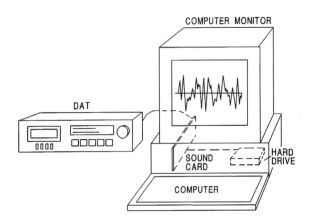

Figure 1.8
Digital Audio Workstation.

3.   You can listen to any part of the recording instantly, and edit the audio program on-screen with a mouse.

4.   Copy the edited recording onto a DAT tape.

Computer hard-disk (HD) recording can be either 2-track or multitrack. If you want to record more than two tracks at once, you need several sound cards, or a card with multiple inputs. Some software and cards let you mix the multitrack HD recording with an external mixer. Otherwise, you mix with your mouse by adjusting controls that appear on your computer screen.

A mixer/multitrack HD recorder combines both functions in a single package. So does a mixer/MiniDisc recorder.

MIDI Sequencing

MIDI sequencing offers sound quality rivaling the best pro studios. Figure 1.9 shows the steps in this process.

1.   Play music on a MIDI controller: a keyboard, drum pads, wind controller, and so on.

2.   As you play the controller, it sends a MIDI signal from its MIDI OUT connector. This signal is a string of numbers—computer code—that tells which keys you pressed, when you pressed them, how fast you pressed them, and so on. In other words, the MIDI signal represents your performance gestures. It's not an audio signal.

3.   The MIDI signal goes to a synthesizer or sound module. In these devices are tone generators which create musical sounds, like a bass, piano, drums, and so on. The MIDI signal plays the tone generators. You hear

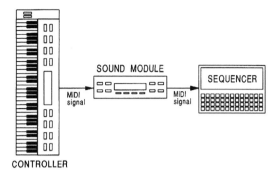

Figure 1.9
MIDI sequencing system.

them with speakers or headphones. The MIDI signal could also go to a drum machine or sampler. They contain samples, which are recordings in computer memory of single notes played by real musical instruments. The MIDI signal plays the samples.

4.   The MIDI signal also goes to a sequencer, a group of computer memory chips which record the MIDI signal. A sequencer might be a circuit in a synthesizer, a stand-alone box, or a program running on a computer. The sequence of notes can be saved on a floppy or hard disk.

5.   When you play the sequencer recording, it activates the sound generators or samples to play the same notes you played. You can edit the sequencer recording. For example, fix incorrect notes. Change the instrument sounds without having to re-do the performance. Or change the tempo without changing the pitch.

Using a device called a tape synchronizer, you can sync a MIDI sequencer with a multitrack recorder. First, record synth and drum-machine parts with the sequencer. Then while listening to the sequence play, record vocals and acoustic instruments with the recorder. The tape synchronizer ensures that the sequencer and recorder stay together in time.

MIDI/digital-audio software lets you record sequences and digital audio on hard disk. First record MIDI sequences onto hard disk, then add several audio tracks: lead vocal, sax solo, or whatever. All these elements are synchronized.

No matter what type of recording you do, each stage contributes to the sound quality of the finished tape. A bad-sounding master tape can be caused by any weak link: low-quality microphones, bad mic placement, improperly set mixer controls, and so on. A good-sounding tape results when you get every stage right. This book will help you reach that goal.

2

EQUIPPING
YOUR STUDIO

It's every musician's dream. You want to set up a home recording system—one with good-quality sound, yet affordable. With today's easy-to-use sound tools, you can do just that. This chapter is a guide to equipment for a recording studio: what it does, and what it costs.

First we'll look at equipment for a multitrack tape studio, then a MIDI studio. Next we'll explain cables and connections. Finally we'll suggest some simple ways to improve your studio's acoustics.

Multitrack Tape Studio

Here is a list of needed equipment:

- A recorder-mixer, or a separate multitrack recorder and mixer
- A 2-track recorder to record the final mix (cassette deck, open-reel deck, or DAT recorder)
- Microphones and mic stands
- A monitoring system (two speakers and a power amplifier, and/or headphones)
- Effects (reverb, delay, compression, etc.)
- Cables
- Rack and patch bay (optional)
- Blank tape
- Miscellaneous equipment

Recorder-Mixer

A recorder-mixer is a small, portable unit combining a mixer with a multi-track recorder (Figure 2.1). The recorder may be cassette, MiniDisc, or hard disk. Currently, cassette records 4 or 8 tracks, MiniDisc records 4 tracks, and hard disk records 4 to 8 tracks (or more if multiple units can be linked).

You plug microphones and electric musical instruments into the mixer, which amplifies their signals and routes them to tape tracks. The 4-track recorder can record up to 4 tracks, each with a different instrument on it. After recording the tracks, you mix or combine them to 2-channel stereo, and record the mix on a separate 2-track recorder. The recording made on that machine is the final product.

Prices range from $259 to $1899 suggested retail. As price increases, you get cleaner, crisper sound, more inputs, more features, and more tracks. The sections that follow cover the features briefly; a detailed explanation is given in Chapter 12.

Recorder-Mixer Features

4 or 8 Tracks

An 8-track machine is more convenient to use than a 4-track machine. When you record a band on a 4-track unit, you often must combine several instruments on each track. As a result, you can't re-adjust the level, tone or

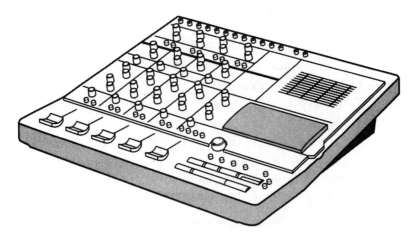

Figure 2.1 A recorder-mixer.

effects of each instrument independently within a recorded track. In most cases, eight tracks is enough to record each instrument on its own track. Then you can control the sound of each instrument independently.

Overdubbing

All recorder-mixers have the overdubbing feature. While listening to tracks already recorded, you play along with them and record a new part on an unused track. For example, suppose you've already recorded bass and drums and you want to add guitar. You listen to a headphone mix of the bass track, drum track, and your guitar signal. While the bass and drum tracks play, you play your guitar along with them, and record the guitar on an unused track.

Simultaneous Recording On All Tracks

Simultaneous recording is useful especially for recording a live performance. When you record in a studio, you often can record one or two tracks at a time. But when you record a live performance, you need to record all the tracks at once. Not all recorder-mixers let you do this.

Punch-In/Out

Use the punch-in and punch-out functions to fix mistakes. As the tape is playing, punch into record mode just before the mistake, play a new correct part that is recorded, and punch out of record mode when you're finished. Most recorder-mixers accept a foot switch so you can punch-in with your foot while playing your instrument.

Bouncing Tracks

When you bounce (or **ping-pong**) tracks, you mix two or three tracks together and record the result on an unused track. Then you can erase the original tracks, freeing them for recording more instruments. This way you can record up to nine tracks with a 4-track machine. All recorder-mixers permit bouncing.

Tape Speed

Tape speed is the rate at which tape moves past the record/playback head. In cassette recorders, two speeds are available: 1-7/8 inches-per-second (ips) and/or 3-3/4 ips. On machines running at 1-7/8 ips, you can play standard commercial cassettes, but recordings made at 3-3/4 ips sound crisper and clearer. (Reel-to-reel units record at 7-1/2, 15, or 30 ips.)

Pitch Control

Pitch control lets you adjust the tape speed up or down so you can match the pitch of recorded tracks to the pitch of new instruments to be recorded. All recorder-mixers have this feature.

EQ (Equalization)

Equalization means tone control. The simplest units have no equalization; you're stuck with the sound you get from your microphones. Others have a single 3-band graphic equalizer—sliding controls that affect bass, midrange, and treble. Most inexpensive units include a bass and treble control, one set per input. Fancier recorder-mixers have **sweepable** or **semi-parametric** EQ, which lets you continuously vary the frequency you want to adjust. This type of EQ offers the most control over the tone quality of each instrument you're recording.

Effects Loop (Aux Loop)

The effects loop is a set of connectors (labeled "send" and "return") for hooking up an external effects unit, such as a reverb or delay device. Effects add a professional polish to your productions.

If your recorder-mixer has no effects loops, you can still record music, but without any effects. It sounds rather dead and plain. A unit with one effects loop lets you add one type of effect; a unit with two effects loops lets you add two for even more sonic interest. A stereo effects loop lets you hear effects in stereo, if your effects device is a stereo unit. Some recorder-mixers have digital reverb built in so that you don't need to buy an external reverb unit.

Noise Reduction

A noise reduction circuit reduces the tape hiss that is inherent in cassette tape recording. Dolby C or dbx work best; Dolby B is less effective. But all help you make clean, noise-free recordings.

Autolocate

Autolocate stores cue-point locations in memory, and shuttles the tape to these locations. You could use this feature to move tape repeatedly between the two preset points, such as the beginning and end of a punch-in.

Return-to-zero is also called **zero stop** or **memory rewind.** When you enable return-to-zero, the tape rewinds to a preset point that you mark "000" on the tape counter. This feature makes it easy to practice a punch-in or a mix repeatedly.

Solenoid Switches

When you push one of your tape-motion controls, the action can be either mechanical or solenoid operated. Solenoid switching is gentler to the tape, and often includes logic which protects the tape from rapid changes in tape motion.

XLR-Type Balanced Inputs

An XLR-type microphone input connector looks like three small holes arranged in a triangle. If your mixer has such inputs, you can run long mic cables without picking up hum. This type of connector is found only in high-end units. Most recorder-mixers use 1/4" phone **jacks** (receptacles) for mic inputs, which are adequate for small studios.

Insert Jacks (Access Jacks)

Insert jack connectors let you plug in a compressor (or other signal processor) in line with an input signal. They make it easy to modify the sound of a single instrument or vocal. Only the more expensive units have this feature.

Sync Track

A sync track (usually the highest numbered track) records a synchronization tone from a MIDI sequencer. The tone synchronizes tape tracks with sequencer tracks. In this way, you can have your sequencer play synthesized bass, drums, and keyboards while your multitrack recorder plays vocals and acoustic instruments—all in sync.

This is a great way to get extra tracks for little cost. Plus, when you do your mixdown, all the synth tracks play "live" (from the synth outputs) rather than from tape. The result is a cleaner mix.

Any track on any multitrack recorder can record the sync tone. But a dedicated sync track has its own input connector for the sync tone, and allows noise reduction to be switched off for more reliable recording.

The sync track also is used for SMPTE (Society of Motion Picture and Television Engineers) time code (explained in Appendix B).

Mixer and Multitrack Tape Recorder

For highest quality, use a separate mixer and a multitrack recorder.

A mixer (Figure 2.2) is an electronic device that mixes and routes the signals from your microphones and tape tracks. Most mixers have an elaborate control panel. You plug mics and electric instruments into the mixer, which amplifies their signals. While recording, you use the mixer controls to

15

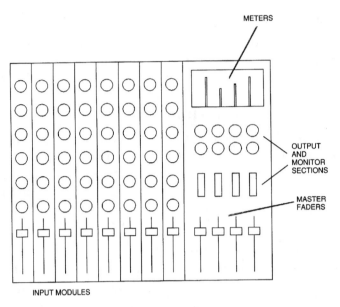

Figure 2.2 A mixer.

send those signals to the desired tape tracks, and set recording levels. During mixdown, the mixer combines (mixes) the tracks to stereo. You can adjust the sound quality of each track. A large, complex mixer is called a **mixing console** or **board**. Mixers are explained in more detail in Chapter 12.

A multitrack tape recorder (Figure 2.3) can record many tracks at once, or one at a time. Usually each track on tape contains the signal of a different instrument. After recording the tracks, you combine them to 2-channel stereo with your mixer. Record the stereo mix on a separate 2-track recorder.

You have a choice of analog or digital multitrack. Analog has more tape hiss, distortion, and wow & flutter (unsteady pitch) than digital. But with analog tape, cymbals sound a little more sweet or gentle, and tape distortion adds a fat, warm sound. Analog multitrack uses 1/4" or wider tape on reels; digital multitrack uses open-reel tape, S-VHS or Hi-8 video cassettes, or a hard disk.

A Modular Digital Multitrack (MDM) is an 8-track digital recorder that records on a video cassette (Figure 2.4). It offers very clean sound and long recording time (40 minutes or 2 hours) but costs about $3500. You can connect several MDMs together to add more tracks.

A multitrack hard-disk recorder records on a hard disk. It offers random access (instant access to any spot without having to fast forward or rewind). However, recording time is about 40 minutes or less, and the hard

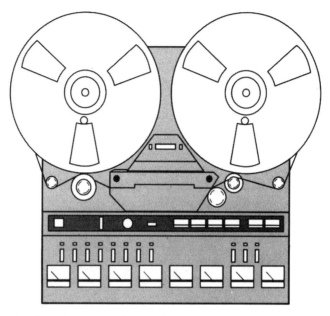

Figure 2.3 An 8-track open-reel recorder.

Figure 2.4
A Modular Digital Multi-
track (MDM).

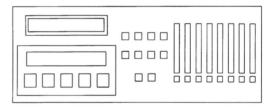

disk is usually not removable. You have to use the same disk over and over—erase old projects to record new ones. Some units, however, offer removable hard drives.

2-Track Recorder

A 2-track recorder is another major component. It records the stereo mix of the tape tracks off your mixer's channel 1 and 2 outputs. The 2-track recorder can be a cassette deck, open-reel deck, or DAT (Digital Audio Tape) recorder. Some people even use a hi-fi VCR.

Cassette decks cost the least. Open-reel decks (Figure 2.5) cost more but have higher sound quality and permit editing. **Editing** is the cutting and splicing of recording tape to remove unwanted noises, to change the

Figure 2.5   A 2-track open-reel recorder.

sequence of songs, or to combine parts of two or more different takes. The open-reel recorder is covered in detail in Chapter 9.

A DAT recorder also gives high sound quality. A DAT tape cannot be edited unless you copy from one DAT recorder to another, or use a Digital Audio Workstation (DAW). DAT and DAW are covered in Chapter 10.

Blank Recording Media

For your recorder you need some blank media to record on. Use the brand suggested by the recorder manufacturer. Cassette: Brand-name metal or chrome cassette tape. Open-reel multitrack: High-output, low-noise tape of width 1/4 inch, 1/2 inch, 1 inch, or 2 inch. MDM: S-VHS video cassettes or Hi-8 video cassettes. MiniDisc multitrack: MD data disc.

Microphones

Good mics are essential for quality sound—and you get what you pay for. If you experiment with various types of microphones, you find big differences in fidelity among them. Two microphones costing at least $125 to $250 each are recommended. Although $250 may seem like a lot of money for a microphone, you can't skimp here and expect to get quality sound.

Any distortion or weird tone quality in the microphone may be difficult or impossible to remove later. It's false economy to use a cheap mic.

You may be able to borrow some good microphones, or use the ones you normally use for P.A. Your ears should tell you if the fidelity is adequate for your purpose. Some people are happy to get any sound on tape; others settle for nothing less than professional sound quality.

How many mics and mic inputs do you need? It depends on the instruments you want to record. If you want to mic a drum set, you might need 8 mics and 8 mic inputs, mixing those to 1 or 2 tracks. On the other hand, if you use MIDI instruments, you might need only one good mic for vocals and acoustic instruments. You can use one mic on several different instruments and vocals if you overdub them one at a time.

Probably the most useful types for home recording are the cardioid condenser mic and cardioid dynamic mic. The cardioid pickup pattern helps reject room acoustics for a tighter sound. The condenser type is commonly used on cymbals, acoustic instruments, and studio vocals; dynamics are used typically on drums and electric amps. (For more information on microphones, see Chapter 6.)

If you plan to record solo instruments or musical ensembles in stereo with two mics out front, you need two condenser mics of the same model number, or a stereo microphone. See Chapter 19 for details.

Phantom-Power Supply

A phantom-power supply powers the circuits in condenser mics. It uses the same cable as the mic's audio signal. You can omit the supply if your condenser mic has a battery, or if your mixer supplies phantom power.

Mic Preamp

This device amplifies a mic signal up to a higher voltage, called "line level," which is needed by mixers and tape recorders. A stand-alone mic preamp provides a little cleaner sound than the mic preamps built into a mixer, but costs much more. Studios on a budget can do without it.

Direct Box

A direct box is a useful accessory for recorder-mixers with balanced XLR-type mic inputs. A direct box is a small device that connects an electric instrument (guitar, bass, synth) to a mixer mic input. It lets you record electric instruments directly into your mixer without a microphone. You can buy a direct box for as little as $50.

A useful feature in a direct box is a filter, which simulates the tone quality of a guitar amplifier/speaker. (See Chapter 8 for more information.)

Low cost recorder-mixers have 1/4" phone-jack inputs. In this case, simply use a short guitar cord between your instrument and mixer input.

A direct box or guitar cord picks up a very clean sound, which may be undesirable for electric guitar. If you want to pick up the distortion of the guitar amp, use a microphone instead. Or try a direct box that plugs into the external speaker jack on your guitar amp. It picks up the amp distortion, and filters it (reduces the treble) to make it sound more like a guitar speaker.

Monitor System

The monitor system lets you hear what you're recording and mixing. You can use a pair of high-quality headphones, or a pair of loudspeakers and a power amplifier. The power amplifier strengthens the mixer's signal so it can drive loudspeakers. The speakers should be accurate, high-fidelity types costing at least $200 each. Your home stereo might be good enough to serve, but skimping on a monitor system is not a smart move.

Nearfield studio monitor speakers (described in Chapter 4) are small, bookshelf-type speakers that are placed about 3 feet apart and 3 feet from you. Also available are powered monitors with built-in amplifiers.

If your monitor speakers are in the same room as your microphones, the mics pick up the sound of the speakers. This causes feedback or a muddy sound. In this case, it's better to monitor with headphones while recording.

If you're recording only yourself, one set of headphones is enough. But if you're recording another musician, you both need headphones. Many recorder-mixers have two headphone jacks for this purpose.

If you want to overdub several people at once, you need headphones for all of them. For example, if you're overdubbing three harmony vocalists, each one needs headphones to hear previously recorded tracks to sing with.

To connect all these headphones, you could build a headphone junction box: an aluminum or plastic box that contains several headphone jacks. These are wired to a cable coming from your mixer's headphone jack. Or you could use a splitter cable, which makes two jacks out of one.

Effects

Effects such as reverberation, echo, and chorus can add sonic excitement to a recording. They are produced by devices called **signal processors** (Figure 2.6).

The most essential effect is **reverberation**, a slow decay of sound such as you hear just after you shout in an empty gymnasium ("HELLO-O-O-o-

Figure 2.6   Signal processor.

o-o . . ."). Reverberation adds a sense of space; it can put your music in a concert hall, a small club, or a cathedral. This effect is usually produced by a digital reverb unit, available for $200 and up.

Another popular effect is **echo**, a repetition of a sound ("HELLO hello hello"). It's made by a **delay unit**, which also provides other effects such as chorus, doubling, and flanging.

The **compressor** is normally used as an automatic volume control for vocals. A compressor keeps the vocal track at a more even volume, making it easier to hear throughout a mix. Home-studio units start around $125.

A **multi-effects processor** combines several effects in a single box. These effects can be heard one at a time or several at once. You can even customize the sounds by pushing buttons to change the presets. (See Chapter 11 for more information on effects.)

Cables and Connectors

Once you have all this hardware, you need several types of cables to carry signals from one component to another. These are covered in detail later in this chapter.

Rack and Patch Bay

A rack is an enclosure to mount signal processors and other equipment in. A patch bay or patch panel in a rack is a group of connectors that are wired to equipment inputs and outputs. Neither are essential, but they are convenient.

Miscellaneous Equipment

Other equipment for your home studio includes power outlet strips, mic pop filters, masking tape and a pen to label inputs and cables, head-cleaning fluid and cotton swabs to clean the tape heads, tape head demagnetizer, MIDI equipment stands, session forms, recording tape, cleaning tape, splicing block and splicing tape, adapters, pen and paper, and so on.

MIDI Studio Equipment

MIDI is covered in detail in Chapter 16. Here are some components in a typical MIDI studio:

MIDI Controller

A MIDI controller is a musical-performance device such as a keyboard, drum pads, etc., that puts out a MIDI signal when you play it. A MIDI signal is a string of numbers that tells which notes you played, when you played them, and so on.

Synthesizer

A synthesizer (synth) is a piano-style keyboard instrument that simulates the sound of real musical instruments, or generates original sounds. When you play a synth, it puts out a MIDI signal.

Sequencer

A sequencer is a device, or a computer program, that records the MIDI signal of your performance into computer memory for later editing and playback. A sequence is a computer file of the notes you played and their timing.

Sampler

A sampler is a device that makes digital recordings of single notes, or short musical phrases, of real musical instruments. The sampler stores the samples in computer memory or on floppy disks. Using a MIDI controller or a sequencer, you trigger the sampler to play notes.

Sound Module

A sound module is a device that plays pre-recorded samples when triggered by a MIDI controller or sequencer. It does not record samples.

Drum Machine

A drum machine is a device that simulates a drummer. It's a sequencer that records a drum performance done on its built-in pads. Each pad plays a different drum sample. When you play back your recorded performance, the samples play. This simulates a drum set and percussion.

Line Mixer

A line mixer is a small mixer that combines the signals of synths, sound modules, samplers, and drum machines. A line mixer has no mic preamps. If you want to use a mic with your MIDI studio, you also need a mic preamp, a mixer with mic preamps, or a DAT recorder with a mic preamp.

MIDI Interface

A MIDI Interface is a circuit card that plugs into a slot in your computer. It converts a MIDI signal into computer data, and vice versa. You use the computer to record and edit sequences of musical notes.

MIDI Tape-Sync Box

A MIDI tape-sync box is a device that synchronizes a multitrack tape recorder with a sequencer. You record MIDI-instrument parts with the sequencer, record vocals and acoustic instruments with the tape recorder, and sync them together with the tape-sync box. Some recorder-mixers have MIDI tape sync already built in.

Digital Audio Workstation (DAW)

A DAW is a system for editing DAT tapes. It includes a circuit card you plug into your computer, and editing software.

MIDI/Digital-Audio Recording System

This system includes a computer, sound card, MIDI interface, and software. It lets you record MIDI sequences (from a MIDI controller) and digital audio tracks (from a mic) on a hard disk. The sequences and digital audio play at the same time, in sync. Both can be edited.

Some Studio Systems

Now that we've defined the equipment, let's put together several different systems and see what they might cost. To save money when you assemble one of these systems, you can buy used equipment, borrow from your stereo or your band's P.A., or rent equipment.

Recording System for Classical Music

Suppose you want to record a classical music ensemble. With this system (Figure 2.7), you can tape an orchestra, symphonic band, choir, pipe organ, string quartet, duets and soloists. You also can record small folk or jazz groups if you record in a good sounding room and balance the musicians by their placement around the mics.

1 DAT recorder	$700
2 high quality condenser mics	$500
1 stereo phantom-power supply (if not in DAT)	$125
1 or 2 mic stands and booms	$80
2 50-foot mic cables	$50
1 high quality headphone	$80
Total	$1535

OPTIONS: Use two battery-powered boundary mics on the stage floor ($120 to $360/pair) and omit the mic stands. These mics will work for small musical ensembles such as a quartet or small choir. A pair of mic preamps can give a cleaner sound than the mic preamps built into the DAT recorder.

Notebook Studio

Use this system to document your musical ideas, work out musical arrangements, or play your song ideas to your band. Sound quality is only fair, but it's good enough for the purpose. If you want to learn the basics of multitrack recording without spending a lot, this is the way to go.

1 4-track recorder-mixer	$450
1 dynamic or condenser mic	$100
1 mic stand and boom	$40
1 mic cable	$12
1 high quality headphone	$80
Total	$682

Figure 2.7 Recording system for classical music.

4-Track Studio for Audition Tapes

This system is good enough to make audition tapes for club owners and music publishers, but not quite good enough to make demos for record companies.

1 4-track recorder-mixer	$600
1 cassette deck	$200
1 digital reverb	$200
5 mics @ $130 each	$650
5 mic stands and booms	$200
1 direct box	$50
6 mic cables	$72
1 high quality headphone	$80
2 stereo patch cords	$10
Total	$2062

4-Track Studio for Demo Tapes

With this system (Figure 2.8), the sound quality is high enough to make demo tapes to send to record companies. If your band has simple instrumentation, you could even record commercial tapes.

1 4-track recorder-mixer	$1000
1 cassette deck	$200
1 multi-effects processor	$400
4 dynamic mics @ $130 each	$520
2 condenser mics @ $250 each	$500
1 stereo phantom-power supply	$125
6 mic stands and booms	$240

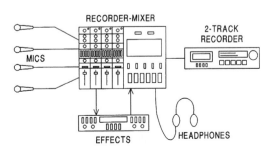

Figure 2.8
4-track studio for demo tapes.

1 direct box	$50
7 mic cables @ $12 each	$84
1 high quality headphone for monitoring	$80
4 musicians' headphones for overdubs	$160
2 Nearfield monitor speakers	$400
1 power amplifier	$500
2 speaker cables	$5
4 stereo patch cords	$20
Total	$4284

OPTIONS: For higher quality, substitute a DAT recorder ($700 and up) for the cassette deck. You might also use a 4-track MiniDisc recorder. To control the dynamics of the lead vocal, add a compressor ($80 and up).

8-Track Studio for Demos and Cassette Albums

This setup uses the same equipment as above, except you substitute an 8-track cassette recorder-mixer ($1899). Add an 8-channel snake for $200. Total cost is $5383.

8-Track Digital Studio

With this system (Figure 2.9), you can record albums for release on cassette or CD. You substitute an MDM and mixer for the cassette recorder-mixer or use a hard disk recorder-mixer.

1 8-track modular digital multitrack (MDM)	$3500
1 8-in, 4-out mixer	$800
1 DAT recorder	$700
Other equipment same as above	$3084
Total	$8084

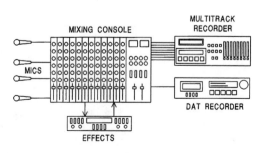

Figure 2.9
8-track digital studio.

OPTIONS: For higher quality, substitute more-expensive condenser mics ($1000 and up). Add more effects. If you want to do your own digital editing, add 2-track computer editing software and a sound card for about $1500, and install it in your home computer. To save cost, substitute an 8-track open-reel recorder for the MDM (cost about $2000, much less if used).

16-Track Digital Studio

This studio is the same, except you add another 8-track MDM ($3500) and substitute a 16-input mixing console ($1100) and 16-channel snake. Total cost is about $12,000.

OPTIONS: Add a remote control for the MDM. Add a rack for housing the effects, MDMs, and patch panel.

MIDI Studio

You can add MIDI equipment to any of the multitrack systems mentioned before. Or you can compose songs using nothing but MIDI devices (Figure 2.10). A simple MIDI studio might include these items:

Synthesizer	$1000
Drum machine	$350
Line mixer	$400
Sequencer	$600
MIDI cables	$50
DAT recorder	$700
1 high-quality headphone	$80
Total	$3180

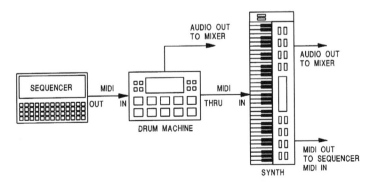

Figure 2.10   A simple MIDI studio.

OPTIONS:

- Musical workstation—a synthesizer/sampling keyboard with a built-in sequencer ($2000).
- Sound module—generates more sounds ($900).
- Tape-sync box ($100)—lets you sync sequencer recordings with a multitrack tape recorder.
- Sequencing software ($150)—used with your home computer as an alternative to a sequencer. Sequencing software provides easier and more extensive editing than a sequencer.
- Sampler—samples your own sounds ($1000).
- Keyboard controller ($400), breath controller, or drum pads.
- Microphone—a quality cardioid condenser mic for vocals and acoustic instruments ($750).
- Notation software ($80–$600)—prints out your performance as a score.
- Editor/librarian software ($200)—edits synth patches and stores them on disk.

MIDI/Digital-Audio Recording System

This system combines MIDI equipment with multitrack hard-disk recording (Figure 2.11). Software lets you do all your recording on hard disk—both MIDI sequences and audio from a microphone or electric guitar. The sequences and audio stay in sync with each other.

Several things happen when you play back what you recorded:

1. The sequence's MIDI signal comes from the MIDI OUT connector in the MIDI interface.
2. The sequence's MIDI signal plays notes through a synth, sound module, and/or drum machine.

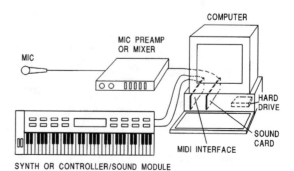

Figure 2.11
MIDI/digital-audio recording system (mixer not shown).

3. The digital audio—a multitrack mix—plays from the sound-card output.

4. A mixer combines the MIDI-equipment audio with the sound-card audio, and you record the mix on DAT.

Other than the computer and hard drive, a small system might include this equipment:

Synth	$1000
Drum machine	$350
Small mixer	$400
MIDI/digital-audio recording software	$400
MIDI interface card	$100
Sound card	$600
DAT recorder	$700
1 high quality microphone	$1000
1 mic stand and boom	$40
1 high-quality headphone	$80
2 Nearfield monitors	$400
1 power amplifier	$350
2 speaker cables	$5
MIDI cables	$50
4 stereo patch cords	$20
Total	$5495

As we've seen, putting together a home studio or project studio needn't cost much. As technology develops, better equipment is available at lower prices. That dream of owning your own studio is within reach.

Setting Up Your Studio

Once you have your equipment, you need to connect it together with cables, and possibly install equipment racks and acoustic treatment. Let's consider each step.

Cables and Equipment Connectors

Cables carry electric signals from one audio component to another. They are usually made of one or two insulated conductors (wires) surrounded by a fine-wire mesh **shield** which reduces hum. Outside the shield is a plastic or rubber insulating jacket.

Cables are either **balanced** or **unbalanced**. A balanced line is a cable that uses two conductors to carry the signal, surrounded by a shield (Figure 2.12). An unbalanced line has a single conductor surrounded by a shield (Figure 2.13). To see how many conductors a cable has, either remove some cable insulation with a wire stripper, or open the connector and look inside.

Recording equipment also has balanced or unbalanced connectors. Be sure your cables match your equipment. Balanced equipment has a 3-pin (XLR-type) connector (Figure 2.14); unbalanced equipment has a 1/4-inch phone jack (Figure 2.15) or phono jack (RCA) connector (Figure 2.16). A jack is a receptacle; a plug inserts into a jack.

Figure 2.12
A 2-conductor shielded, balanced line.

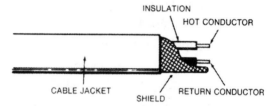

INSULATION
HOT CONDUCTOR
CABLE JACKET
SHIELD
RETURN CONDUCTOR

Figure 2.13
A 1-conductor shielded, unbalanced line.

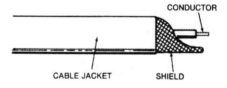

CONDUCTOR
CABLE JACKET
SHIELD

Figure 2.14
A 3-pin XLR-type connector used in balanced equipment.

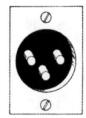

Figure 2.15
A 1/4-inch phone jack used in unbalanced equipment.

Figure 2.16
A phono (RCA) jack used in unbalanced equipment.

The balanced line rejects hum better than an unbalanced line, but an unbalanced line under 10 feet long usually provides adequate hum rejection and costs less.

A cable carries one of these three signal levels or voltages:

- Mic level (about 2 millivolts, or .002 volt)
- Line level (0.316 volt for unbalanced equipment, 1.23 volts for balanced equipment)
- Speaker level (about 20 volts)

The term "0.316 volt" also is known as "–10 dBV"; the term "1.23 volts" also is known as +4 dBu (see Appendix A).

Cable Connectors

There are several types of connectors used in audio. Figure 2.17 shows a 1/4-inch phone plug, used with cables for unbalanced microphones, synthesizers, and electric instruments. The tip terminal is soldered to the cable's center conductor; the sleeve terminal is soldered to the cable shield.

Figure 2.18 shows an RCA or phono plug, used to connect unbalanced line-level signals. The center pin is soldered to the cable's center conductor; the cup terminal is soldered to the cable shield.

Figure 2.19 shows a 3-pin pro' audio connector (XLR-type). It is used with cables for balanced mics and balanced recording equipment. The

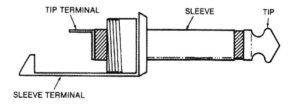

Figure 2.17
A 1/4-inch phone plug.

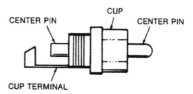

Figure 2.18
An RCA (phono) plug.

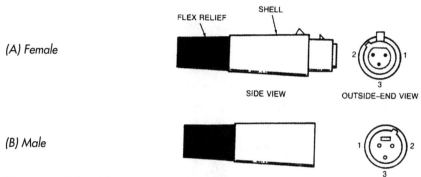

(A) Female

(B) Male

Figure 2.19 XLR-type connectors.

female connector (with holes, Figure 2.19 A) plugs into equipment outputs. The male connector (with pins, Figure 2.19 B) plugs into equipment inputs. Pin 1 is soldered to the cable shield; pin 2 is soldered to the "hot" red or white lead, and pin 3 is soldered to the remaining lead. This wiring applies to both female and male connectors.

Figure 2.20 shows a stereo phone plug, used with stereo headphones and with some balanced line-level cables. For headphones, the tip terminal is soldered to the left-channel lead; the ring terminal is soldered to the right-channel lead, and the sleeve terminal is soldered to the common lead. For balanced line-level cables, the sleeve terminal is soldered to the shield; the tip terminal is soldered to the hot red or white lead, and the ring terminal is soldered to the remaining lead.

Some mixers have **insert** jacks that are stereo phone jacks; each jack accepts a stereo phone plug. Tip is the **send** signal to an audio device input; ring is the **receive** signal from the device output, and sleeve is ground.

If you have unbalanced microphone inputs (1/4-inch diameter holes) on your recorder or mixer, use a balanced cable from mic to input. This reduces hum. Solder the shield and black lead to the long ground lug (sleeve terminal) on the phone plug. Solder the white or red lead to the small center lug (tip terminal) on the phone plug (Figure 2.21).

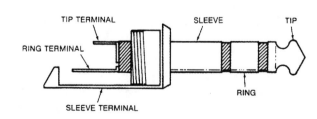

Figure 2.20
A stereo phone plug.

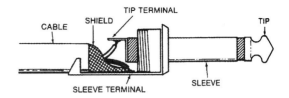

Figure 2.21
Wiring a balanced mic cable to an unbalanced 1/4-inch phone plug.

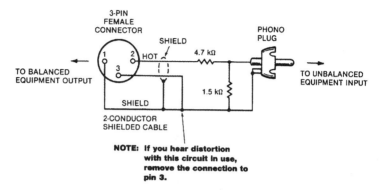

NOTE: If you hear distortion with this circuit in use, remove the connection to pin 3.

Figure 2.22   A circuit for matching a balanced +4 dBu output to an unbalanced −10 dBV input. It reduces the level 12 dB.

When you connect balanced equipment to unbalanced equipment, use the circuit shown in Figure 2.22 to match their levels. Balanced equipment operates at a relatively high level or voltage (called +4 dBu or +4); unbalanced equipment operates at a lower level (called −10 dBV or −10).

Cable Types

Cables are also classified according to their function. In a studio, you'll use five types of cables: mic cables, guitar cords, patch cords, MIDI cables, and speaker cables.

A mic cable is usually 2-conductor shielded—it has two wires to carry the signal—surrounded by a fine-wire cylinder or shield that reduces hum pickup. On one end of the cable is a connector that plugs into the microphone, usually a female XLR-type. On the other end is either a 1/4-inch phone plug or a male XLR-type connector that plugs into your mixer.

Rather than running several mic cables to your recorder-mixer, you might consider using a **snake**: a box with multiple mic connectors, all wired to a thick multiconductor cable (Figure 2.23). A snake is especially convenient if you're running long cables to recording equipment in a separate room.

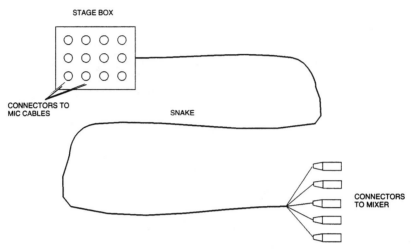

Figure 2.23 A stage box/snake.

A guitar cord is made of 1-conductor shielded cable with a 1/4-inch phone plug on each end. Use it to record instruments direct—the electric guitar, electric bass, synthesizer, and drum machine.

Patch cords connect your recorder-mixer to external devices: an effects unit, stereo cassette deck, and power amplifier. An unbalanced patch cord is made of 1-conductor shielded cable with either a 1/4-inch phone plug or a phono (RCA) connector on each end. A stereo patch cord is two patch cords joined together. Professional balanced equipment is interconnected with 2-conductor shielded cable having a female XLR on one end and a male XLR on the other. Professional patch bays use balanced cables with a stereo phone plug on each end.

A MIDI cable uses a 5-pin DIN connector on each end of a 2-conductor shielded cable. The cable connects between one component's MIDI OUT (or MIDI THRU) connector, and another component's MIDI IN connector.

A speaker cable connects the power amp to each loudspeaker. Speaker cables are normally made of lamp cord (zip cord). To avoid wasting power, speaker cables should be as short as possible, and should be heavy gauge (between 12 and 16 gauge). Number 12 gauge is thicker than 14; 14 is thicker than 16.

Rack/Patch Panel

You might want to mount your signal processors in a **rack**, a wooden or metal enclosure with mounting holes for equipment (Figure 2.24). You also might want to install a **patch panel** or **patch bay**: a group of connectors that are wired to equipment inputs and outputs. Using a patch panel and patch cords, you can change equipment connections easily. You also can bypass

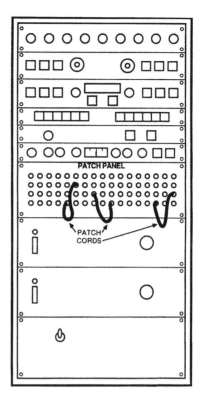

Figure 2.24
A rack and patch panel.

or patch around defective equipment. Note that patch bays increase the chance of hum pickup slightly because of the additional cables and connectors. Figure 2.25 shows some typical patch panel assignments.

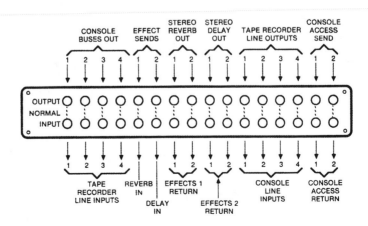

Figure 2.25   Some typical patch panel assignments.

Equipment Connections

The instruction manuals of your equipment tell how to connect each component to the others. In general, use cables as short as possible to reduce hum, but long enough to be able to make changes.

Be sure to label all your cables on both ends according to what they plug into, for example, TRACK 6 OUT or REVERB IN. If you change connections temporarily, or the cable becomes unplugged, you'll know where to plug it back in.

Typically, you follow this procedure to connect equipment (Figure 2.26):

1. Plug audio equipment and electric musical instruments into outlet strips fed from the same circuit breaker.
2. Connect mics and direct boxes to mic cables.
3. Connect mic cables either to the snake junction box, or directly into mixer mic inputs. Connect the snake connectors into mixer mic inputs.
4. Set the output volume of synthesizers and drum machines about 3/4 up. Connect synthesizers and drum machines to mixer line inputs. If this causes hum, use a direct box. If you lack a direct box, and the

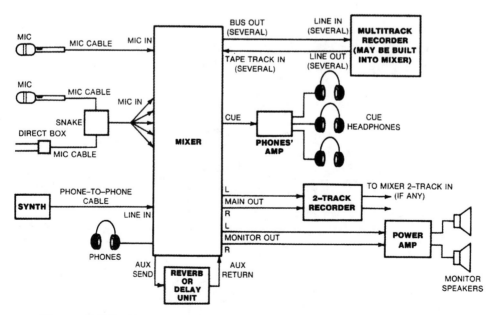

Figure 2.26 Equipment connections.

instrument and mixer both have 3-prong power cords, try cutting or unsoldering the cable shield at the instrument end of the cable.

5. Connect the mixer main outputs to a 2-track recorder.
6. Connect the 2-track recorder outputs to the mixer 2-track inputs (if any).
7. Connect the mixer monitor outputs to the power-amp inputs.
8. Connect the power-amp outputs to loudspeakers.
9. Connect the mixer aux-send connectors to effects inputs.
10. Connect the effects outputs to the mixer aux-return or bus-in connectors.
11. If you're using a separate mixer and multitrack recorder, connect mixer bus 1 to recorder track 1 IN; connect bus 2 to track 2, and so on. Also connect the recorder's track 1 OUT to the mixer's tape-track 1 IN; connect the track 2 OUT to the mixer's tape-track 2 in, and so on.
12. As an alternative, connect insert jacks to multitrack inputs and outputs. At each insert plug, connect the tip (send) terminal to a track input; connect the ring (receive) terminal to the same track's output.
13. If you have several headphones for musicians, connect the cue output to a small amplifier to drive their headphones. Or if the mixer's headphone signal is powerful enough, connect it to a box with several headphone jacks wired in parallel.

Acoustic Treatment

You might want to place some material in the studio to absorb sound reflections. This treatment reduces the reverberation in the room, and gives a clearer recorded sound.

For budget or improvised studios, the acoustic treatment has to be limited. Try surrounding the instrument and its mic with thick blankets or sleeping bags hung a few feet away. Carpet the floor and nail some convoluted (mattress) foam to wood-paneled walls. You can also try some sound absorbing cylinders, made commercially.

Add absorbers spaced evenly around the walls until your recordings sound reasonably dry (free of audible room reverberation). More on acoustic treatments in Chapter 3.

NOTE: By using close miking, direct boxes, and overdubs, you might be able to make good recordings in a room without any acoustic treatment.

This chapter covered the equipment and connectors for a recording studio. The rest of this book explains each piece of equipment and how to use it for best results.

3

SOUND, SIGNALS, AND STUDIO ACOUSTICS

When you make a recording, you deal with at least two kinds of invisible energy: sound waves and electrical signals. They carry the musicians' message. If you know the properties of sound and signals, you will understand what you're doing as you manipulate them in the recording process.

Sound Wave Creation

To produce sound, most musical instruments vibrate against air molecules, which pick up the vibration and pass it along as sound waves. When these vibrations strike your ears, you hear sound. To illustrate how sound waves are created, imagine a vibrating speaker cone in a guitar amp. When the cone moves out, it pushes the adjacent air molecules closer together. This forms a **compression.** When the cone moves in, it pulls the molecules farther apart, forming a **rarefaction.** As shown in Figure 3.1, the compressions have a higher pressure than normal atmospheric pressure; the rarefactions have a lower pressure than normal.

These disturbances pass from one molecule to the next in a springlike motion—each molecule vibrates back and forth to pass the wave along. The sound waves travel outward from the sound source at 1,130 feet per second, the speed of sound.

At some receiving point, such as an ear or a microphone, the air pressure varies up and down as the disturbances pass by. Figure 3.2 is a graph showing how sound pressure varies with time—the "wave" motion. The high point of the graph is called a **peak**; the low point is called a **trough.** The horizontal center line of the graph is normal atmospheric pressure.

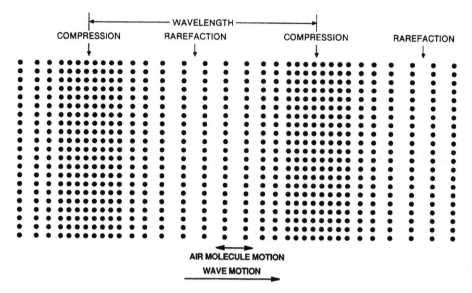

Figure 3.1 A sound wave.

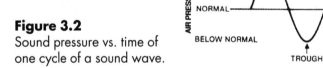

Figure 3.2
Sound pressure vs. time of
one cycle of a sound wave.

Characteristics of Sound Waves

Figure 3.3 shows three waves in succession. One complete vibration from normal to high to low pressure and back to the starting point is called one **cycle.** The time it takes to complete one cycle—from the peak of one wave to the next—is called the **period** of the wave. One cycle is one period long.

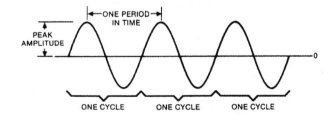

Figure 3.3
Three cycles of a wave.

Amplitude

The height of the wave is its **amplitude**. Loud sounds have high amplitudes (big pressure changes); quiet sounds have low amplitudes (small pressure changes).

Wavelength

When a sound wave travels through the air, the physical distance from one peak (compression) to the next is called a **wavelength** (Figure 3.1). Low-pitched sounds have long wavelengths (several feet); high-pitched sounds have short wavelengths (a few inches or less).

Frequency

The sound source (in this case, the guitar amp loudspeaker) vibrates back and forth many times a second. The number of cycles completed in one second is called **frequency.** The faster the speaker vibrates, the higher the frequency of the sound. Frequency is measured in **hertz** (Hz), which stands for cycles per second. One thousand hertz is called "one kilohertz," abbreviated kHz.

The higher the frequency, the higher the perceived pitch of the sound. Low frequency tones have a low pitch (like low E on a bass, which is 41 Hz). High frequency tones have a high pitch (like four octaves above middle-C, or 4186 Hz). Doubling the frequency raises the pitch one **octave.**

Children can hear frequencies from 20 Hz to 20 kHz, and most adults with good hearing can hear up to 15 kHz or higher. Each musical instrument produces a range of frequencies, say, 41 Hz to 9 kHz for a string bass, or 196 Hz to 15 kHz for a violin.

Phase and Phase Shift

The **phase** of any point on the wave is its degree of progression in the cycle—the beginning, the peak, the trough, or anywhere between. Phase is measured in degrees, with 360 degrees being one complete cycle. The beginning of a wave is 0 degrees; the peak is 90 degrees (1/4 cycle), and the end is 360 degrees. Figure 3.4 shows the phase of various points on the wave.

If there are two identical waves traveling together, but one is delayed with respect to the other, there is a **phase shift** between the two waves. The more delay, the more phase shift. Phase shift is measured in degrees. Figure 3.5 shows two waves separated by 90 degrees (1/4 cycle) of phase shift. The dashed wave lags the solid wave by 90 degrees.

When there is a 180-degree phase shift between two identical waves, the peak of one wave coincides with the trough of another. If these two waves are combined, they cancel out. This phenomenon is called **phase cancellation.**

41

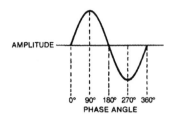

Figure 3.4
The phase of various points on a wave.

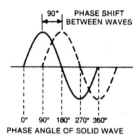

Figure 3.5
Two waves that are 90 degrees out-of-phase.

Suppose you have a signal with a wide range of frequencies, like the singing voice. If you delay this signal and combine it with the original undelayed signal, some frequencies will be 180 degrees out-of-phase and will cancel. This makes a hollow, filtered tone quality.

Here's an example of how this can happen. Suppose you're recording a singer/guitarist with one mic near the singer and another mic near the guitar. Both mics pick up the singer. The singer's mic is close to the mouth, and you hear it with no delay in the signal. The guitar mic is farther from the mouth, so its voice signal is delayed. When you mix the two mics, you often hear a funny tone quality caused by phase cancellations between the two mics.

Suppose you're recording a stage play with a mic on a short stand on the floor. The mic picks up the direct sound from the actors, but it also picks up delayed reflections off the floor. Direct and delayed sounds combine at the mic, causing phase cancellations. You hear it as a hollow, filtered sound that changes when the actor walks while talking.

Harmonics

The type of wave shown in Figure 3.2 is called a **sine wave.** It is a pure tone of a single frequency, like a signal from a tone generator. In contrast, most musical tones have a complex waveform, which has more than one frequency component. All sounds are combinations of sine waves of different frequencies and amplitudes. Figure 3.6 shows sine waves of three frequencies combined to form a complex wave.

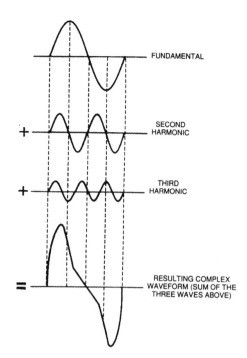

Figure 3.6
Adding fundamental and
harmonic waveforms to
form a complex waveform.

The lowest frequency in a complex wave is called the **fundamental frequency.** It determines the pitch of the sound. Higher frequencies in the complex wave are called **overtones** or **upper partials.** If the overtones are multiples of the fundamental frequency, they are called **harmonics.** For example, if the fundamental frequency is 200 Hz, the second harmonic is 400 Hz, and the third harmonic is 600 Hz.

The harmonics and their amplitudes help determine the tone quality or **timbre** of a sound, and help to identify the sound as a trumpet, piano, organ, voice, etc. **Noise** (such as tape hiss) contains a wide band of frequencies and has an irregular, non-repeating waveform.

Envelope

Another characteristic that identifies a sound is its envelope. When a note sounds, it doesn't last forever. It rises in volume, lasts a short time, then falls back to silence. This rise and fall in volume of one note is called the note's **envelope.** The envelope connects the peaks of successive waves that make up a note. Each musical instrument has a different envelope.

Most envelopes have four sections: **attack, decay, sustain,** and **release** (see Figure 3.7). During the attack, a note rises from silence to its maximum

THE LIBRARY
GUILDFORD COLLEGE
of Further and Higher Education

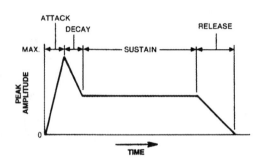

Figure 3.7
The four sections of the envelope of a note.

volume. Then it decays from maximum to some mid-range level. This middle level is the sustain portion. During release, the note falls from its sustain level back to silence.

Percussive sounds, such as drum hits, are so short that they have only a rapid attack and decay. Other sounds, such as organ or violin notes, last longer. They have slower attacks and longer sustains. Guitar plucks and cymbal crashes have quick attacks and slow releases. They hit hard then fade out slowly.

You can shorten a guitar string's decay or ringing by damping the string with the side of your hand. You can press a blanket against a kick drum head to damp the decay and get a tighter sound.

Signal Characteristics of Audio Devices

When a mic converts sound to electricity, this electricity is called the **signal.** It has the same frequency and the same amplitude changes as the incoming sound wave.

When this signal passes through an audio device, the device may alter the signal. It might change the level of some frequencies, or add unwanted sounds that are not in the original signal. Let's look at some of these effects.

Frequency Response

The frequency response of an audio device is the range of frequencies it reproduces at an equal level (within a tolerance, such as +/– 3 dB). A frequency-response graph shows the level vs. frequency. Signal level is measured in dB, while frequency in measured in Hz (hertz).

In Figure 3.8, the frequency response is 50 Hz to 12,000 Hz +/– 3 dB. That means the audio device passes all frequencies from 50 Hz to 12,000 Hz at a nearly equal level—within 3 dB. It reproduces low sounds and high

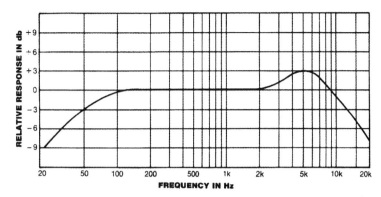

Figure 3.8 An example of a frequency response.

sounds equally well. The response is down 3 dB at 50 Hz and 12,000 Hz, and is up 3 dB at 5,000 Hz.

If the response is the same at all frequencies in the specified range, the graph forms a horizontal straight line, and so is called a **flat frequency response** (see Figure 3.9).

If an audio component has a flat frequency response, it passes all frequencies in the audio band without changing their relative levels. You get out the same amount of bass and treble that went in. A flat response does not affect the incoming sound.

When you turn a bass or treble knob on your stereo or your mixer, you're changing the frequency response. If you turn up the bass, the low frequencies rise in level. If you turn up the treble, the high frequencies are

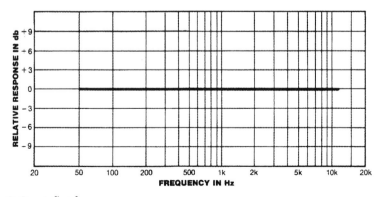

Figure 3.9 A flat frequency response.

emphasized. The ear interprets these effects as changes in tone quality—warmer, brighter, thinner, duller, and so on.

As shown in the right side of Figure 3.8, the response at high frequencies might roll off or decline. This shows that the upper harmonics are weak. The result is a dull sound. If the response rolls off at low frequencies (as in the left side of Figure 3.8), the fundamentals are weakened and the result is a thin sound. If a rolloff is above or below the range of frequencies that an instrument makes, you can't hear the rolloff (except maybe as harshness).

Usually, the wider the frequency response, the more natural and real the recording sounds. A wide, flat response gives accurate reproduction. A frequency response of 200 Hz to 8,000 Hz (+/– 3 dB) is narrow (poor fidelity); 80 Hz to 12,000 Hz is wider (better fidelity), and 20 Hz to 20,000 Hz is widest (best fidelity).

Also, the flatter the frequency response, the greater the fidelity or accuracy. A response deviation of +/–3 dB is good; +/–2 dB is better, and +/–1 dB is excellent.

The frequency response of an audio device might be made non-flat on purpose. For example, you might cut low frequencies with an equalizer to reduce breath pops from a microphone. Also, a microphone may sound best with a non-flat response—such as boosted high frequencies which add presence and sizzle.

Noise

Every audio component produces noise—a rushing sound like wind in trees. Noise in a recording is undesirable unless it's part of the music. You can make noise less audible by keeping the signal level relatively high. If the signal level in an audio device is very low, you have to turn up the listening volume in order to hear the signal well. Turning up the volume of the signal also turns up the volume of the noise, so you hear noise along with the signal. If the signal level is high to start with, you don't have to turn up the listening level so high and the noise remains in the background.

Distortion

If you turn up the signal level too high, the signal distorts. Then you hear a gritty, grainy sound. This type of distortion is sometimes called **clipping** because the peaks and troughs of the waveform are clipped off so they are flattened. To hear distortion, simply record a signal at a very high recording level (with the meters going well into the red area) and play it back.

Optimum Signal Level

You want the signal level high enough to cover up the noise, but low enough to avoid distortion. Every audio component works best at a certain optimum signal level, and this is usually indicated by a "0" on a meter or lights that show the signal level.

Figure 3.10 shows the range of signal levels in an audio device. At the bottom is the **noise floor** of the device—the level of noise it produces even with no signal. At the top is the **distortion level**—the point at which the signal distorts and sounds grungy. In between is a range in which the signal sounds clean. Try to maintain the signal around the 0 point on the average.

Signal-to-Noise Ratio

The level difference in dB between the signal level and the noise floor is called the **signal-to-noise ratio** (S/N) (Figure 3.10). The higher the signal-to-noise ratio, the cleaner the sound. A signal-to-noise ratio of 50 dB is fair, 60 dB is good, and 70 dB or greater is excellent.

To illustrate signal-to-noise ratio, imagine a person yelling a message over the sound of a train. The message being yelled is the signal; the noise is the train. The louder the message, or the quieter the train, the greater the signal-to-noise ratio. And the greater the signal-to-noise ratio, the clearer the message.

Headroom

The level difference in dB between the normal signal level and the distortion level is called **headroom** (Figure 3.10). The greater the headroom, the

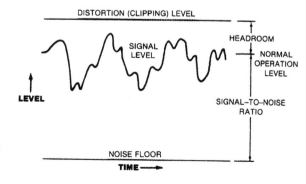

Figure 3.10
The range of signal levels in an audio device.

greater the signal level the device can pass without running into distortion. If an audio device has a lot of headroom, it can pass high-level peaks through without clipping them.

Behavior of Sound in Rooms

Because most music is recorded in rooms, you need to understand how room surfaces affect sound.

Echoes

Musical instruments make sound waves that travel outward in all directions. Some of the sound travels directly to your ears (or to a microphone) and is called **direct sound.** The rest strikes the walls, ceiling, floor, and furnishings of the recording room. At those surfaces, some of the sound is absorbed, some is transmitted through the surface, and the rest is reflected back into the room.

Because sound waves take time to travel (about 1 foot per millisecond), the reflected sound reaches you after the direct sound. The reflection repeats the original sound after a short delay. If the sound is delayed about 50 msec or more, we call it an echo (Figure 3.11). In some concert halls we

(A) Echo formation.

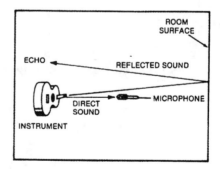

(B) Intensity vs. time of direct sound and its echoes.

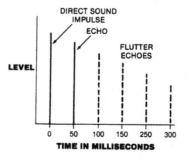

Figure 3.11
Echoes.

48

hear single echoes; in small rooms we often hear a short, rapid succession of echoes called flutter echoes. You can detect them by clapping your hand next to a wall. Flutter echoes happen when sound bounces back and forth between two parallel walls.

Reverberation

Sound reflects many times from all the surfaces in the room. These reflections sustain the sound of each note the musician plays. This persistence of sound in a room after the original sound has stopped is called **reverberation** (reverb). For example, reverberation is the sound you hear just after you shout in an empty gymnasium. The sound of your shout stays in the room and gradually dies away (decays).

Reverb is hundreds of echoes that gradually get quieter. The echoes follow each other so rapidly that they merge into a single continuous sound. Eventually, the room surfaces completely absorb the echoes. The timing of the echoes is random, and the echoes increase in number as they decay. Figure 3.12 shows how reverberation develops in a recording room.

Reverberation is a continuous fade-out of sound ("HELLO-O-O-o-o"), while an echo is a discrete repetition of a sound ("HELLO hello hello hello").

Reverberation time (RT60) is the time it takes for reverb to decay 60 dB. Too long a reverb time makes a recording sound distant, muddy, and

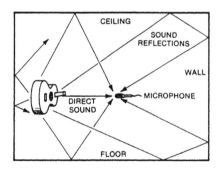

(A) Multiple sound reflections create reverberation.

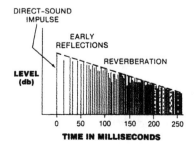

(B) Intensity vs. time of direct sound, early reflections, and reverberation.

Figure 3.12
Reverberation.

washed-out. That's why pop-music recordings are made in a fairly "dead" or non-reverberant studio, which has an RT60 of about 0.4 second or less. In contrast, classical music is recorded in "live," reverberant concert halls (RT60 about 1 to 3 seconds) because we want to hear reverb with classical music—it's part of the sound.

Reverberation comes to you from every direction because it is a pattern of many sound reflections off the walls, ceiling, and floor. Since we can tell where sounds come from, we can distinguish between the direct sound of an instrument coming to us from a single location and the reverberation coming to us from everywhere else. So we can ignore the reverb and concentrate on the sound source. In fact, we normally are not aware of reverberation.

But record an instrument in that room and play back the recording. You'll hear a lot more reverb than what you heard live. What's going on? The reverb you recorded on tape is not all around you. Instead, it's all up front between the speakers. So it's more audible; you can't discriminate against the reverb spatially. To reduce the amount of reverb in your recordings, you need to place mics close to instruments, and maybe add some sound-absorbing materials to the room.

Leakage

Sound from an instrument travels to the nearest mic, and also "leaks" into the mics intended for other instruments. This overlap of an instrument's sound into the mic of another instrument is called **leakage** (or **bleed** or **spill**).

It's very important to minimize leakage—to ensure that each mic picks up only its intended instrument. Suppose that you're miking a drum set and an acoustic piano. As the musicians play, you monitor what the mics pick up. When you turn up just the drum mics, the drums sound close-up or "tight." But when you also turn up the piano mic, the drum sound becomes distant or muddy. That's because the piano mic picks up drum leakage at a distance.

How to Tame Echoes, Reverb, and Leakage

Echoes, reverb, and leakage can make your recordings sound mushy and distant. There are two ways to prevent these problems: with recording techniques and with acoustic treatment.

Controlling Room Problems with Recording Techniques

Sometimes you can make clean recordings in an ordinary room, such as a club, living room, or basement, if you follow these suggestions:

- Mike close. Place each mic 1 to 6 inches from each instrument or voice. Then the mics will hear more of the instruments and less of the room reflections. You might want to use mini mics, which attach directly to instruments.
- Use directional mics—cardioid, supercardioid, or hypercardioid— which reject room acoustics.
- Record bass guitar and synth directly with a guitar cord or a direct box. Since you omit the microphone, you pick up no room acoustics. To get a good sound when recording electric guitar direct, record off the effects boxes or use a guitar amp simulator.
- Space the musicians fairly far apart—about 12 feet if possible.
- First record the loud instruments—drums, bass, and electric guitar. Then overdub the quiet instruments—acoustic guitars, sax, piano, vocals.
- Overdub instruments one at a time rather than recording them all at once. You'll pick up a much cleaner sound. However, this loses the emotional interaction that occurs when all the musicians play together.
- Record in a large room. This lets you spread the musicians farther apart, and weakens the wall reflections.

Controlling Room Problems with Acoustic Treatment

When should you add some acoustic treatment to your recording room, or build a studio?

- You clap your hands next to a wall and you hear flutter echoes.
- Your studio is a very live environment, such as a garage or concrete block basement.
- Your recording room is very small.
- You hear outside noises in your recordings.
- You want the freedom to mike several feet away.

If these conditions apply, check out the following suggestions on upgrading the acoustics of your studio.

In a pop-music studio, it's best to keep the reverb time fairly short: about 0.4 second. Since reverberation is caused by sound reflections off room surfaces, any surface that is highly sound-absorbent helps to reduce echoes, reverb, and leakage.

To absorb high frequencies, use porous materials such as fiberglass insulation, acoustic tile, foam plastic, carpeting, and curtains. If possible, space these materials several inches from the wall. The spacing helps them absorb mid-bass frequencies.

To absorb low frequencies, you can make bass traps. Construct them of flexible surfaces such as wood paneling or linoleum nailed to studs. Wood paneling, and cavities such as closets or air spaces behind couches absorb bass, too.

Start with just a little absorption behind the musician you're recording. The room will still be pretty live. You can add more absorbers, a few at a time, until your recordings sound as dead as you wish.

You don't want the room to be completely absorbent (dead) because such an environment is stifling. In such a room, musicians may feel they are playing in a vacuum—they get no reinforcement from sounds reflecting from nearby walls. Some reflections are beneficial, not only for the musicians' comfort, but also for the "air" and liveliness they add to the recorded sound.

Here's a list of simple acoustic treatments:

- To prevent flutter echoes, put some absorbent material (foam or blankets) on one or both walls, or make the walls non-parallel.
- Open closet doors, and place couches and books a few inches from the walls.
- Carpet the floor.
- Hang canvas or blankets from the ceiling in deep folds.
- Make a big pillow by wrapping 6-inch thick fiberglass insulation in muslin to contain the fibers. Tape or nail several of these pillows on the wall.
- Hang thick curtains, comforters, or sleeping bags on the walls. Better yet, hang them at least 2 feet from the walls.
- Attach open-cell acoustic-foam wedges on or near the walls. Even convoluted foam for mattresses is better than nothing. The thicker the foam, the better the low frequency absorption. Four-inch foam on the wall absorbs frequencies from about 400 Hz up.
- In a basement studio, nail acoustic tile to the ceiling joists, with fiberglass insulation in the air space between tiles and ceiling.
- For bass trapping, make some panel absorbers by nailing 1/4-inch and 1/8-inch thick plywood panels to 2-inch furring strips (battens). Put fiberglass insulation in the air space behind the panel (Figure 3.13).

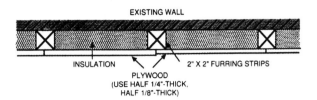

Figure 3.13
One type of bass trap: a panel absorber.

Cover about half the wall area in this manner. Or buy some tubular bass traps, such as those made by Acoustical Sciences Corp.

NOTE: You may not need any bass traps if you don't put any bass into the room. For example, don't turn up the bass guitar amp—just record the bass direct and have the musicians wear headphones to hear the bass.

• Try making some wide-range absorbers. Get some pressed fiberglass board (Owens-Corning Type 703, 3 lb/cu ft), 2- or 4-inch thick. Nail the board to 2x6 studs, spaced 4 feet apart on the existing wall. Put fiberglass insulation in the air space. Place the board in patches around the room rather than all together.

Controlling Room Resonance

If you play an amplified bass guitar through a speaker in a room, and do a bass run up the scale, you may hear some notes that boom out in the room. The room is resonating at those frequencies. These resonant frequencies, which are strongest below 300 Hz, are called **room modes** or **normal modes.** They occur in patterns called standing waves. Room modes can give a tubby or boomy coloration to musical instruments.

Room resonances are worst in a cubical room. They are less of a problem if the room's length, width, and height are not multiples of each other. Here are some recommended ratios of room dimensions:

Height	Width	Length
1	1.14	1.39
1	1.17	1.47
1	1.26	1.41
1	1.28	1.54
1	1.45	2.10
1	1.47	1.70
1	1.60	2.33
1	1.62	2.62

As an example, let's try the top ratio. If the ceiling height is 10 feet, the room width should be 11.4 feet and the length should be 13.9 feet.

Try to record in a large room because the room resonance frequencies are likely to be below the musical range. Use bass traps to absorb room resonances. Contrary to popular opinion, non-parallel walls don't prevent standing waves.

Making a Quieter Studio

The following tips will keep noises out of your recordings:

- Turn off appliances and telephones while recording.
- Pause for ambulances and airplanes to pass.
- Close windows.
- Close doors and seal with towels.
- Remove small objects that can rattle or buzz.
- Weather-strip doors all around, including underneath. (Leave the doors open for ventilation when not recording.)
- Replace hollow doors with solid doors.
- Block openings in the room with thick plywood and caulking.
- Put several layers of plywood and carpet on the floor above the studio, and put insulation in the air space between the studio ceiling and the floor above.
- Place microphones close to instruments and use directional microphones. This won't reduce noise in the studio, but it will reduce noise picked up by the microphones.
- When building a new studio, you might want to make the walls of plastered concrete block because massive walls reduce sound transmission. Or make the walls of gypsum board and staggered studs. Nail gypsum board to 2x4 staggered studs on 2x6 footers as seen in Figure 3.14. Staggering the studs prevents sound transmission through the studs. Fill the airspace between walls with insulation.

Figure 3.14
Staggered-stud construction to reduce noise transmission.

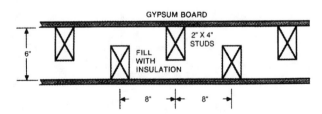

The ideal home-recording room for pop music is a large, well-sealed room with optimum dimensions. It should have some soft surfaces (carpet, acoustic-tile ceiling, drapes, couches), and some hard vibrating surfaces (wood paneling or gypsum-board walls on studs). The room is in a quiet neighborhood. Your recording room may not need the acoustic treatments presented here—do some trial recordings to find out. If you do need some adjustments, however, the suggestions in this chapter should point you in the right direction.

4

MONITORING

One of the most exciting moments in recording comes when the finished mix is played over the studio monitor speakers. The sound is so clear you can hear every detail, and so powerful you can feel the deep bass throbbing in your chest.

You use the monitor system to listen to the output signals of the console or the tape recorders. It consists of the console monitor mixer, the power amplifiers, loudspeakers, and the listening room. Each power amplifier boosts the electrical power of a console signal to a sufficient level to drive a loudspeaker. The speaker converts the electrical signal into sound, and the listening room acoustics affect the sound from the speaker.

A quality monitor system is a must if you want your mixes to sound good. The power amp and speakers tell what you're doing to the recorded sound. According to what you hear, you adjust the mix and judge your mic techniques. Clearly, the monitor system affects the settings of many controls on your mixer, as well as your mic selection and placement. And all those settings affect the sound you're putting on tape. So, using inadequate monitors can result in a poor sounding product coming out of your studio.

It's important to use accurate speakers that have a flat frequency response. If your monitors are weak in the bass, you will tend to boost the bass in the mix until it sounds right over those monitors. But when that mix is played over speakers with a flatter response, it will sound too bassy because you boosted the bass on your mixer. So, using monitors with weak bass results in bassy recordings; using monitors with exaggerated treble results in dull recordings, and so on. In general, colorations in the monitors will be inverted in your mixdown tape.

That's why it's so important to use an accurate monitor system—one with a wide, smooth frequency response. Such a system lets you hear exactly what's on the tape.

Speaker Requirements

The requirements for an accurate studio monitor are these:

- **Wide, smooth frequency response.** To ensure accurate tonal reproduction, the on-axis response of the direct sound should be +/– 4 dB or less from 40 Hz to 15 kHz. The low-frequency response of a small monitor speaker should extend to at least 70 Hz.

- **Uniform off-axis response.** The high frequency output of a speaker tends to diminish off-axis. Ideally the response at 30 degrees off-axis should be only a few dB down from the response on-axis. That way, a producer and engineer sitting side-by-side will hear the same tonal balance. Also the tonal balance will not change as the engineer moves around at the console.

- **Good transient response.** This is the ability of the speaker to accurately follow the attack and decay of musical sounds. If a speaker has good transient response, bass guitar sounds tight, not boomy. Drum hits have sharp impact. Some speakers are designed so that the woofer and tweeter signals are aligned in time. This aids transient response.

- **Clarity and detail.** You should be able to hear small differences in the sonic character of instruments, and to sort them out in a complex musical passage.

- **Low distortion.** Low distortion is necessary because it lets you listen to the speaker for a long time without your ears hurting. A good spec might be: Total Harmonic Distortion under 3 percent from 40 Hz to 20 kHz at 90 dB SPL.

- **Sensitivity.** Sensitivity is the sound pressure level a speaker produces at 1 meter when driven with 1 watt of pink noise. **Pink noise** is random noise with equal energy per octave. This noise is either band-limited to the range of the speaker or is a 1/3-octave band centered at 1 kHz. Sensitivity is measured in dB/W/m (dB Sound Pressure Level per 1 watt at 1 meter). A spec of 93 dB/W/m is considered high; 85 dB/W/m is low. The higher the sensitivity, the less amplifier power you need to get adequate loudness.

- **High output capability.** This is the ability of a speaker to play loudly without burning out. You often need to monitor at high levels to hear quiet details in the music. Plus, when you record musicians who play loudly in the studio, it can be a letdown for them to hear a quiet playback. So you may need a maximum output of 110 dB SPL.

  This formula calculates the maximum output of a speaker (how loud it can play): dB SPL = 10 log (P) + S where dB SPL is the sound pressure level at 1 meter, P is the continuous power rating of the speaker in watts, S is the sensitivity rating in dB/Watt/meter. For example, if a speaker is rated at 100 watts maximum continuous power, and its sensitivity is 94 dB SPL/watt/meter, its maximum output SPL is 10 log (100) + 94 = 114 dB SPL (at 1 meter from the speaker). The level at 2 meters will be about 4 to 6 dB less.

Nearfield™ Monitors

Many professional recording studios use large monitor speakers which have deep bass. However, they are expensive, heavy, difficult to install, and are affected by the acoustics of the control room.

If you want to avoid this hassle and expense, consider using a pair of **Nearfield monitor speakers** (Figure 4.1). A Nearfield monitor is a small, wide-range speaker typically using a cone woofer and dome-shaped tweeter. You place a pair of them about 3 or 4 feet apart, on stands just behind the console, about 3 or 4 feet from you. Nearfields are far more popular than large wall-mounted speakers.

This technique, developed by audio consultant Ed Long, is called **Nearfield monitoring.** Because the speakers are close to your ears, you hear mainly the direct sound of the speakers and tend to ignore the room acoustics. This way, the speakers tend to sound about the same in any environment, so you may not need to acoustically treat your control room. Plus, Nearfield monitors sound very clear and provide sharp stereo imaging. They are normally used without equalization. Some units have bass or treble tone controls built in.

Nearfield monitors are designed for close listening—they have enough bass to sound full when placed far from walls. Although most

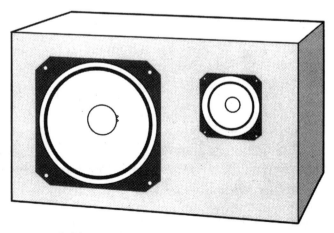

Figure 4.1 A Nearfield monitor speaker.

Nearfields lack deep bass, they can be supplemented with a subwoofer to reproduce the complete audio spectrum. Or you can check the mix occasionally with headphones that have deep bass.

Some Nearfields are in a **satellite-subwoofer** format. The two satellite speakers are small units, typically including a 4-inch woofer and 3/4-inch dome tweeter. The satellites are too small to produce deep bass, but that is handled by the subwoofer. It is a single cabinet with one or two large woofer cones. Typically, the subwoofer (sub) produces frequencies from 100 Hz down to 40 Hz or below. Since we do not localize sounds below about 100 Hz, all the sound seems to come from the satellite speakers. The sub-satellite system is more complicated to set up than two larger speakers, but offers deeper bass.

Powered Monitors

Some monitors have a power amplifier built in. You feed them a line-level signal (labeled MONITOR OUT) from your mixing console. Some powered monitors are bi-amplified: they have one amplifier for the woofer and one for the tweeter. The advantages of bi-amplification are:

- Distortion components caused by clipping the woofer power amplifier will not reach the tweeter, so there is less likelihood of tweeter burnout if the amplifier clips. In addition, clipping distortion in the woofer amplifier is made less audible.
- Intermodulation distortion is reduced.
- Peak power output is greater than that of a single amplifier of equivalent power.
- Direct coupling of amplifiers to speakers improves transient response—especially at low frequencies.
- Bi-amping reduces the inductive and capacitive loading of the power amplifier.
- The full power of the tweeter amp is available regardless of the power required by the woofer amp.

The Power Amplifier

If your monitor speakers are not powered, you need a power amplifier (Figure 4.2). It boosts your mixer's line-level signal to a higher power in order to drive the speakers.

How many watts of power do you need? The monitor speaker's data sheet gives this information. Look for the specification called "Recommended amplifier power." A power amp of 50 watts per channel continuous

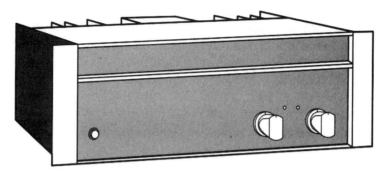

Figure 4.2 A power amplifier.

is about the minimum for Nearfield monitors; 150 watts is better. Too much power is better than too little, because an underpowered system is likely to clip or distort. This creates high frequencies that can damage tweeters.

A good monitor power amp has distortion under 0.05 percent at full power. It should have a high damping factor—at least 100—to keep the bass tight. The amp should be reliable. Look for separate level controls for left and right channels. The amplifier should have a clip or peak light that flashes when the amp is distorting.

Speaker Cables and Polarity

When you connect the power amp to the speakers, use good wiring practice. Long or thin cables waste amplifier power by heating. So put the power amp(s) close to the speakers and use short cables with thick conductors—at least 16 gauge. The low resistance of these cables help the power amplifier to damp the speaker motion and tighten the bass.

If the two speakers are wired in opposite polarity, one speaker's cone moves out while the other speaker's cone moves in. This causes vague stereo imaging, weak bass, and a strange sense of pressure on your ears. Be sure to wire the speakers in the same polarity as follows: In both channels, connect the amplifier positive (+ or red) terminal to the speaker positive (+ or red) terminal. Setting the correct polarity is also called "speaker phasing."

Control-Room Acoustics

The acoustics of the control room affect the sound of the speakers. Sound waves leaving the speaker strike the room surfaces. At those surfaces, some frequencies are absorbed, while other frequencies are reflected. At your ears, the sound waves reflected from the room surfaces combine with the direct sound. Reflections that arrive within 20 to 65 milliseconds after the direct sound blend with the direct sound and affect the tonal balance you hear.

Suppose the walls are covered with carpet so that they absorb only the high frequencies. Then the walls will reflect mainly the low frequencies. When you listen to a speaker playing in such a room, you hear the direct sound from the speaker plus the bassy wall reflections. The combined sound will be bass-heavy. Now suppose the walls are made of wood paneling mounted on studs. Such a vibrating surface absorbs lows and reflects highs. The sound you hear probably will be thin and overly bright.

Clearly, the room surfaces should reflect (or absorb) all frequencies about equally to avoid coloring the sound of the speakers. Equal absorption (+/− 25 percent) from about 250 Hz to 4000 Hz is usually adequate. As described in Chapter 3, you can use flexible panels to absorb lows, in combination with fibrous materials or foam to absorb highs. Or use thick fibrous material spaced from the wall and ceiling.

Room resonances or **standing waves** can cause some notes to blare out and cause other notes to disappear. Be sure to control these resonances as suggested in Chapter 3.

Room acoustics also affect the decay-in-time of the sound coming from the speakers. Whenever the speakers play a note that ends suddenly, the sound of that note continues to bounce around the room. This causes echoes and reverberation that prolong the sound. This long decay of sound is not part of the recording. So the control room should be relatively **dead**— that is, it should have a short reverberation time. A typical living room has a reverb time of about 0.4 second; the control room should too, so the engineer will hear about the same amount of room reverb that a home listener will hear. A totally dead room is uncomfortable to listen in.

Using Nearfield monitors makes the room acoustics less important, but it still helps to treat the room acoustics. To prevent sound reflections from the wall behind the speakers, apply muslin-covered fiberglass insulation or acoustic foam. This treatment improves the monitors' sound. Stereo imaging and depth are greatly improved, the sound is clearer, and the frequency response is flatter. The treatment will reduce boominess and ringing, and make transients sharper. Also, your recordings will translate better to other speakers.

If your control room is separate from the studio, the control room should be built to keep out sound from the studio. You want to hear only the sound from the monitors, not the live sound from the musicians. In a home studio, you can achieve isolation simply by putting the control room equipment in a room far removed from the studio, with the doors closed.

A control room built next to the studio needs good isolation. Use double-wall construction with staggered studs (Figure 4.3). Put fiberglass insulation between the two walls. The door between the two rooms should be solid wood and should be weather-stripped all around—including underneath. Use a double-pane window (mounted in rubber) between the control room and studio.

Speaker Placement

Mount the speakers at ear height so the mixer doesn't block their sound. To prevent sound reflections off the mixer, place the speakers on stands behind the meter bridge, rather than putting them on top. For best stereo imaging, align the speaker drivers vertically and mount the speakers symmetrically with respect to the side walls. Place the two speakers as far apart as you're sitting from them; aim them toward you, and sit exactly between them (Figure 4.4).

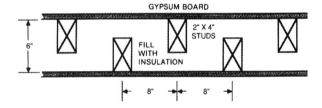

Figure 4.3
Staggered-stud construction to reduce noise transmission.

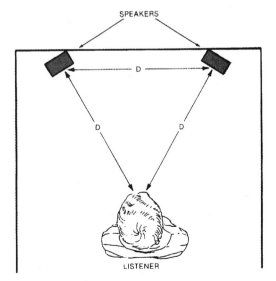

Figure 4.4
The recommended speaker/listener relationship for best stereo imaging.

Try to position the monitors several feet from the nearest wall. Wall reflections can degrade the frequency response and stereo imaging. The closer to the wall the speakers are, the more bass you hear.

Using the Monitors

You've treated the room acoustics, and you connected and placed the speakers as described earlier. Now it's time to adjust the stereo balance.

1. Play a mono musical signal and assign it to channels 1 and 2 in your mixer.
2. Adjust the channel 1 and 2 master faders so that the signal reads the same on the channel 1 and 2 meters.
3. Place the two speakers the same distance from you.
4. Sit at the mixer exactly midway between the speakers, and listen to the image of the sound between the speaker pair. You should localize it midway between the monitors—that is, straight ahead.
5. If necessary, center the image by adjusting the left or right volume control on your power amp. (Note: An off-center listener will hear the image shifted toward one side.)

When you do a mixdown, try to keep the listening level around 85 dB SPL—a typical home listening level. As discovered by Fletcher and Munson, we hear less bass in a program that is played quietly than in the same program played loudly. If you mix a program while monitoring at, say, 100 dB SPL, the same program will sound weak in the bass when heard at a lower listening level—which is likely in the home. So, programs meant to be heard at 85 dB SPL should be mixed and monitored at that level.

Loud monitoring also exaggerates the frequencies around 4 kHz. A recording mixed loud may sound punchy, but the same recording heard at a low volume will sound dull and lifeless.

Here's another reason to avoid extreme monitor levels: Loud sustained sound can damage your hearing or cause temporary hearing loss at certain frequencies. If you must do a loud playback for the musicians (who are used to high SPLs in the studio), protect your ears by wearing earplugs or leaving the room.

You can get a low-cost sound level meter from Radio Shack. Play a musical program at 0 VU on the mixer meters and adjust the monitor level to obtain an average reading of 85 dB SPL on the sound level meter. Mark the monitor-level setting.

Before doing a mix, you may want to play some familiar records over your monitors to remind yourself what a good tonal balance sounds like.

Listen to the amount of bass, midrange, and treble, and try to match those in your mixes. But listen to several records since they vary.

While mixing, monitor the program alternately in stereo and mono to make sure there are no out-of-phase signals that cancel certain frequencies in mono. Also beware of center-channel buildup: Instruments or vocals that are panned to center in the stereo mix sound 3 dB louder when monitored in mono than they do in stereo. That is, the balance changes in mono—the center instruments are a little too loud. To prevent this, don't pan tracks hard left and hard right. Bring in the side images a little so they will be louder in mono.

You'll mix the tracks to sound good on your accurate monitors. But also check the mix on small inexpensive speakers to see whether anything is missing or whether the mix changes drastically. Make sure that bass instruments are recorded with enough edge or harmonics to be audible on the smaller speakers. It's a good idea to make a cassette copy of the mix and give it to the client to audition in a car, boom box, or compact stereo.

Headphones

Quality headphones are a low-cost alternative to loudspeakers for a home studio. You may find that headphones provide adequate isolation if the music you're recording is quiet.

Many home studios have the mixer and monitors in the same room as the musicians. In this case, you monitor with headphones while recording and overdubbing, then monitor with speakers when you mix. If you're monitoring as the musicians are playing, block out their sound by using closed-cup headphones or in-the-ear earphones.

Compared to speakers, headphones have several advantages:

- They cost much less.
- There is no coloration from room acoustics.
- The tone quality is the same in different environments.
- They are convenient for on-location monitoring.
- It's easy to hear small changes in the mix.
- Transients are sharper due to the absence of room reflections.

The disadvantages of headphones are:

- They become uncomfortable after long listening sessions.
- Cheap headphones have inaccurate tone quality.
- Headphones don't project bass notes through your body.

- The bass response varies due to changing headphone pressure.
- The sound is in your head rather than out front.
- You hear no room reverberation, so you may add in too much or too little artificial reverb.
- It's difficult to judge the stereo spread. Over headphones, panned signals tend to sound not as far off center as the same signals do heard over speakers. The same is true of stereo recordings made with a coincident pair of mics.

Even though headphones may not sound like speakers, you can do your mixes over headphones to match commercial records heard over those same headphones. Then your mixes should sound commercial over speakers, too.

The Cue System

The cue system is a monitor system for musicians to use as they're recording. It includes the cue or monitor knobs in your mixer, a small amplifier, a headphone connector box, and headphones. Musicians often can't hear each other well in the studio, but they can listen over headphones to hear each other in a good balance. Also, they can listen to previously recorded tracks while overdubbing.

Headphones for a cue system should be durable and comfortable. They should be closed-cup to avoid leakage into microphones. This is an ideal situation; open-air phones may work well enough. Also, the cue phones should have a smooth response to reduce listening fatigue, and should play loud without burning out. Make sure they are all the same model so each musician hears the same thing. A built-in volume control is convenient.

A suggested cue system is shown in Figure 4.5. Connect a power amp to the cue or monitor output of your mixer. The amp drives several

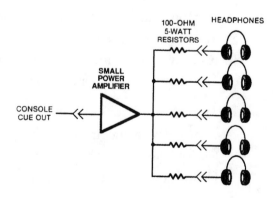

Figure 4.5
A cue system.

resistor-isolated headphones, which are in parallel. Sometimes you can get by with four headphone jacks in a small metal box, wired in parallel and connected to your mixer's headphone jack.

Although some consoles can provide several independent cue mixes, the ideal situation is to set up individual cue mixers near each musician. Then they can set their own cue mix and listening level. The inputs of these mixers are fed from the console output buses.

Suppose a vocalist sings into a microphone and hears that mic's signal over the cue headphones. If the singer's voice and the headphone's sound are opposite in polarity, the voice partially cancels or sounds funny in the 'phones. Make sure that the voice and headphones are the same polarity.

Here's how. While talking into a mic and listening to it on head-phones, reverse the ground and signal leads to the headphones connector. The position that gives the fullest, most solid sound in the headphones is correct. All the headphones in your studio should be the same model, so that everyone will hear with correct polarity.

Summary

Ultimately, what you hear from the monitors influences your recording techniques and affects the quality of your recordings. So take the time to plan and adjust the control room acoustics. Choose and place the speakers carefully. Monitor at proper levels and listen on several systems. You'll be rewarded with a monitor system you can trust.

5

HUM PREVENTION

You patch in a piece of audio equipment, and there it is—HUM! It's a low-pitched tone or buzz. This annoying sound is a tone at 60 Hz (50 Hz in Europe) and multiples of that frequency.

What's causing it? How can you get rid of it? This chapter answers those questions.

Grounding Definitions

To understand the information in this chapter, you need a few definitions related to grounding:

Grounding: Connecting pieces of electronic equipment to ground.

Ground: A point designated as a zero-voltage reference. Voltages and signals in the system are measured relative to this ground point.

Earth ground or **physical ground:** A connection to moist dirt usually made through a copper ground rod.

Power ground (Safety ground): An earth ground connection to equipment chassis made through the power cord. If you look at a modern AC outlet, you'll see three holes. The U-shaped hole is the safety ground. This terminal connects by a long wire to the power company's earth ground: a copper rod driven in the earth or a building's underground steel structure.

Many electronic devices have 3-wire power cords; the round ground pin on the cord is connected to the equipment chassis. A **chassis** is a metal housing that surrounds the circuitry in a piece of audio equipment. When you plug a device's 3-prong power cord into an outlet, the chassis of the device is connected to the safety ground. If a short circuit accidentally occurs between the chassis and a hot power line, the chassis current will flow to the safety ground rather than through someone touching the chassis, preventing shocks.

AC mains: The 60 Hz, 120/240V AC power wiring supplied by the power company (sometimes a type called 3-phase).

Phase: One part or one leg of a 3-phase power wiring.

Preventing Hum Pickup in Audio Cables

One source of hum is your studio's power wiring, which radiates hum fields that vibrate at 60 Hz or higher. Your audio cables can pick up these hum fields. Let's explain how.

Electrostatic Interference

Power lines act as one plate of a capacitor, while the conductors in audio cables act as the other plate. An oscillating electrostatic field is set up between these two plates, causing hum to be transmitted (coupled) from the power lines to the cable conductors. An electrostatic field couples best at high frequencies, and so is heard as a buzz primarily made of the upper harmonics of 60 Hz.

To block hum fields, each audio cable has a shield—a cylinder of foil or fine wires that surrounds the signal wires in the center of the cable. A shield is a conductive enclosure around signal-carrying conductors, used to keep out electrostatic hum fields and radio frequency interference (RFI). Metal racks, equipment chassis, and microphone handles also are shields.

A shield works only when connected to a ground, such as the ground terminal of a chassis-mounted connector. That way, hum currents induced in the shield go to ground instead of getting into your audio. The ground provides a drain path for shield charges caused by the electrostatic fields.

If a cable shield is broken (usually at a connector), you might hear hum because there is no path to ground for the induced hum currents. So if you hear hum, check for broken shield connections in your cables. Also, use good-quality cables that have a high percentage of shielding (90 percent or better). The greater the shield coverage, the better it rejects hum.

Magnetic Interference

Power lines and transformers also act as electromagnets, radiating magnetic lines of force that oscillate at 60 Hz and its harmonics. These lines of force "cut" through the conductors in audio cables, causing the conductors to generate electricity at 60 Hz and its harmonics. Magnetic fields couple best at low frequencies, and so are usually heard as a low tone at 60 Hz. A magnetic hum field is directional, so you can detect it by rotating the device that is picking up hum. If the hum level varies, the hum is induced magnetically.

A shield must be made of a magnetic material (such as steel) to block magnetic hum fields. This shield need not be grounded unless you also want to use it for electrostatic shielding.

There are a number of steps you can take to avoid magnetic interference:

- Avoid using fluorescent lights in the studio.
- Install your equipment at least several inches (or feet) from the large power transformers in power amplifiers.
- Extension cords, power cords, and lighting cables radiate hum fields that audio cables can pick up. So separate your audio cables from those other types by at least one foot. If you must cross those cables, do so at right angles and space them vertically; this reduces the coupling between cables.

Coiled AC cords radiate more hum than straight cords. You might want to shorten unused AC cable to avoid coiling it.

CAUTION: This may void your warranty or UL approval.

Balanced and Unbalanced Lines

The kind of cables you use—balanced or unbalanced—affects the amount of hum pickup. A balanced line is a cable made of two conductors that carry the signal, surrounded by a shield (Figure 5.1). The shield does not carry the signal; it is connected to ground and keeps hum fields out of the conductors. Balanced equipment uses 3-pin pro audio connectors (XLR type). Professional audio equipment uses balanced connectors. Pin 1 is ground or shield, pin 2 is audio in-polarity (hot), and pin 3 is audio return (cold).

In a balanced line, the two conductors pick up equal amounts of hum interference. The cable plugs into a balanced input, which is sensitive to the voltage difference between the two conductors. Because there is little or no difference in hum voltage between the two conductors, hum picked up by the cable is not amplified, or is canceled in the process.

Figure 5.1
Balanced line.

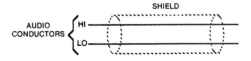

Twisted-pair cable picks up less magnetically induced hum than non-twisted-pair cable. This is because twisted conductor pairs occupy the same point in space on the average, so they are the same distance from the hum source and receive equal hum interference. This equal interference is then canceled by the balanced input circuitry of the equipment.

An **unbalanced line** is a cable that uses a single conductor surrounded by a shield (Figure 5.2). An unbalanced cable has either an RCA (phono) plug or a 1/4-inch 2-conductor phone plug. Unbalanced equipment has RCA (phono) jacks, or has 1/4-inch phone jacks. Home stereos and semipro recording equipment use unbalanced connectors.

Both the conductor and the shield carry the audio signal, so the shield isn't as effective in blocking electrostatic hum as it is in a balanced line. Also, the inner conductor and outer shield are at different impedances to ground, so they pick up different amounts of hum interference. This difference in induced hum voltage is amplified by the equipment the cable is plugged into.

Unbalanced lines pick up more hum than balanced lines. However, if you use unbalanced lines less than 10 feet long, hum is usually inaudible. You can get by with unbalanced lines in a small home studio where everything is close together. But if you need to run long mic cables (over 25 feet), consider getting a mixer with balanced (XLR-type) mic inputs to reject hum.

When connecting balanced equipment to unbalanced equipment, you may want to add a 1:1 audio isolation transformer at the unbalanced input or output (Figure 5.3). This allows most of the interconnecting cable to be balanced.

Figure 5.2
Unbalanced line.

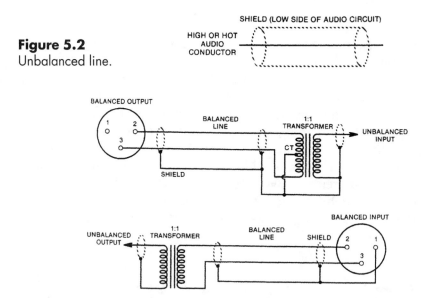

Figure 5.3 Transformer connections between balanced and unbalanced equipment.

Preventing "Dirty" AC Power

The AC power from wall outlets is seldom a clean sine wave. Distortion, spikes, RFI, and "hash" appear on the waveform and can cause buzzes in your audio. To ensure clean power:

- Avoid silicon control rectifier (SCR) light dimmers. Instead, use multi-way incandescent bulbs to vary the studio lighting levels.
- Install a power conditioner, AC isolation transformer, or line filter between the AC power outlets and the audio-equipment power cords.
- Use a balanced AC power supply. It cancels hum by using balanced AC power from a center-tapped power transformer. Instead of one 120V line and one 0V line, it has two 60V lines. They are in phase with each other, and sum to 120V. But they are connected to the center-tap ground out of phase (one is +60V; the other is –60V). Any hum and noise on the grounding system cancels out.

Preventing Ground Loops

A major cause of hum is the **ground loop**. It is the circuit loop that is formed when equipment is connected to ground though more than one path. It occurs when two pieces of audio gear are connected to each other through a shield and also through the AC safety ground.

Figure 5.4 shows a ground loop. Two equipment chassis are connected to two separate safety grounds by their AC cords. Also, the equipment chassis

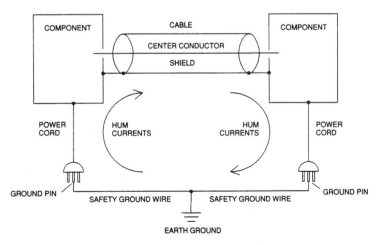

Figure 5.4 A ground loop.

are connected together by the shield of the audio cable. The shield and safety-ground wires form a ground loop. A ground loop also can be created between two cable shields connected to the same piece of equipment.

Here is a real world example of a ground loop situation. Suppose you're recording a synthesizer. The synth and your mixer have 3-prong power cords that connect to the safety ground. The synth is plugged into an AC outlet across the room, and your mixer is plugged into a nearby outlet. When you connect the synth to your mixer with a guitar cord, and monitor the signal, you hear hum.

Figure 5.4 shows what's happening. Chances are that the outlets are fed from different circuit breakers, so the outlets are at different ground voltages. When you plug your synth and mixer into these separated out-lets, and connect the equipment together with a guitar cord, the difference in ground voltages makes a 60 Hz hum current flow between the synth and mixer. That's a ground loop. There are several ways to prevent ground loops, and we'll look at each one.

Isolate the Grounds

One solution is to use a transformer-isolated direct box between the synth and your mixer's balanced mic input. Unlike a guitar cord, the transformer passes the signal without connecting the two chassis together. On the direct box, flip the ground-lift switch to the position where you hear the least hum.

A similar solution works for two pieces of unbalanced gear. If you hear hum after connecting them with a cable, try wiring a 1:1 isolation transformer between them (see Figure 5.5).

Connect Everything to the Same Ground

Another cure for ground loops is to plug all your equipment into one or more outlet strips fed from the same circuit breaker. That way, the ground voltage for all the equipment is about the same, so no hum current can flow

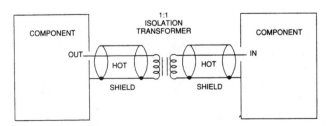

Figure 5.5 Wiring an isolation transformer between two unbalanced devices.

between their chassis. Plug the electric musical instruments and recording gear into the same strips. Use thick extension cords for distant equipment.

There still may be a slight voltage difference between components because their power supplies reflect different voltages onto their chassis. A balanced AC power supply can eliminate this problem.

Before you plug in all those power cords, make sure that the sum of the equipment fuse ratings does not exceed the amperage rating for that circuit. In most cases, a single 20-amp breaker will handle a small home studio.

Break the Loop

Suppose that the synth and mixer are plugged into the same outlet strip. If you hear hum when you connect them together, unsolder or cut the shield at one end of the cable between them. This breaks the ground loop. The safety-ground leads, instead of the shield, serve as the signal return path. If you still hear hum, use a direct box.

Some people try to prevent ground loops by putting a 3-to-2 adapter on each power cord. This breaks the loop by removing the safety ground. It's a quick fix, but it creates a serious shock hazard and is not recommended.

What if you connect two unbalanced devices that have 2-prong power cords? Leave the shield connected at both ends because it carries the signal. Ground loops aren't normally a problem with this type of equipment. The orientation of two-prong AC cords in non-polarized outlets makes a difference. For each piece of audio gear, rotate the two-prong power cord in its outlet to find the minimum hum position.

In balanced line-level cables, you can break the loop by cutting or unsoldering the cable shield in one connector. Then no hum current can flow between equipment. The shield still drains electrostatic interference to ground through its single ground connection. Connect the shield only in the female XLR connector. Leave the shield unsoldered in the male connector— cut it short and shrink-tubed it so that it doesn't short to other contacts.

CAUTION: Don't lift (remove) the shield unless you are sure that the connected end is to a chassis ground, not a signal ground.

To disconnect the shield temporarily—as in on-location work—insert a ground-lift adapter (Figure 5.6) in the male cable connector.

If you hear radio stations or other type of RFI with a lifted-ground cable, solder a 0.01 uF capacitor between the unconnected shield and pin 1 of the XLR-type connector.

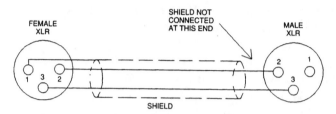

Figure 5.6   A ground-lift adapter for balanced line-level cables.

In some situations, the shield **should** be connected to pin 1 on both ends of the cable:

- In mic cables (so that the mic housing will be grounded).
- When you use balanced AC power.
- When pin 1 of the equipment connector is grounded to the equipment chassis (check this with an ohmmeter).

Connect Shields to Grounded Chassis

The following method is the recommended way to prevent ground loops in an audio system with balanced equipment. Connect the cable shield only to the equipment chassis on both ends of the cable. Ground all equipment chassis by the power-cord safety ground.

If pin 1 is tied (connected) only to the chassis (not to signal ground), solder the cable shield to pin 1 in the connectors that plug into that component. If pin 1 is tied to the signal ground on an internal PC board, wire the XLR-type connector as shown in Figure 5.7.

Keep the Loop Area Small

A ground loop can be created between two cables connected to the same piece of equipment. If these two cables are widely separated, they form a big loop or circle that can pick up radiated hum fields like an antenna.

To reduce this kind of hum pickup, make the loop area small. Keep your equipment close together and use short audio cables. When running send-and-return cables to a signal processor, tape these cables together. When you run left-channel and right-channel cables for a stereo hookup, use a single stereo cable.

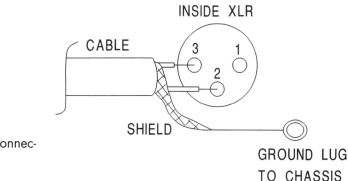

Figure 5.7
Wiring an XLR-type connector to tie the shield to chassis.

Prevent Ground Loops in Racks

A rack is a grounded metal cabinet used to hold audio equipment and patch panels. The equipment bolts onto rack channels, which are vertical metal strips with holes in them. Inside the rack is an AC outlet strip to power the rack equipment. Also inside the rack, near the bottom, is a **rack ground** securely bonded to the rack.

Try to put all unbalanced equipment in a single rack to shorten the cables between them. There are two opposite approaches to mounting unbalanced equipment in a rack. One is to isolate the chassis from the rack; the other is to ground the chassis to the rack.

According to the "isolate" approach, a ground loop can occur when two chassis of unbalanced equipment contact each other through a rack. To prevent this, put unbalanced equipment in a wooden rack with wooden rack channels, and keep the chassis separated (insulated) from each other with electrical tape. If you must use a rack with metal rails, isolate all the unbalanced rack equipment from the rack (and each other) by using electrical tape, nylon mounting bolts, and nylon washers. You can buy isolation tabs made for this purpose.

In contrast, some engineers prefer to ground all chassis to the rack channels, which are tied to safety ground. That way, equipment with "wall wart" power supplies will be grounded.

There are two opposite approaches to grounding at a patch panel:

1. Connect corresponding pairs of input and output jack sleeves together, but isolate the shields from the rack (Figure 5.8). An example of a corresponding pair is "Tape Track 7 Out" and "Mixer Tape Input 7."

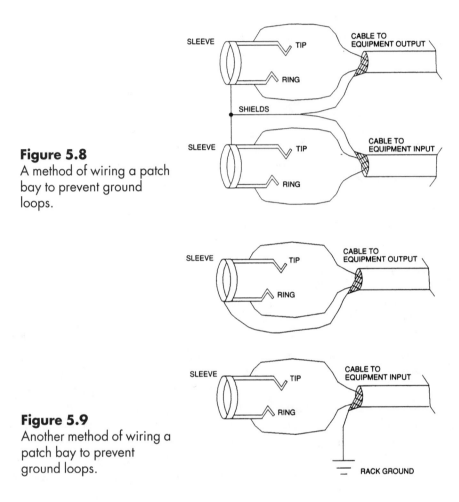

Figure 5.8
A method of wiring a patch bay to prevent ground loops.

Figure 5.9
Another method of wiring a patch bay to prevent ground loops.

2. Don't connect anything to sleeves of patch bay jacks wired to equipment inputs. In addition, connect input cable shields to the rack ground bus (Figure 5.9).

Prevent Accidental Ground Loops

Ground loops can happen when a connector shell or a snake box touches a metal surface. Inside an XLR-type cable connector is a ground lug that contacts the metallic connector shell. If the ground lug is soldered to pin 1 (the shield's pin), the shell is connected to the shield through the ground lug and pin 1. Ground loops may occur if the shell touches a grounded metallic surface. So don't wire the ground lug to pin 1.

If you locally ground a microphone snake box, this will create a ground loop. So don't ground the snake box except through its cable shield. Make sure the snake box doesn't touch any metal surfaces.

Each audio cable should have an insulating rubber or plastic jacket to prevent ground loops. If the shield is exposed, it can contact grounded metallic surfaces at more than one point, creating a ground loop.

Reducing Hum from Mics and Electric Guitars

Microphones and electric guitars are sensitive to hum pickup because they produce low-level signals. These signals need a lot of amplification, which also amplifies any hum picked up by these devices and their cables.

Microphone Hum

Use these tips to minimize microphone hum pickup:

- Use low-impedance microphones (150-600 ohms), which pick up less hum than high-impedance microphones.
- Use microphones with balanced outputs (3-pin connectors), which pick up less hum than unbalanced mics (hot conductor plus shield).
- Use a balanced cable from the mic to the input. If you have unbalanced mic inputs on your recorder or mixer, solder the shield and pin-3 lead to the phone plug sleeve or ground terminal; solder the pin-2 lead to the plug's hot or tip terminal.
- If the mic cable still picks up hum, unbalance the cable through a 1:1 transformer that is plugged directly into the input (refer to Figure 5.3).
- If hum pickup is severe with a dynamic microphone, use a dynamic mic with a humbucking coil built in. Or change to a condenser mic.
- Use twisted-pair mic cable to reduce pickup of magnetically induced hum. The more shield coverage, the less pickup of electrostatically induced hum. **Braided shield** generally offers the best coverage; **double-spiral wrapped** is next best; and **spiral-wrapped** is worst.
- Use star-quad cables such as those made by Canare, Mogami, or Belden.
- Routinely check mic cables to make sure the shield is connected at both ends.

- Check that the mic-connector set screw is securely screwed clockwise into the mic handle. This set screw is in the handle near the connector.

- For outdoor work, tape over cracks between connectors to keep out dust and rain.

Electric Guitar Hum

Electric guitars are high-impedance, unbalanced devices, which makes them very susceptible to hum. Try the following suggestions to reduce hum associated with electric guitars:

- Replace or repair guitar cords that have broken shields. Use only high-quality cords with metal-jacket plugs.

- Flip the polarity switch (if present) on the guitar amp to the lowest hum position.

- Flip the ground-lift switch on each direct box to the lowest hum position.

- Have the guitarist turn up the volume of the guitar all the way, and then turn down the gain on the guitar amp.

- Have the guitarist move around or turn around (rotate) to find a spot with minimum hum pickup.

- Replace any defective tubes in the guitar amp. If the power-supply filter capacitors in the guitar amp are corroded, replace them. This replacement should be done by an authorized technician.

- Use guitars with humbucking pickups, or install modern humbuckers in older guitars.

- Use a quieter amplifier.

Reducing Radio Frequency Interference (RFI)

RFI is heard as buzzing, clicks, radio programs, or "hash" in the audio signal. It's caused by CB transmitters, computers, lightning, radar, radio and TV transmitters, industrial machines, auto ignitions, stage lighting, and other sources. Many of the following techniques are the same used to reduce hum from other sources. To reduce RFI:

- Use wide copper straps or braids, rather than wires, for ground connections, to reduce the high ground resistance caused by **skin effect**— the tendency of RF signals to travel on the outside of a conductor.

- Install high-quality RFI filters in the AC power outlets. The cheap types available from local electronics shops are generally ineffective.
- Physically separate the lighting power wiring from the audio cables.
- Avoid SCR dimmers—instead, use multiwatt incandescent bulbs to vary the studio lighting levels.
- Use enclosed equipment racks. The metal enclosure acts as a shield.
- Avoid long ground leads and unbalanced lines that are over ten feet long.

Also, for each unbalanced mic input, consider connecting a 0.01 uF capacitor between the hot terminal and ground (if the mixer lacks such a capacitor). Then, if the shell of the mic-input connector isn't grounded to the mixer chassis, connect a 0.001 uF Mylar capacitor between the shell terminal and the mixer chassis (if the mixer lacks such a capacitor).

Long speaker cables can act as an RF antenna. If you suspect the cables to be the source of the RFI, bypass the RF to ground at the power-amp speaker terminals. Connect a 0.01 uF to 0.03 uF disk capacitor between one speaker lead and the amplifier chassis ground. Use one capacitor per channel. Do the same for the "ground" side of the amp's speaker terminals, if they are above chassis ground.

In connectors that have the shield floating (NOT tied to pin 1), connect the shield to pin 1 through a 0.01 uF capacitor. This bypasses the RFI to ground without connecting the audio signal to ground. Also, for balanced lines in mic junction boxes, solder a 0.01 uF capacitor between pins 1 and 2, and between pins 1 and 3. You also could do this at the mixer at each balanced mic input.

Special Considerations for On-Location Work

In a remote recording job, you're outside the controlled environment of the studio. You need to take special precautions with power distribution, electric guitar grounding, and interconnecting multiple sound systems.

Power-Distribution System

A touring sound company should carry its own single-phase power distribution system because the building ground wiring on-location is unreliable. See Figure 5.10 for a suggested AC power distribution system.

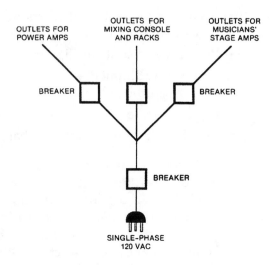

Figure 5.10
An AC power distribution system for a touring sound system.

Electric Guitar Grounding

At times, electric guitar players can receive a shock when they touch their guitar and a mic simultaneously. This occurs when the guitar amp is plugged into an electrical outlet on stage, and the mixing console (to which the mics are grounded) is plugged into a separate outlet across the room. As stated before, these two power points may be at widely different ground voltages, so a current can flow between the grounded mic housing and the grounded guitar strings.

> CAUTION: Electric guitar shock is especially dangerous when the guitar amp and the console are on different phases of the AC mains.

It helps to power all instrument amps and audio gear from the same AC distribution outlets. That is, run a heavy extension cord from a stage outlet back to the mixing console (or vice versa). Plug all the power-cord ground pins into grounded outlets. That way, you prevent shocks and hum at the same time.

If you're picking up the electric guitar direct, use a transformer-isolated direct box and set the ground-lift switch to the minimum-hum position. Using a neon tester or voltmeter, measure the voltage between the electric-guitar strings and the metal grille of the microphones. If there is a voltage, flip the polarity switch on the amp. Use foam windscreens for additional protection against shocks.

Interconnecting Multiple Sound Systems

When you record a concert, three mixers are commonly in use: house, monitor, and recording. On stage, a mic splitter takes the signal from each mic and sends it three ways to the three mixers (see Figure 5.11).

To avoid ground loops between the three systems, ground the mic-cable shields to only one mixer. At the splitter, use the ground-lift switches to disconnect the shields going to the other two mixers. Ground to the mixer that provides the least hum. You might need to power two of the mixers from outlets near the other mixer. That way, the three mixers share a common ground. If you encounter an unknown system where balanced audio cables might be grounded at both ends, use some cable ground-lift adapters (Figure 5.5) to float (remove) the extra ground connection at equipment inputs.

Often, a radio station or video crew will take an audio feed from a studio's mixing console. In this case, you can prevent a hum problem by using a console with transformer-isolated inputs and outputs. Or you can use a 1:1 audio isolation transformer between the console and the feeds.

Special Considerations for Small Studios

If you have a small studio with fewer than 20 power cords in use, you can use the wall-outlet power ground to ground your audio equipment. Plug your equipment into 3-wire grounded outlet strips powered by the same wall-outlet circuit breaker. If your equipment has 2-prong power cords, you probably won't have any ground-related problems if you connect the equipment as described earlier. In particular, review the section on preventing ground loops.

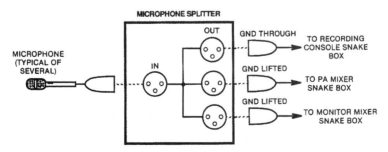

Figure 5.11 Using a mic splitter to feed three sound systems.

Special Considerations for Large Studios

The following suggestions are for large professional installations using balanced audio lines. If you have a small studio, you may want to skip this section or just read it for the advice it contains.

AC Power Wiring

Hum prevention starts with properly designed power wiring. Sound-system equipment should be powered on its own circuit separate from others, such as those used for lighting or air-conditioning, which can put noise spikes on the AC line.

Listed below are four ways to power the audio system. The first is most effective but most expensive; the last is least effective but least expensive:

- Power the audio system from its own power transformer on a telephone pole outside the studio.
- Get power from an independent breaker box.
- Have an electrician put the audio system on a different phase of the incoming AC mains than the phase(s) other equipment is connected to.
- Power the audio from its own circuit breaker.

In any case, all the audio equipment (including guitar amps in the studio or on stage) should be on the same phase of the power line to prevent hum.

Ground Wiring

Even if your wall outlets have a safety ground, you can't always trust these grounds to have a low-resistance connection to earth ground. If you suspect that the earth ground at the electrical service entrance is insufficient, have an electrician replace it.

Large studios often require special wiring of the safety ground to prevent ground loops. Although studio power wiring must meet electrical code requirements, the ground wiring may be nonstandard; so confer with the electrician to make sure the work is done as described in this section and also meets code.

If you're installing a system in a building or home which is already wired for AC power, use a balanced AC power supply to power the studio. Plug all the 3-prong power cords into outlet strips that are fed from the balanced supply. This eliminates ground loop problems.

If you're installing new power wiring, use AC outlets with isolated-ground terminals. That is, use special wall outlets that float (isolate) the ground terminal from the wall box; the wall box is grounded through the conduit. Such outlets are available from electrical supply houses.

Typical AC outlets with modern wiring contain three wires: hot (black), neutral (white), and ground (bare or green). The ground wire goes to the U-shaped hole in the outlet. Substitute your own ground wire for the bare or green one by following this procedure:

1. Have an electrician run a low-resistance (No. 10 gauge) insulated wire from each outlet's ground receptacle back to the ground bus bar: a copper plate in the bottom of the circuit-breaker box. Try to keep the lengths of all the ground leads about equal for equal resistance to earth ground. These ground wires should not short to the conduit, or you get ground loops.
2. Plug the rack equipment into the rack's AC outlet strip.
3. Plug the 3-prong AC power cord of each piece of equipment (and rack outlet strip) into an isolated ground AC outlet.

You may want to enclose the power wiring in grounded metal conduit to prevent hum radiation from the power lines into audio circuits.

A proper grounding scheme resembles a tree (Figure 5.12). The earth ground is the roots, the safety-ground wires are the big branches, and the rack safety-ground wires are the small branches. This is also called a **star** or **single-point** grounding system.

Summary

For unbalanced equipment in a small studio, and using short connecting cables, these are the most important points to remember about hum prevention:

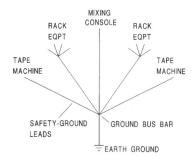

Figure 5.12
Recommended layout of safety-ground wires.

- Plug all equipment into outlet strips powered by the same breaker.
- Put unbalanced equipment in a single rack, isolated from the rack and each other (say, by using a wooden rack). Some engineers prefer instead to ground all chassis to the rack rails.
- Use balanced mic cables if possible.

For balanced equipment in a large installation:

- Put audio equipment on a separate power feed.
- Use AC isolation transformers or AC line filtering if necessary.
- Consider installing a balanced AC power supply, such as made by Equi=Tech or Furman.
- In each balanced line-level cable, connect the shield to pin 1 in the female XLR; disconnect the shield from pin 1 in the male XLR. Leave the shield connected on both ends if the equipment chassis are tied to pin 1, or if you use balanced AC power.
- As an alternative, ground cable shields only to equipment chassis as shown in Figure 5.7.
- If you're installing new power wiring, use isolated-ground outlets. Connect an insulated low-resistance wire from each outlet's ground receptacle to a single ground point. Plug all the rack equipment into the same AC outlet strip in the rack. Plug all the power cords into 3-prong outlets.

Hum-Fixing Checklist

Even if your system is wired properly, a hum or buzz may appear when you make a connection. Follow these tips to stop the hum:

- Check cables and connectors for broken leads.
- Remove audio cables from your devices and monitor each device by itself. It may be defective.
- Use a direct box instead of a guitar cord between instrument and mixer.
- Try a 1:1 isolation transformer between the two devices.
- Add ground-lift adapters to line-level balanced cables at the male XLR end. **Caution:** Lifted shields can act as an RF antenna.
- Try the connection shown in Figure 5.7.
- Make sure that the snake box is not touching metal.

- Tighten the mic-connector screws.
- Try another mic.
- Flip AC polarity switches or ground-lift switches.
- Have the guitarist move or aim in a different direction.
- Route mic cables away from power cords; separate them vertically where they cross.

By following all these tips, you should be able to connect audio equipment without introducing any hum. Good luck!

6

MICROPHONES

What microphone is best for recording an orchestra? What's a good snare mic? Should the microphone be a condenser or dynamic, omni or cardioid? You can answer these questions more easily once you know the types of microphones and understand their specs.

First, it always pays to get a high-quality microphone—which costs at least $125. The mic is a source of your recorded signal. If that signal is noisy, distorted, or tonally colored, you'll be stuck with those flaws through the whole recording process. Better get it right up front. Even if you have a MIDI studio and get all your sounds from samples or synthesizers, you still might need a good microphone for sampling, or to record vocals, sax, acoustic guitar, and so on.

A microphone is a **transducer**—a device that changes one form of energy into another. Specifically, a mic changes sound into an electrical signal. Your mixer amplifies and modifies this signal.

Transducer Types

Mics for recording can be grouped into three types depending on how they convert sound to electricity: dynamic, ribbon, or condenser. A dynamic mic capsule, or transducer, is shown in Figure 6.1. A coil of wire attached to a diaphragm is suspended in a magnetic field. When sound waves vibrate the diaphragm, the coil vibrates in the magnetic field and generates an electrical signal similar to the incoming sound wave. Another name for a dynamic mic is moving-coil mic, but this term is seldom used.

In a ribbon mic capsule, a thin metal foil or ribbon is suspended in a magnetic field (Figure 6.2). Sound waves vibrate the ribbon in the field and generate an electrical signal.

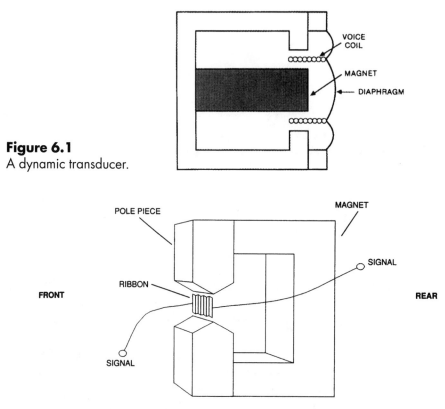

Figure 6.1
A dynamic transducer.

Figure 6.2 A ribbon transducer.

A **condenser** or **capacitor** mic capsule has a conductive diaphragm and a metal backplate placed very close together (Figure 6.3). They are charged with static electricity to form two plates of a capacitor. When sound waves strike the diaphragm, it vibrates. This varies the spacing between the plates. In turn, this varies the capacitance and generates a signal similar to the incoming sound wave.

Two types of condenser mics are true condenser and electret condenser. In a **true condenser** mic (externally biased mic), the diaphragm and backplate are charged with a voltage from a circuit built into the mic. In an **electret condenser** mic, the diaphragm and backplate are charged by an electret material, which is in the diaphragm or on the backplate. Electrets and true condensers can sound equally good, although some engineers prefer true condensers which tend to cost more.

A condenser mic needs a power supply to operate, such as a battery or phantom power supply. **Phantom power** is 12 to 48 volts DC applied to pins

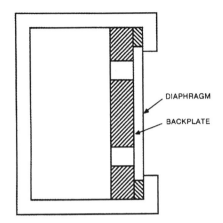

Figure 6.3
A condenser transducer.

2 and 3 of the mic connector through two equal resistors. The microphone receives phantom power and sends audio signals on the same two conductors. Many mixing consoles supply phantom power at their mic input connectors. You simply plug the mic into the mixer to power it. Dynamics and ribbons need no power supply. You can plug these types of mics into a phantom supply without damage, unless either signal conductor is accidentally shorted to the mic housing.

Because of its lower diaphragm mass and higher damping, a condenser mic responds faster than a dynamic mic to rapidly changing sound waves (transients).

Figure 6.4 shows a cutaway view of a typical dynamic mic and condenser mic.

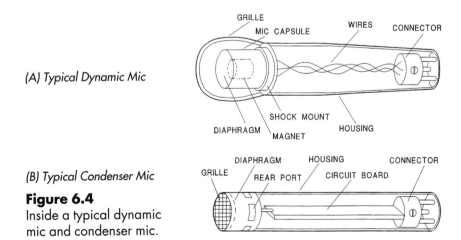

(A) Typical Dynamic Mic

(B) Typical Condenser Mic

Figure 6.4
Inside a typical dynamic mic and condenser mic.

General Traits of Each Transducer Type

Condenser

- Wide, smooth frequency response
- Detailed sound, extended highs
- Omni type has excellent low-frequency response
- Transient attacks sound sharp and clear
- Preferred for acoustic instruments, cymbals, studio vocals
- Can be miniaturized

Dynamic

- Tends to have rougher response, but still quite usable
- Rugged and reliable
- Handles heat, cold, and high humidity
- Handles high volume without distortion
- Preferred for guitar amps and drums
- If flat response, can take the "edge" off woodwinds and brass

Ribbon

- Prized for its warm, smooth tone quality
- Delicate
- Complements digital recording

There are exceptions to the tendencies listed above. Some dynamics have a smooth, wide-range frequency response. Some condensers are rugged and handle high SPLs. It depends on the specs of the particular mic.

Polar Pattern

Microphones also differ in the way they respond to sounds coming from different directions. An **omnidirectional** microphone is equally sensitive to sounds arriving from all directions. A **unidirectional** mic is most sensitive to sound arriving from one direction—in front of the mic—but softens sounds entering the sides or rear of the mic. A **bidirectional** mic is most sensitive to sounds arriving from two directions—in front of and behind the mic—but rejects sounds entering the sides.

There are three types of unidirectional patterns: cardioid, supercardioid, and hypercardioid. A mic with a **cardioid** pattern is sensitive to sounds arriving from a broad angle in front of the mic. It is about 6 dB less sensitive at the sides, and about 15 to 25 dB less sensitive in the rear. The supercardioid pattern is 8.7 dB less sensitive at the sides and has two areas of least pickup at 125 degrees away from the front. The hypercardioid pattern is 12 dB less sensitive at the sides and has two areas of least pickup at 110 degrees away from the front.

To hear how a cardioid pickup pattern works, talk into a cardioid mic from all sides while listening to its output. Your reproduced voice is loudest when you talk into the front of the mic, and softest when you talk into the rear.

The supercardioid and hypercardioid reject sound from the sides more than the cardioid—they are more directional. But they pick up more sound from the rear than the cardioid does.

A microphone's polar pattern is a graph of its sensitivity vs. the angle at which sound comes into it. The polar pattern is plotted on polar graph paper. Sensitivity is plotted as distance from the origin. Figure 6.5 shows various polar patterns.

Traits of Different Polar Patterns

Omnidirectional

- All-around pickup
- Most pickup of room reverberation
- Not much isolation unless you mike close
- Low sensitivity to pops (explosive breath sounds)
- Low handling noise
- No up-close bass boost (proximity effect)
- Extended low-frequency response in condenser mics (great for pipe organ or bass drum in an orchestra or symphonic band)
- Lower cost in general

Unidirectional (Cardioid, Supercardioid, Hypercardioid)

- Selective pickup
- Rejection of room acoustics, background noise, and leakage
- Good isolation—good separation between tracks
- Up-close bass boost (except in mics that have holes in the handle)
- Better gain-before-feedback in a sound-reinforcement system
- Coincident or near-coincident stereo miking (explained in Chapter 7)

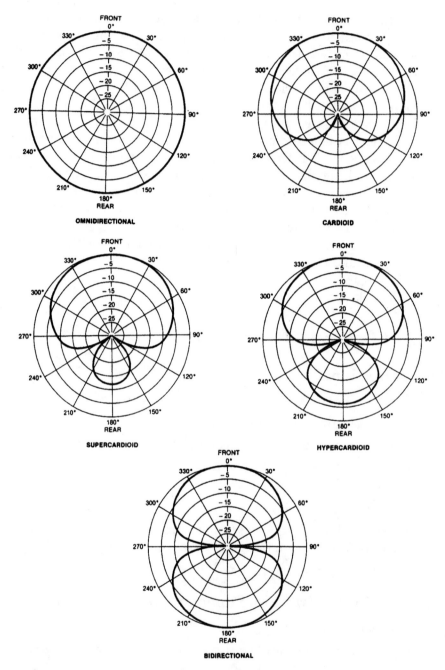

Figure 6.5 Various polar patterns. Sensitivity is plotted vs. angle of sound incidence.

Cardioid

- Broad-angle pickup of sources in front of the mic
- Maximum rejection of sound approaching the rear of the mic

Supercardioid

- Maximum difference between front hemisphere and rear hemisphere pickup (good for stage-floor miking)
- More isolation than a cardioid
- Less reverb pickup than a cardioid

Hypercardioid

- Maximum side rejection in a unidirectional mic
- Maximum isolation—maximum rejection of reverberation, leakage, feedback, and background noise

Bidirectional

- Front and rear pickup, with side sounds rejected (for across-table interviews or two-part vocal groups, for example)
- Maximum isolation of an orchestral section when miked overhead
- Blumlein stereo miking (two bidirectional mics crossed at 90 degrees)

In a good mic, the polar pattern should be about the same from 200 Hz to 10 kHz. If not, you'll hear off-axis coloration: the mic will have a different tone quality on and off axis. Small-diaphragm mics tend to have less off-axis coloration than large-diaphragm mics.

You can get either the condenser or dynamic type with any kind of polar pattern (except bidirectional dynamic). Ribbon mics are either bidirectional or hypercardioid. Some condenser mics come with switchable patterns. Note that the shape of a mic does not indicate its polar pattern.

If a mic is **end addressed**, you aim the end of the mic at the sound source. If a mic is **side addressed**, you aim the side of the mic at the sound source. Figure 6.6 shows a typical side-addressed condenser mic with switchable polar patterns.

Boundary mics that mount on a surface have a pattern that is either half-omni (hemispherical), half-supercardioid, or half-cardioid (like an apple sliced in half through its stem). The boundary mounting increases the directionality of the mic, thus reducing pickup of room acoustics.

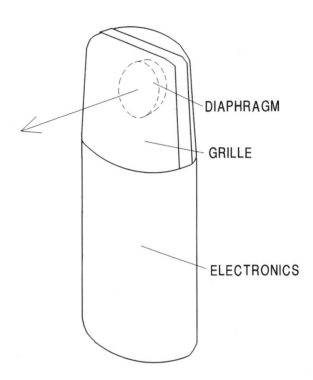

Figure 6.6
A typical multi-pattern mic
that is side-addressed.

Frequency Response

As with other audio components, a microphone's frequency response is the range of frequencies that it will reproduce at an equal level (within a tolerance, such as +/–3 dB). The following is a list of sound sources and the microphone frequency response that is adequate to record the source with high fidelity. A wider-range response works, too.

Most instruments: 80 Hz to 15 kHz

Bass instruments: 40 Hz to 9 kHz

Brass and voice: 80 Hz to 12 kHz

Piano: 40 Hz to 12 kHz

Cymbals and some percussion: 300 Hz to 15 or 20 kHz

Orchestra or symphonic band: 40 Hz to 15 kHz

If possible, use a mic with a response that rolls off below the lowest fundamental frequency of the instrument you're recording. For example, the frequency of the low-E string on an acoustic guitar is about 82 Hz. A mic used on the acoustic guitar should roll off below that frequency to avoid picking up low-frequency noise such as rumble from trucks and air conditioning. Some mics have a built-in low-cut switch for this purpose. Or you can filter out the unneeded lows at your mixer.

A **frequency-response curve** is a graph of the mic's output level in dB at various frequencies. The output level at 1 kHz is placed at the 0 dB line on the graph, and the levels at other frequencies are so many dB above or below that reference level.

The shape of the response curve suggests how the mic sounds at a certain distance from the sound source. (If the distance is not specified, it's probably 2 to 3 feet.) For example, a mic with a wide, flat response reproduces the fundamental frequencies and harmonics in the same proportion as the sound source. So a flat-response mic tends to provide accurate, natural reproduction at that distance.

A rising high end or a "presence peak" around 5 to 10 kHz sounds more crisp and articulate because it emphasizes the higher harmonics (Figure 6.7). Sometimes this type of response is called "tailored" or "contoured." It's popular for guitar amps and drums because it adds punch and emphasizes attack. Some microphones have switches that alter the frequency response.

Most unidirectional and bidirectional mics boost the bass when used within a few inches of a sound source. You've heard how the sound gets bassy when a vocalist sings right into the mic. This low-frequency boost related to close mic placement is called the **proximity effect**, and it's often plotted on the frequency-response graph. Omni mics have no proximity effect; they sound the same at any distance.

The warmth created by proximity effect adds a pleasing fullness to drums. In most recording situations, though, the proximity effect lends an unnatural boomy or bassy sound to the instrument or voice picked up by the mic. Some mics—multiple-D or variable-D types—are designed to reduce it. These types have holes or slots in the mic handle. Some mics have

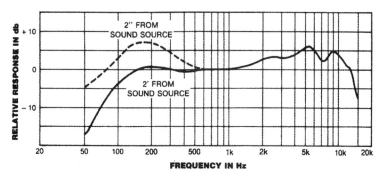

Figure 6.7 An example of the frequency response of a microphone with proximity effect and a presence peak around 5 kHz.

a bass-rolloff switch to compensate for the bass boost. Or you can roll off the excess bass with your mixer's equalizer until the sound is natural. By doing so, you also reduce low-frequency leakage picked up by the microphone.

Note that mic placement can greatly affect the recorded tone quality. A flat-response mic does not always guarantee a natural sound because mic placement has such a strong influence. Tonal effects of mic placement are covered in Chapter 7.

Impedance (Z)

This spec is the mic's effective output resistance at 1 kHz. A mic impedance between 150 and 600 ohms is low; 1000 to 4000 ohms is medium, and above 25 kilohms is high.

Always use low-impedance mics. If you do, you can run long mic cables without picking up hum or losing high frequencies. The input impedance of a mixer mic input is about 1500 ohms. If it were the same impedance as the mic, about 250 ohms, the mic would "load down" when you plug it in. Loading down a mic makes it lose level, distort, or sound thin. To prevent this, a mic input has an impedance much higher than that of the microphone. But it's still called a low-Z input.

Maximum SPL

To understand this spec, first we need to understand SPL. Sound Pressure Level is a measure of the intensity of a sound. The quietest sound we can hear, the threshold of hearing, is 0 dB SPL. Normal conversation at 1 foot measures about 70 dB SPL; painfully loud sound is above 120 dB SPL.

If the maximum SPL spec is 125 dB SPL, the mic starts to distort when the instrument being miked is putting out 125 dB SPL at the mic. A maximum SPL spec of 120 dB is good, 135 dB is very good, and 150 dB is excellent.

Dynamic mics tend not to distort, even with very loud sounds. Some condensers are just as good. Some have a pad you can switch in to prevent distortion in the mic circuitry. Since a mic pad reduces signal-to-noise ratio, use it only if the mic distorts.

Sensitivity

This spec tells how much output voltage a mic produces when driven by a certain sound pressure level (SPL). A high-sensitivity mic puts out a stronger signal (higher voltage) than a low-sensitivity mic when both are exposed to an equally loud sound.

A low-sensitivity mic needs more mixer gain than a high-sensitivity mic. More gain usually results in more noise.

When you record quiet music at a distance (classical guitar, string quartet), use a mic of high sensitivity to override mixer noise. When you record loud music or mike close, sensitivity matters little because the mic signal level is well above the mixer noise floor. That is, the signal-to-noise ratio is high. Listed below are typical sensitivity specs for three transducer types:

Condenser: 5.6 mV/Pa (high sensitivity)

Dynamic: 1.8 mV/Pa (medium sensitivity)

Ribbon or small dynamic: 1.1 mV/Pa (low sensitivity)

The louder the sound source, the higher the signal voltage the mic puts out. A very loud instrument, such as a kick drum or guitar amp, can cause a microphone to generate a signal strong enough to overload the mic preamp in your mixer. That's why most mixers have pads or input-gain controls—to prevent preamp overload from hot mic signals.

Self-Noise

Self-noise or **equivalent noise level** is the electrical noise or hiss a mic produces. It's the dB SPL of a sound source that would produce the same output voltage that the noise does.

Usually the self-noise spec is A-weighted. That means the noise was measured through a filter that makes the measurement correlate more closely with the annoyance value. The filter rolls off low and high frequencies to simulate the frequency response of the ear.

An A-weighted self-noise spec of 18 dB SPL or less is excellent (quiet); a spec around 28 dB SPL is good, and a spec around 35 dB SPL is fair—not good enough for quality recording. Because a dynamic mic has no active electronics to generate noise, it has very low self-noise (hiss) compared to a condenser mic. So most spec sheets for dynamic mics do not specify self-noise.

Signal-to-Noise Ratio

This is the difference in dB between the mic's sensitivity and its self-noise. The higher the SPL of the sound source at the mic, the higher the S/N ratio. Given an SPL of 94 dB, a S/N spec of 74 dB is excellent; 64 dB is good. The higher the S/N ratio, the cleaner (more noise-free) is the signal, and the greater is the reach of the microphone.

"Reach" is the clear pickup of quiet, distant sounds due to high S/N. Reach is not specified in data sheets because any mic can pick up a source at any distance if the source is loud enough. For example, even a cheap mic can reach several miles if the sound source is a thunderclap.

Polarity

The polarity spec relates the polarity of the electrical output signal to the acoustic input signal. The standard is "pin 2 hot." That is, the mic produces a positive voltage at pin 2 with respect to pin 3 when the sound pressure pushes the diaphragm in (positive pressure).

Be sure that your mic cables do not reverse polarity. On **both** ends of each cable, the wiring should be pin 1 shield, pin 2 red, pin 3 white or black. Or the wiring on both ends should be pin 1 shield, pin 2 white, pin 3 black. If some mic cables are correct polarity and some are reversed, and you mix their mics to mono, the bass may cancel.

Special Microphone Types

The following sections describe three types of recording mics used for special purposes: the **boundary microphone, miniature microphone,** and **stereo microphone.**

Boundary Microphone

Boundary mics are designed to be used on surfaces. Tape them to the underside of a piano lid, or tape them to the wall for pickup of room ambience. They can be used on hard baffles between instruments, or on panels to make the mics directional. A boundary mic uses a mini condenser mic capsule mounted very near a sound-reflecting plate or boundary. Due to this construction, the mic picks up direct sound and reflected sound at the same time—in-phase at all frequencies. So you get a smooth response free of phase cancellations. A conventional mic near a surface sounds colored; a boundary mic on a surface sounds natural.

One example of a boundary mic is the Crown Pressure Zone Microphone (PZM) series (Figure 6.8). The claimed benefits are a wide, smooth frequency response free of phase cancellations, excellent clarity and "reach," and the same tone quality anywhere around the mic. The polar pattern is half-omni or hemispherical. Some boundary mics have a half-cardioid or half-supercardioid polar pattern. They work great on a conference table, or near the front edge of a stage floor to pick up drama or musicals.

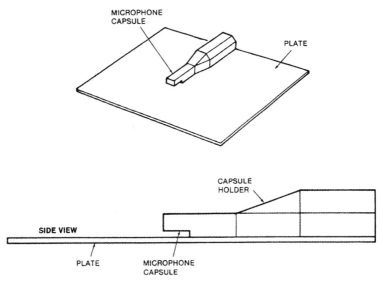

Figure 6.8   Typical PZM construction.

Miniature Microphone

Mini condenser mics can be attached to drum rims, flutes, horns, acoustic guitars, and so on. Their tone quality is about as good as larger studio microphones and the price is relatively low. With these tiny units you can mike a band in concert without cluttering the stage with boom stands (Figure 6.9). Or you can mike a whole drum set with two or three of these. Although you

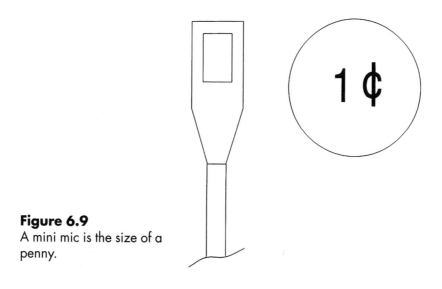

Figure 6.9
A mini mic is the size of a penny.

lose individual control of each drum in the mix, the cost is low and the sound is quite good with some bass and treble boost. Compared to large mics, mini mics tend to have more noise (hiss) in distant-miking applications.

Stereo Microphone

A stereo microphone combines two directional mic capsules in a single housing for convenient stereo recording (Figure 6.10). Simply place the mic a suitable distance and height from the sound source, and you'll get a stereo recording with little fuss.

Because there is no spacing between the mic capsules, there also is no delay or phase shift between their signals. Coincident stereo microphones are mono-compatible—the frequency response is the same in mono and stereo—because there are no phase cancellations if the two channels are combined.

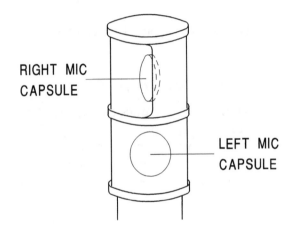

RIGHT MIC
CAPSULE

LEFT MIC
CAPSULE

Figure 6.10
A stereo microphone.

Microphone Selection

Table 6.1 is a guide to choosing a mic based on your requirements.

Table 6.1 Mic application guide

Requirements	Characteristic
Natural, smooth tone quality	Flat frequency response
Bright, present tone quality	Rising high-frequency response

Table 6.1 Mic application guide *(continued)*

Requirements	Characteristic
Extended lows	Omni condenser or dynamic with extended low-frequency response
Extended highs (detailed sound)	Condenser
Reduced "edge" or detail	Dynamic
Boosted bass up close	Single-D cardioid
Flat bass response up close	Omnidirectional, multiple-D cardioid, or single-D cardioid with bass rolloff
Reduced pickup of leakage, feedback, and room acoustics	Unidirectional, or omni up close
Enhanced pickup of room acoustics	Omnidirectional, or unidirectional farther away
Miking close to a surface, even coverage of moving sources or large sources, inconspicuous mic	Boundary mic or miniature mic
Coincident or near-coincident stereo (see Chapter 18)	Unidirectional mics or stereo mic
Extra ruggedness	Moving coil (dynamic)
Reduced handling noise	Omni, or unidirectional with shock mount
Reduced breath popping	Omni, or unidirectional with pop filter
Distortion-free pickup of very loud sounds	Condenser with high maximum SPL spec, or dynamic
Noise-free pickup of quiet sounds	Low self-noise, high sensitivity

Let's try an application example. Suppose you want to record a grand piano playing with several other instruments. You need the microphone to reduce leakage. Table 6.1 recommends a unidirectional mic or an omni mic up close. For this particular piano, you also want a natural sound, for which the table suggests a mic with a flat response. You want a detailed sound, so a condenser mic is the choice. A microphone with all these characteristics is a flat-response, unidirectional condenser mic. If you're miking close to a surface (the piano lid), a boundary mic is recommended.

Now suppose you're recording an acoustic guitar on stage, and the guitarist roams around. This is a moving sound source, for which the table recommends a mini mic attached to the guitar. Feedback and leakage are not a problem because you're miking close, so you can use an omni mic. Thus, a mini omni condenser mic is a good choice for this application.

For a home studio, a suggested first choice is a cardioid condenser mic with a flat frequency response. This type of mic is especially good for studio vocals, cymbals, percussion, and acoustic instruments. Remember that the mic needs a power supply to operate, such as a battery or phantom power supply. Your second choice of microphone for a home studio is a cardioid dynamic microphone with a presence peak in the frequency response. This type is good for drums and guitar amps.

I recommend cardioid over omni for a home studio. The cardioid pattern rejects the leakage, background noise and room reverb often found in home studios. An omni mic, however, can do that too if you mike close enough. Also, omni mics tend to provide a more natural sound at lower cost, and have no proximity effect.

Mic Accessories

There are many devices used with microphones to route their signals or to make them more useful, including **pop filters, stands** and **booms, shock mounts, cables** and **connectors, stage boxes** and **snakes,** and **splitters.**

Pop Filters

A much needed accessory for a vocalist's microphone is a pop filter or windscreen. It usually is a foam "sock" which you put over the mic.

Why is it needed? When a vocalist sings a word starting with "p," "b," or "t" sounds, a turbulent puff of air is forced from the mouth. A microphone placed close to the mouth is hit by this air puff, resulting in a thump or little explosion called a **pop.** The windscreen reduces this problem.

Some microphones have pop filters or ball-shaped grilles built in. The best type of pop filter is a nylon screen in a hoop placed a few inches from the mic. You can also reduce pop by placing the mic above or to the side of the mouth, or by using an omni mic.

Stands and Booms

Stands and booms hold the microphones and let you position them as desired. A mic stand has a heavy metal base that supports a vertical pipe. At the top of the pipe is a rotating clutch that lets you adjust the height of a smaller telescoping pipe inside the large one. The top of the small pipe has a standard 5/8-inch 27 thread, which screws into a mic stand adapter.

A boom is a long horizontal pipe that attaches to the vertical pipe. The angle and length of the boom are adjustable. The end of the boom is threaded to accept a mic stand adapter, and the opposite end is weighted to balance the weight of the microphone.

Shock Mounts

A shock mount holds a mic in a resilient suspension to isolate the mic from mechanical vibrations, such as bumps to the mic stand and floor thumps. Many mics have an internal shock mount that isolates the mic capsule from its housing; this reduces handling noise as well as stand thumps.

Cables and Connectors

Mic cables carry the electrical signal from the mic to the mixing console or tape recorder. With low-impedance mics, you can use hundreds of feet of cable without hum pickup or high-frequency loss. Some mics have a permanently attached cable for convenience and low cost; others have a connector in the handle to accept a separate mic cable. The second method is preferred for serious recording because if the cable breaks, you have to repair or replace only the cable, not the whole microphone.

Mic cables are made of one or two insulated conductors surrounded by a fine-wire mesh shield to keep out electrostatic hum. If you hear a loud buzz when you plug in a microphone, check that the shield is securely soldered in place.

After acquiring a microphone, you may need to wire its 2-conductor shielded cable to a 3-pin audio connector. Here are the solder connections:

Pin 1: Shield

Pin 2: "Hot" lead (usually red or white)

Pin 3: "Cold" lead (usually black)

If the mic output is 3-pin balanced, but your recorder or mixer mic input is an unbalanced phone jack, a different wiring is needed:

Phone-plug tip: Hot lead

Phone-plug long ground lug: Shield and cold lead

Wind your mic cables onto a large spool, which can be found in the electrical section of hardware stores. Plug the cables together as you wind them.

Snakes

It is messy and time-consuming to run mic cables from several mics all the way to a mixer. Instead, you can plug all your mics into a stage box with several connectors (Figure 2.23). The snake—a thick multi-conductor cable—carries the signals to the mixer. At the mixer end, the cable divides into several mic connectors that plug into the mixer.

Splitters

When you record a band in concert, you might need to feed each mic's signal to your recording mixer and to the band's PA and monitor mixers. A mic splitter does the job. It has one input for each microphone and two or three isolated outputs per microphone to feed each mixer.

Summary

We talked about some mic types, specs, and accessories. You should have a better idea about what kind of microphone to choose for your own applications.

Mic manufacturers are happy to send you free catalogs and application notes. Your dealers may have this literature. You can get the company addresses from them or from the World Wide Web.

Remember, you can use any microphone on any instrument if it sounds good to you. Just try it and see if you like it. To make high quality recordings, though, you need good mics with a smooth, wide-range frequency response, low noise, and low distortion.

7

MICROPHONE TECHNIQUE BASICS

Suppose you're going to mike a singer, or a sax, or a guitar. Which mic should you choose? Where should you place it?

Your mic technique has a powerful effect on the sound of your recordings. In this chapter we'll look at some general principles of miking that apply to all situations. Chapter 8 covers common mic techniques for specific instruments.

Which Mic Should I Use?

Is there a "right" mic to use on a piano, a kick drum, or a guitar amp? No. Every microphone sounds different, and you choose one that gives you the sound you want. Still, it helps to know about two main characteristics of mics that affect the sound: frequency response and polar pattern.

Most condenser mics have an extended high-frequency response—they reproduce sounds up to 15 or 20 kHz. This makes them great for cymbals, or other instruments that need a detailed sound, such as acoustic guitar, strings, piano, and voice. Dynamic moving-coil microphones have a response good enough for drums, guitar amps, horns, and woodwinds. Loud drums and guitar amps sound dull if recorded with a flat-response mic; a mic with a presence peak (a boost around 5 kHz) gives more edge or punch.

The polar pattern of a mic affects how much leakage and ambience it picks up. Leakage is unwanted sound from instruments other than the one the mic is aimed at. Ambience is the acoustics of the recording room, its early reflections and reverb. The more leakage and ambience you pick up, the more distant the instrument sounds.

An omni mic picks up more ambience and leakage than a directional mic when both are the same distance from an instrument. So an omni tends to sound more distant. To compensate, you have to mike closer with an omni.

How Many Mics?

The number of mics you need varies with what you're recording. If you want to record an overall acoustic blend of the instruments and room ambience, use just two microphones or a stereo mic (Figure 7.1). This method works great on an orchestra, symphonic band, choir, string quartet, pipe organ, small folk group, or a piano/voice recital. Stereo miking is covered in detail later in this chapter.

To record a pop-music group, you mike each instrument or instrumental section. Then you adjust the mixer volume control for each mic to control the balance between instruments (Figure 7.2).

To get the clearest sound, don't use two mics when one will do the job. Sometimes you can pick up two or more sound sources with one mic (Figure 7.3). You could mike a brass section of four players with one mic on four players, or with two mics on every two players.

Picking up more than one instrument with one mic has a problem: during mixdown, you can't adjust the balance among instruments recorded on the same track. You have to balance the instruments before recording them. Monitor the mic, and listen to see if any instrument is too quiet. If so, move it closer to the mic.

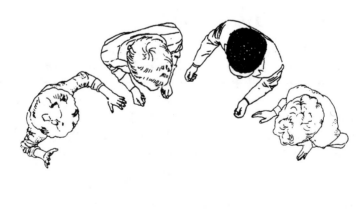

Figure 7.1 Overall miking of a musical ensemble with two distant microphones.

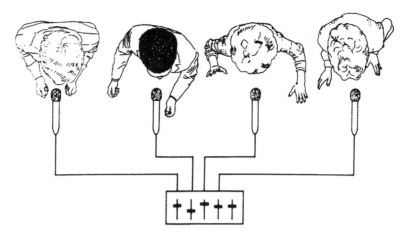

Figure 7.2 Individual miking with multiple close mics and a mixer.

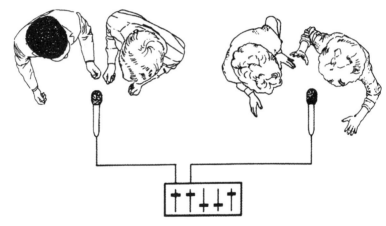

Figure 7.3 Multiple miking with several sound sources on each microphone.

How Close Should I Place the Mic?

Once you've chosen a mic for an instrument, how close should the mic be? Mike a few inches away to get a tight, present sound; mike farther away for a distant, spacious sound. (Try it to hear the effect.) The farther a mic is, the more it picks up ambience, leakage, and background noise. So mike close to reject these unwanted sounds. Mike farther away to add a live, loose, airy feel to overdubs of drums, lead-guitar solos, horns, etc.

Close miking sounds close; distant miking sounds distant. Here's why. If you put a mic close to an instrument, the sound at the mic is loud. So you need to turn up the mic gain on your mixer only a little to get a 0 VU recording level. And since the gain is low, you pick up very little reverb, leakage, and background noise (Figure 7.4 A).

If you put a mic far from an instrument, the sound at the mic is quiet. You'll need to turn up the mic gain a lot to get a 0 VU recording level. And since the gain is high, you pick up a lot of reverb, leakage, and background noise (Figure 7.4 B).

If the mic is very far away—maybe 10 feet—it's called an ambience mic. It picks up mostly room echoes and reverb. A popular mic for ambience is a boundary microphone taped to the wall. You mix it with the usual close mics to add a sense of space. Use two for stereo. When you record a live concert, you might want to place ambience mics over the audience, aiming at them from the front of the hall, to pick up the crowd reaction and the hall acoustics.

Classical music is always recorded at a distance (about 4 to 20 feet away) so that the mics will pick up reverb from the concert hall. It's a desirable part of the sound.

Reducing Leakage

Suppose you're close-miking a drum set and a piano at the same time (Figure 7.5). When you listen to the drum mics alone, you hear a close, clear sound. But when you mix in the piano mic, that nice, tight drum sound degrades into a distant, muddy sound. That's because the drum sound leaked into the piano mic. The piano mic picked up a distant drum sound from across the room.

(A) A close microphone picks up mainly direct sound, resulting in a close sound quality.

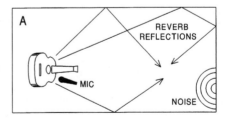

(B) A distant microphone picks up mainly reflected sound, resulting in a distant sound quality.

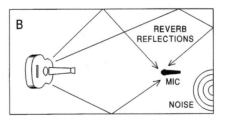

Figure 7.4

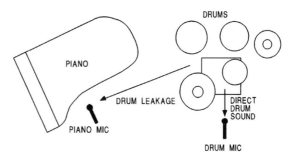

Figure 7.5
Example of leakage. The piano mic picks up leakage from the drums—changing the close drum sound to distant.

There are many ways to reduce leakage:

- Mike each instrument closely. That way the sound level at each mic is high. Then you can turn down the mixer gain of each mic, which reduces leakage at the same time.
- Overdub each instrument one at a time.
- Record direct.
- Filter out frequencies above and below the range of each instrument.
- Use directional mics (cardioid, etc.) instead of omni's.
- Record in a large, fairly dead studio. In such a room, leakage reflected from the walls is weak.
- Put portable walls (goboes) between instruments.

Don't Mike Too Close

Miking too close can color the recorded tone quality of an instrument. If you mike very close, you might hear a bassy or honky tone instead of a natural sound.

Why? Musical instruments are designed to sound best at a distance, at least 1 1/2 feet away. The sound of an instrument needs some space to develop. A mic placed a foot or two away tends to pick up a well balanced, natural sound. That is, it picks up a blend of all the parts of the instrument that contribute to its character or timbre.

Think of a musical instrument as a loudspeaker with a woofer, midrange, and tweeter. If you place a mic a few feet away, it will pick up the sound of the loudspeaker accurately. But if you place the mic close to the woofer, the sound will be bassy. Similarly, if you mike close to an instrument, you emphasize the part of the instrument that the microphone is near. The tone quality picked up very close may not reflect the tone quality of the entire instrument.

Suppose you place a mic next to the sound hole of an acoustic guitar, which resonates around 80 to 100 Hz. A microphone placed there hears this bassy resonance, giving a boomy recorded timbre that does not exist at a greater miking distance. To make the guitar sound more natural when miked close to the sound hole, you need to roll off the excess bass on your mixer, or use a mic with a bass rolloff in its frequency response.

The sax projects highs from the bell, but projects mids and lows from the tone holes. So if you mike close to the bell, you miss the warmth and body from the tone holes. All that's left at the bell is a harsh tone quality. You might like that sound, but if not, move the mic out and up to pick up the entire instrument. If leakage forces you to mike close, change the mic or use EQ.

Usually, you get a natural sound if you put the mic as far from the source as the source is big. That way, the mic picks up all the sound-radiating parts of the instrument about equally. For example, if the body of an acoustic guitar is 18 inches long, place the mic 18 inches away to get a natural tonal balance. If this sounds too distant or hollow, move in a little closer.

Where Should I Place the Mic?

Suppose you have a mic placed a certain distance from an instrument. If you move the mic left, right, up, or down, you change the recorded tone quality. In one spot, the instrument might sound bassy; in another spot, it might sound natural, and so on. So, to find a good position, simply place the mic in different locations—and monitor the results—until you find one that sounds good to you.

Here's another way to do the same thing. Close one ear with your finger, listen to the instrument with the other ear, and move around until you find a spot that sounds good. Put the mic there. Then make a recording and see if it sounds the same as what you heard live. Don't try this with kick drums or screaming guitar amps!

Why does moving the mic change the tone quality? A musical instrument radiates a different tone quality in each direction. Also, each part of the instrument produces a different tone quality. For example, Figure 7.6 shows the tonal balances picked up at various spots near a guitar.

Other instruments work the same way. A trumpet radiates strong highs directly out of the bell, but does not project them to the sides. So a trumpet sounds bright when miked on-axis to the bell and sounds more natural or mellow when miked off to one side. A grand piano miked one foot over the middle strings sounds fairly natural; under the soundboard sounds bassy and dull, and in a sound hole sounds constricted.

It pays to experiment with all sorts of mic positions until you find a sound you like. There is no one right way to place the mics because you place them to get the tonal balance you want.

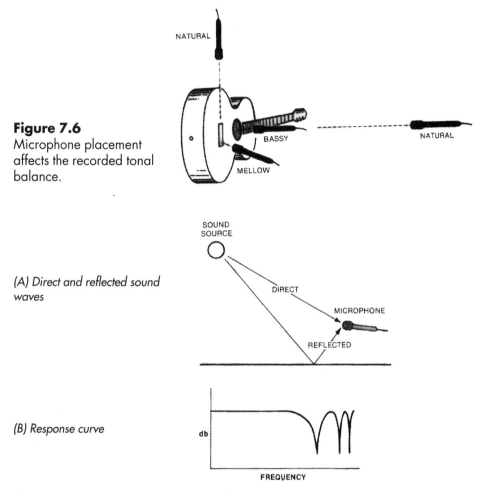

Figure 7.6
Microphone placement affects the recorded tonal balance.

NATURAL

BASSY

MELLOW

NATURAL

(A) Direct and reflected sound waves

SOUND SOURCE

DIRECT

MICROPHONE

REFLECTED

(B) Response curve

db

FREQUENCY

Figure 7.7 A mic placed near a surface picks up direct sound and delayed reflections, giving a comb-filter frequency response.

On-Surface Techniques

Sometimes you're forced to place a mic near a hard reflecting surface. Examples: Recording drama or opera with the mics near the stage floor. Recording an instrument that has hard surfaces around it. Recording a piano with the mic close to the lid. In these cases, you'll often pick up an unnatural, filtered tone quality.

Here's why. Sound travels to the microphone via two paths: directly from the sound source, and reflected off the nearby surface (Figure 7.7). Because of its longer travel path, the reflected sound is delayed compared

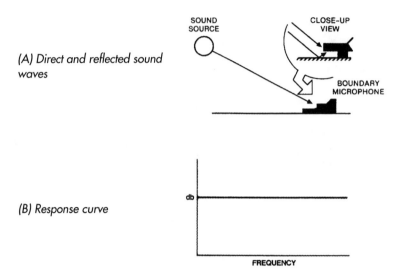

(A) Direct and reflected sound waves

(B) Response curve

Figure 7.8   A boundary mic on a surface picks up direct and reflected sounds in phase.

to the direct sound. The direct and delayed sound waves combine at the mic, which causes phase cancellations of various frequencies. The series of peaks and dips in the response is called a comb-filter effect, and it sounds like mild flanging.

Boundary mics solve the problem. In a boundary mic, the diaphragm is very close to the reflecting surface so that there is no delay in the reflected sound. Direct and reflected sounds add in-phase over the audible range of frequencies, resulting in a flat response (Figure 7.8).

You might tape an omni boundary mic to the underside of a piano lid, to a hard-surfaced panel, or to a wall for ambience pickup. A unidirectional boundary mic works great on a stage floor to pick up drama. A group of these mics will clearly pick up people at a conference table.

The Three-to-One Rule

Let's say you're miking several instruments, each with its own mic. If you place the musicians too close together, the sound will be blurred. But if you spread out the musicians and mike them close, the sound will be clearer.

Specifically, try to space the mics at least three times the mic-to-source distance (as in Figure 7.9). This is called the 3:1 rule. For example, if two mics are each placed 1 foot from their sound sources, the mics should be at least 3 feet apart. This will prevent the blurred, colored sound caused by

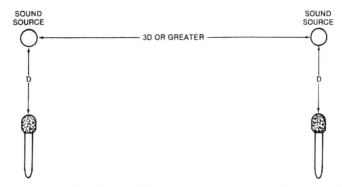

Figure 7.9 The 3:1 rule of microphone placement avoids phase interference between microphone signals.

phase cancellations between mics. The mics can be closer than 3:1 if you use two cardioid mics aiming in opposite directions.

Suppose you're recording a singer/guitarist. There's a mic on the singer and a mic on the acoustic guitar. The vocal picked up by the guitar mic is delayed because the vocal sound travels a longer path to that mic. The two vocal signals in the mix—direct and delayed—interfere with each other and make a hollow sound.

Try these solutions:

- Mike the voice and guitar very close. Roll off the excess bass with your mixer's EQ.
- Place two bidirectional mics so their grilles touch. This gets rid of any delay between their signals. Aim the "dead" side of the vocal mic at the guitar; aim the dead side of the guitar mic at the mouth.
- Use just one mic midway between the mouth and guitar. Adjust the balance by changing the mic's height.

Off-Axis Coloration

Some mics have off-axis coloration: a dull or colored effect on sound sources that are not right in front of the mic. Try to aim the mic at sound sources that put out high frequencies, such as cymbals. When you pick up a large source like an orchestra, use a mic that has the same response over a wide angle. Such a mic has similar polar patterns at middle and high frequencies. Most large-diaphragm mics have more off-axis coloration than smaller mics (3/4-inch or under).

Stereo Mic Techniques

Stereo mic techniques capture the sound of a musical group as a whole, using only two or three microphones. When you play back a stereo recording, you hear phantom images of the instruments in various spots between the speakers. These image locations—left to right, front to back—correspond to the instrument locations during the recording session.

Stereo miking is the preferred way to record classical-music ensembles and soloists. In the studio, you can stereo-mike a piano, drum set cymbals, vibraphone, harmony singers, or other large sound sources.

Goals of Stereo Miking

One goal is accurate localization. That is, instruments in the center of the group are reproduced midway between the two speakers. Instruments at the sides of the group are heard from the left or right speaker. Instruments halfway to one side are heard halfway to one side, and so on.

Figure 7.10 shows three stereo localization effects. Figure 7.10 A shows some instrument positions in an orchestra: left, left-center, center, right-center, right. In Figure 7.10 B, the reproduced images of these instruments are accurately localized between the speakers. The stereo spread, or stage

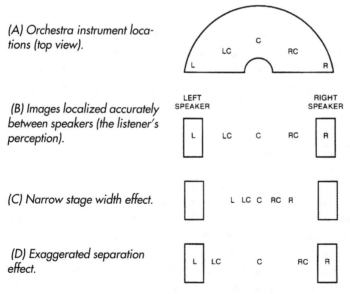

(A) Orchestra instrument locations (top view).

(B) Images localized accurately between speakers (the listener's perception).

(C) Narrow stage width effect.

(D) Exaggerated separation effect.

Figure 7.10 Stereo localization effects.

width, extends from speaker to speaker. (You might want to record a string quartet with a narrower spread.)

If you space or angle the mics too close together, you get a narrow stage width (Figure 7.10 C). If you space or angle the mics too far apart, you hear exaggerated separation (Figure 7.10 D). That is, instruments halfway to one side are heard near the left or right speaker.

To judge stereo effects, you have to sit exactly between your monitor speakers (the same distance from each). Sit as far from the speakers as the spacing between them. Then the speakers appear to be 60 degrees apart. This is about the same angle an orchestra fills when viewed from a typical ideal seat in the audience (say, tenth row center). If you sit off-center, the images shift toward the side on which you're sitting and are less sharp.

Types of Stereo Mic Techniques

To make a stereo recording, you use one of these basic techniques:

- Coincident pair (XY or MS)
- Spaced pair (AB)
- Near-coincident pair (ORTF, etc.)
- Baffled omni pair (sphere, OSS, SASS, PZM wedge, etc.)

Let's look at each technique.

Coincident Pair

With this method, you mount two directional mics with grilles touching, diaphragms one above the other, and angled apart (Figure 7.11). For example, mount two cardioid mics with one grille above the other, and angle them 120 degrees apart. You can use other patterns too: supercardioid, hypercardioid, or bidirectional. The wider the angle between mics, the wider the stereo spread.

How does this technique make images you can localize? A directional mic is most sensitive to sounds in front of the mic (on-axis) and progressively less sensitive to sounds arriving off-axis. That is, a directional mic puts out a high-level signal from the sound source it's aimed at, and produces lower-level signals from other sound sources.

The coincident pair uses two directional mics that are angled symmetrically from the center line (Figure 7.11). Instruments in the center of the group make the same signal from each mic. During playback, you hear a phantom image of the center instruments midway between your speakers. That's because identical signals in each channel produce an image in the center.

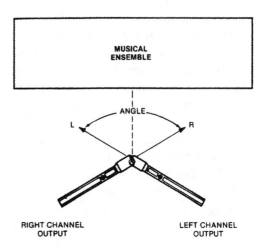

Figure 7.11
Coincident-pair technique.

If an instrument is off-center to the right, it is more on-axis to the right-aiming mic than to the left-aiming mic. So the right mic will produce a higher-level signal than the left mic. During playback of this recording, the right speaker will play at a higher level than the left speaker. This reproduces the image off-center to the right—where the instrument was during recording.

The coincident pair codes instrument positions into level differences between channels. During playback, the brain decodes these level differences back into corresponding image locations. A pan pot in a mixing console works on the same principle. If one channel is 15 to 20 dB louder than the other, the image shifts all the way to the louder speaker.

Suppose we want the right side of the orchestra to be reproduced at the right speaker. That means, the far-right musicians must produce a signal level 20 dB higher from the right mic than from the left mic. This happens when the mics are angled far enough apart. The correct angle depends on the polar pattern. Instruments part-way off center produce interchannel level differences less than 20 dB, so you hear them part-way off center.

Listening tests have shown that coincident cardioid mics tend to reproduce the musical group with a narrow stereo spread. That is, the group does not spread all the way between speakers.

A coincident-pair method with excellent localization is the Blumlein array. It uses two bidirectional mics angled 90 degrees apart and facing the left and right sides of the group.

A special form of the coincident-pair technique is Mid-Side or MS (Figure 7.12). In this method, a cardioid or omni mic faces the middle of the orchestra. A matrix circuit sums and differences the cardioid mic with a bidirectional mic aiming to the sides. This produces left- and right-channel signals. You can remote-control the stereo spread by changing the ratio of

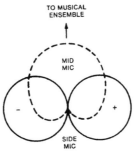

TO MUSICAL
ENSEMBLE

MID
MIC

−   +

SIDE
MIC

RIGHT CHANNEL = MID + SIDE
LEFT CHANNEL = MID − SIDE

Figure 7.12
Mid-Side (MS) Technique.

the mid signal to the side signal. This remote control is useful at live concerts, where you can't physically adjust the mics during the concert. MS
localization can be accurate.

To make coincident recordings sound more spacious, boost the bass
dB (+2 dB at 600 Hz) in the L-R or side signal.

A recording made with coincident mics is mono-compatible. That is,
the frequency response is the same in mono or stereo. Since the mics
occupy almost the same point in space, there is no time or phase difference
between their signals. And when you combine them to mono, there are no
phase cancellations to degrade the frequency response. If you expect that
your recordings will be heard in mono (say, on TV), then you'll probably
want to use coincident methods.

Spaced Pair

Here, you mount two identical mics several feet apart and aim them
straight ahead (Figure 7.13). The mics can have any polar pattern, but omni
is most popular for this method. The greater the spacing between mics, the
greater the stereo spread.

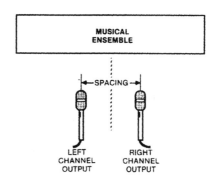

MUSICAL
ENSEMBLE

◄─ SPACING ─►

Figure 7.13
Spaced-pair technique.

LEFT
CHANNEL
OUTPUT

RIGHT
CHANNEL
OUTPUT

How does this method work? Instruments in the center of the group make the same signal from each mic. When you play back this recording, you hear a phantom image of the center instruments midway between your speakers.

If an instrument is off-center, it is closer to one mic than the other, so its sound reaches the closer microphone before it reaches the other one. Both mics make about the same signal, except that one mic signal is delayed compared with the other.

If you send a signal to two speakers with one channel delayed, the sound image shifts off center. With a spaced-pair recording, off-center instruments produce a delay in one mic channel, so they are reproduced off center.

The spaced pair codes instrument positions into time differences between channels. During playback, the brain decodes these time differences back into corresponding image locations.

A delay of 1.2 msec is enough to shift an image all the way to one speaker. You can use this fact when you set up the mics. Suppose you want to hear the right side of the orchestra from the right speaker. The sound from the right-side musicians must reach the right mic about 1.2 msec before the left mic. To make this happen, space the mics about 2 to 3 feet apart. This spacing makes the correct delay to place right-side instruments at the right speaker. Instruments part-way off center make interchannel delays less than 1.2 msec, so they are reproduced part-way off center.

If the spacing between mics is, say, 12 feet, then instruments that are slightly off center produce delays between channels that are greater than 1 msec. This places their images at the left or right speaker. I call this "exaggerated separation" or a "ping pong" effect (Figure 7.10D).

On the other hand, if the mics are too close together, the delays produced will be too small to provide much stereo spread. Also, the mics will tend to emphasize instruments in the center because the mics are closest to them.

To record a good musical balance of an orchestra, you need to space the mics about 10 or 12 feet apart. But then you get too much separation. You could place a third mic midway between the outer pair and mix its output to both channels. That way, you pick up a good balance, and you hear an accurate stereo spread.

The spaced-pair method tends to make off-center images unfocused or hard to localize. Why? Spaced-pair recordings have time differences between channels. Stereo images produced solely by time differences are unfocused. You still hear the center instruments clearly in the center, but off-center instruments are hard to pinpoint. Spaced-pair miking is a good choice if you want the sonic images to be diffuse or blended, instead of sharply focused.

Another flaw of spaced mics: If you mix both mics to mono, you may get phase cancellations of various frequencies. This may or may not be audible.

Spaced mics, however, give a "warm" sense of ambience, in which the concert hall reverb seems to surround the instruments and, sometimes, the listener. Here's why: The two channels of recorded reverb are incoherent; that is, they have random phase relationships. Incoherent signals from stereo speakers sound diffuse and spacious. Since spaced mics pick up reverb incoherently, it sounds diffuse and spacious. The simulated spaciousness caused by the phasiness is not necessarily realistic, but it is pleasant to many listeners.

Another advantage of the spaced pair is that you can use omni mics. An omni condenser mic has deeper bass than a uni condenser mic.

Near-Coincident Pair

In this method, you angle apart two directional mics, and space their grilles a few inches apart horizontally (Figure 7.14). Even a few inches of spacing increases the stereo spread and adds a sense of ambient warmth or air to the recording. The greater the angle or spacing between mics, the greater the stereo spread.

How does this method work? Angling directional mics produces level differences between channels. Spacing mics produces time differences. The level differences and time differences combine to create the stereo effect.

If the angling or spacing is too great, you get exaggerated separation. If the angling or spacing is too small, you'll hear a narrow stereo spread.

A common near-coincident method is the ORTF system, which uses two cardioids angled 110 degrees apart and spaced 7 inches (17 cm) horizontally. Usually this method gives accurate localization. That is, instruments at the sides of the orchestra are reproduced at or very near the speakers, and instruments half-way to one side are reproduced about half-way to one side.

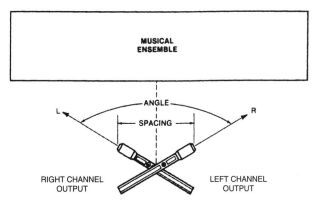

Figure 7.14 Near-coincident pair technique.

Baffled Omni Pair

This method uses two omni mics, usually ear-spaced, and separated by either a hard or soft baffle (Figure 7.15). To create stereo, it uses time differences at low frequencies and level differences at high frequencies. The spacing between mics creates time differences. The baffle creates a sound shadow (reduced high frequencies) at the mic farthest from the source. Between the two channels, there are spectral differences—differences in frequency response.

Comparing Techniques

Coincident pair:

- Uses two directional mics angled apart with grilles touching.
- Level differences between channels produce the stereo effect.
- Images are sharp.
- Stereo spread ranges from narrow to accurate.
- Signals are mono compatible.

Spaced pair:

- Uses two mics spaced several feet apart, aiming straight ahead.
- Time differences between channels produce the stereo effect.
- Off-center images are diffuse.
- Stereo spread tends to be exaggerated unless a third center mic is used, or unless spacing is under 2 to 3 feet.
- Provides a warm sense of ambience.

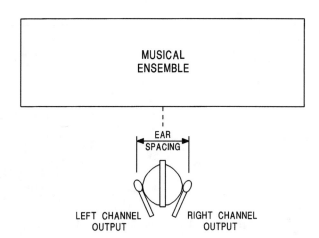

Figure 7.15
Baffled-omni technique.

- Tends not to be mono compatible.
- Good low-frequency response if you use omni condensers.

Near-coincident pair:

- Uses two directional mics angled apart and spaced a few inches apart horizontally.
- Level and time differences between channels produce the stereo effect.
- Images are sharp.
- Stereo spread tends to be accurate.
- Provides a greater sense of air than coincident methods.
- Tends not to be mono compatible.

Baffled omni pair:

- Uses two omni mics, usually ear-spaced, with a baffle between them.
- Level and time differences (spectral differences) produce the stereo effect.
- Images are sharp.
- Stereo spread tends to be accurate.
- Good low-frequency response.
- Good imaging with headphones.
- Provides more air than coincident methods.
- Tends not to be mono compatible.

Hardware

A handy device is a stereo mic adapter or stereo bar. It mounts two mics on a single stand, and lets you adjust the angle and spacing. You might prefer to use a stereo mic instead of two mics. It has two mic capsules in a single housing for convenience.

How to Test Imaging

Here's a way to check the stereo imaging of a mic technique.

1. Set up the stereo mic array in front of a stage.
2. Record yourself speaking from various locations on stage where the instruments will be—center, half-right, far right, half-left, far left. Announce your position.
3. Play back the tape over speakers.

You'll hear how accurately the technique translated your positions, and you'll hear how sharp the images are.

We looked at several mic arrays to record in stereo. Each has its pros and cons. Which method you choose depends on the sonic compromises you're willing to make.

Surround Miking

You may want to mike an ensemble for reproduction over a 5.1 playback system. Start with the usual stereo pair spaced a little wider than usual. Add a center mic aiming straight ahead to feed the center channel. Also add a rear-aiming stereo pair to pick up the hall reverb, which will feed the rear surround speakers. You can place the rear pair of mics immediately behind the front pair, or up to 25 feet behind the front pair. Record all mics on an MDM.

8

MICROPHONE TECHNIQUES

This chapter describes some ways to select and place mics for musical instruments and vocals. These techniques are popular, but they're just suggested starting points. Feel free to experiment.

Before you mike an instrument, listen to it live in the studio, so you know what sound you're starting with. You might want to duplicate that sound through your monitor speakers.

Electric Guitar

Let's start by looking at the chain of guitar, effects, amplifier, and speaker. At each point in the chain where you record, you'll get a different sound (Figure 8.1).

1. The electric guitar puts out an electrical signal that sounds clean and clear.

2. This signal might go through some effects boxes, such as distortion, wah wah, compression, chorus, or stereo effects.

3. Then the signal goes through a guitar amp, which boosts the signal and adds distortion. At the amplifier output (preamp out or external speaker jack), the sound is very bright and edgy.

4. The distorted amp signal is played by the speaker in the amp. Since the speaker rolls off above 4 kHz, it takes the edge off the distortion and makes it more pleasant.

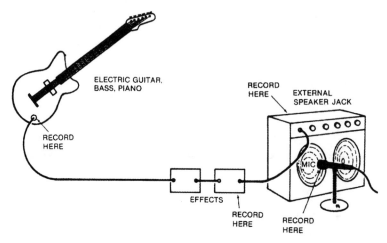

Figure 8.1 Recording an electric guitar.

You can record the electric guitar in many ways (Figure 8.1):

- With a mic in front of the guitar amp
- With a direct box
- Both miked and direct
- Through a signal processor

The song you're recording will tell you what method it wants. Just mike the amp when you want a rough, raw sound with tube distortion and speaker coloration. Rock 'n' Roll or Heavy Metal usually sound best with a miked amp. If you record through a direct box, the sound is clean and clear, with crisp highs and deep lows. That might work for quiet jazz or R&B. Use whatever sounds right for the particular song you're recording.

First, try to kill any hum you hear from the guitar amp. Turn up the guitar's volume and treble controls so that the guitar signal overrides hum and noise picked up by the guitar cable. Ask the guitarist to move around, or rotate, to find a spot in the room where hum disappears. Flip the polarity switch on the amp to the lowest hum position. To remove buzzes between guitar notes, try a noise gate, or ask the player to keep their hands on the strings.

Miking the Amp

Small practice amplifiers tend to be better for recording than large, noisy stage amps. If you use a small one, place it on a chair to avoid picking up sound reflections from the floor (unless you like that effect).

A common mic for the guitar amp is a cardioid dynamic type with a "presence peak" in its frequency response (a boost around 5 kHz). The cardioid pattern reduces leakage (off-mic sounds from other instruments). The dynamic type handles loud sounds without distorting, and the presence peak adds "bite." Of course, you can use any mic that sounds good to you.

As a starting point, try miking the amp about an inch from a speaker cone, slightly off-center—where the cone meets the dome. The closer you mike the amp, the bassier the tone. The farther off-center the mic is, the duller the tone.

Often, distant miking sounds great when you overdub a lead guitar solo played through a stack of speakers in a live room. Try a boundary mic on the floor or on the wall, several feet away.

Recording Direct

Now let's look at recording direct (also known as direct injection or DI). The electric guitar produces an electrical signal which you can plug into your mixer. You bypass the mic and guitar amp, so the sound is clean and clear. Just remember that amp distortion is desirable in some songs.

Mixer mic inputs tend to have an impedance (Z) around 1500 ohms. But a guitar pickup is several thousand ohms. So if you plug a high-Z electric guitar directly into a mic input, the input will load down the pickup and give a thin sound.

To get around this loading problem, use a direct box between the guitar and your mixer (Figure 8.2). The DI box has a high-Z input and low-Z output, thanks to a built-in transformer or circuit.

Some direct boxes let you record off the amp's external-speaker jack to pick up distortion. The sound at that jack is really annoying—edgy and sizzly. To compensate, some DI boxes include a circuit which filters the highs out of that signal to simulate the effect of the speaker.

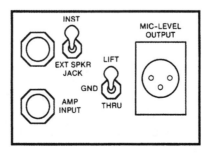

Figure 8.2
Typical direct box.

The direct box should have a ground-lift switch to prevent ground loops and hum. Set it to the position where you monitor the least hum. You might try a mix of direct sound and miked sound.

Electric Guitar Effects

If you want to record the guitarist's special effects, connect the output of the effects boxes into the direct-box input. Many players have a rack of signal processors that creates their unique sound, and they just give you their direct feed. Be open to their suggestions, and diplomatic about changing the sound. If they are studio players, they often have a better handle on effects than you might as the engineer.

You might want a "fat" or spacious lead-guitar sound. Here are some ways to get it:

- Send the guitar signal through a digital delay set to 10 to 20 milliseconds. Pan guitar left, delay right. Adjust levels for nearly equal loudness from each speaker. (Watch out for phase cancellations in mono.)
- Send the guitar signal through a pitch-shifter, set for just a few cents of pitch bending. Pan guitar left, pitch-shifted guitar right.
- Record two guitarists playing identical parts, and pan them left and right. This works great for rhythm-guitar parts in Heavy Metal.
- Double the guitar. Have the player re-record the same part on an unused track while listening to the original part. Pan the original part left and pan the new part right.
- Add stereo reverb or stereo chorus.

Some guitar processors add many effects to an electric guitar, such as distortion, EQ, chorus, and compression. You simply plug the electric guitar into the processor, adjust it for the desired sound, and record the signal direct. You wind up with a fully produced sound with a minimum of effort.

Electric Bass

BWAM, dik diddy bum. Do your bass tracks sound that clear? Or are they more muddy, like, "Bwuh, dip dubba duh"? Here how to record the electric bass so it's clean and easy to hear in a mix.

As always, first you work on the sound of the instrument itself. Put on new strings if the old ones sound dull. Adjust the pickup screws (if any) for equal output from each string. Also adjust the intonation and tuning.

Usually, you record the electric bass direct for the cleanest possible sound. A direct pickup gives deeper lows than a miked amp, but the amp gives more midrange punch. You might want to mix the direct and miked sound. Use a condenser or dynamic mic with a good low-frequency response, placed 1 to 6 inches from the speaker.

When mixing a direct signal and a mic signal, make sure they are in-phase with each other. To do this, set them to equal levels and reverse the polarity of the direct signal or the mic signal. The polarity that gives the most bass is correct.

Have the musician play some scales to see if any notes are louder than the rest. You might set a parametric equalizer to soften these notes, or use a compressor.

The bass guitar should be fairly constant in level (a dynamic range of about 6 dB) to be audible throughout the song, and to avoid saturating the tape on loud peaks. To do this, run the bass guitar through a compressor. Set the compression ratio to about 4:1; set the attack time fairly slow (8 to 20 milliseconds) to preserve the attack transient, and set the release time fairly fast (1/4 to 1/2 second). If the release time is too fast, you get harmonic distortion.

EQ can make the bass guitar more clear. Try cutting around 200 to 400 Hz. Don't boost the extreme lows. A boost at 1500 to 5000 Hz adds edge or slap, and a boost at 700 to 800 Hz adds "growl" and harmonic clarity.

Here are some ways to make the bass sound clean and well defined:

- Record the bass direct.
- Use no reverb or echo on the bass.
- Have the bass player turn down the bass amp in the studio, just loud enough to play adequately. This reduces muddy-sounding bass leakage into other mics.
- Better yet, don't use the amp. Instead, have the musicians monitor the bass (and each other) with headphones.
- Have the bass player try new strings or a different guitar. Some guitars are better for recording than others.
- Ask the bass player to use the treble pickup near the bridge.
- Be sure to record the bass with enough edge or harmonics so the bass will be audible on small, cheap speakers.
- Try a bass-guitar signal processor such as the Bass Rockman.

If the bass part is full and sustained, it's probably best to go for a mellow sound without much pluck. Let the kick drum define the rhythmic pattern. But if both the bass and kick are rhythmic and work independently, then you should hear the plucks. Listen to the song first, then get a bass sound appropriate for the music. A sharp, twangy timbre is not always right for a ballad; a full, round tone will get lost in a fusion piece.

Often, a musician plays bass lines on a synth or sound module. The module is triggered from a keyboard, a sequencer, or a bass guitar plugged into a pitch-to-MIDI converter. Connect the module output to your mixer line in.

Two effects boxes for the electric bass are the octave box and the bass chorus. The octave box takes the bass signal and drops it an octave in pitch by dividing the bass signal's fundamental frequency in two. You put 82 Hz in; you get 41 Hz out. This gives an extra deep, growly sound. So does a 5-string bass.

A bass chorus gives a wavy, shimmering effect. Like a conventional chorus box, it detunes the signal and combines the detuned signal with the direct signal. Also, it removes the lowest frequencies from the detuned signal, so that the chorus effect doesn't thin out the sound.

Synthesizer, Drum Machine, Electric Piano

For the most clarity, you usually DI a synth, MIDI sound module, drum machine, or electric piano. Set the volume on the instrument about 3/4 up to get a strong signal. Try to get the sound you want from patch settings rather than EQ.

Plug the instrument into a phone jack input on your mixer, or use a direct box. If you connect to a phone jack and hear hum, you probably have a ground loop. Here are some fixes:

- Power your mixer and the instrument from the same outlet strip. If necessary, use a thick extension cord between the outlet strip and the instrument.

- Use a direct box instead of a guitar cord, and set the ground-lift switch to the position where you monitor the least hum.

- To reduce hum from a low-cost synth, use battery power instead of an AC adapter.

A synth can sound dry and sterile. To get a livelier, funkier sound, you might run the synth signal into a power amp and speakers, and mike the speakers a few feet away.

If the keyboard player has several keyboards plugged into a keyboard mixer, you may want to record a premixed signal from that mixer's output. Record both outputs of stereo keyboards.

You don't need to record MIDI instruments to multitrack. Just use your sequencer and sync it to tape (see Chapter 16).

Leslie Organ Speaker

This glorious device has a rotating dual-horn on top for highs and a woofer on the bottom for lows. Only one horn of the two makes sound; the other is for weight balance. The swirling, grungy sound comes from the phasiness and Doppler effect of the rotating horn, and from the distorted tube electronics that drive the speaker. Here are a few ways to record it (Figure 8.3):

- In mono: Mic the top and bottom separately, 3 inches to 2 feet away. Aim the mics into the louvers. In the top mic's signal, roll off the lows below 150 Hz.

- In stereo: Record the rotating horn in stereo with a mic on either side. Problem: The horn will sound like it's rotating twice as fast, because the mics will pick up the horn twice per rotation.

- In stereo: Record the top horn with a stereo mic or a pair of mics out front. Put a mic with a good low end on the bottom speaker, and pan it to center.

When you record the Leslie, watch out for wind noise from the rotating horn, and buzz from the motor. Mike farther away if you monitor these noises.

Drum Set

The first step is to make the drums sound good live in the studio. If the set sounds poor, you'll have a hard time making it sound great in the control room!

You might put the drum set on a riser 1 1/2 feet high to reduce bass leakage and to provide better eye contact between the drummer and the rest of the band. To reduce drum leakage into other mics, you could surround the set with goboes: padded thick-wood panels about 4 feet tall. For more isolation, place the set in a drum booth, a small padded room with windows. It's also common to overdub the set in a live room.

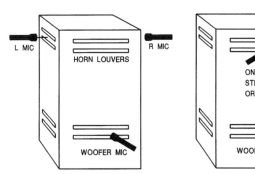

Figure 8.3
Miking a Leslie organ speaker.

Tuning

One secret of creating a good drum sound lies in careful tuning. It's easier to record a killer sound if you tune the set to sound right in the studio before miking it.

First let's consider drum heads. Plain heads have the most ring or sustain, while heads with sound dots or hydraulic heads dampen the ring. Thin heads are best for recording because they have a crisp attack and long sustain. Old heads become dull, so use new heads.

When you tune the toms, first take off the heads and remove the damping mechanism which can rattle. Put just the top head on and hand-tighten the lugs. Then, using a drum key, tighten opposite pairs of lugs one at a time, one full turn. After you tighten all the lugs, repeat the process, tightening one-half turn. Then press on the head to stretch it. Continue tightening a half-turn at a time until you reach the pitch you want. You'll get the most pleasing tone when the heads are tuned within the range of the shell resonance.

To reduce ugly overtones, try to keep the tension the same around the head. While touching the center of the head, tap with a drumstick on the head near each lug. Adjust tension for equal pitch around the drum. If you want a downward pitch bend when the head is struck, loosen one lug.

Keep the bottom head off the drum for the most projection and the broadest range of tuning. In this case, pack the bottom lugs with felt to prevent rattles. But you may want to add the bottom head for extra control of the sound. Projection is best if the bottom head is tighter than the top head, say, tuned a fourth above the top head. There will be a muted attack, an "open" tone, and some note bending. If you tune the bottom head looser than the top, the tone will be more "closed," with good attack.

With the kick drum (bass drum), a loose head gives lots of slap and attack, and almost no tone. The opposite is true for a tight head. Tune the head to complement the style of music. For more attack, use a hard beater.

Tune the snare drum with the snares off. A loose batter head or top head gives a deep, fat sound. A tight batter head sounds bright and crisp. With the snare head or bottom head loose, the tone is deep with little snare buzz, while a tight snare head yields a crisp snare response. Set the snare tension just to the point where the snare wires begin to "choke" the sound, then back off a little.

Damping and Noise Prevention

Usually the heads should ring without any damping. But if the toms or snare drum ring too much, put some plastic damping rings on them. Or tape some gauze pads, tissues, or folded handkerchiefs to the edge of the heads. Put masking tape on three sides of the pad so that the untaped edge is free to vibrate and dampen the head motion. Don't overdo the damping, or the drum set will sound like cardboard boxes.

Oil the kick-drum pedal to prevent squeaks. Tape rattling hardware in place.

Sometimes a snare drum buzzes in sympathetic vibration with a bass-guitar passage or a tom-tom fill. Try to control the buzz by wedging a thick cotton wad between the snares and the drum stand. Or tune the snare to a different pitch than the toms.

Drum Miking

Now you're ready to mike the set. For a tight sound, place a mic near each drum head. For a more open, airy sound, use fewer mics or mix in some room mics placed several feet away. Typical room mics are omni condensers or boundary mics. Figure 8.4 shows typical mic placements for a rock drum set. Let's look at each part of the kit.

Snare

The most popular type of mic for the snare is a cardioid dynamic with a presence peak. The cardioid pattern reduces leakage; its proximity effect boosts the bass for a fatter sound. The presence peak adds attack. You might prefer a cardioid condenser for its sharp transient response.

Bring the mic in from the front of the set on a boom. Place the mic about one inch in from the rim, one or two inches above the head (Figure 8.5). Angle the mic down to aim where the drummer hits. Or attach a mini condenser mic to the side of the snare drum so it "looks at" the top head over the rim.

Some engineers mike both the top and bottom heads of the snare drum, with the microphones in opposite polarity. A mic under the snare drum gives a zippy sound; a mic over the snare drum gives a fuller sound.

Whenever the hi-hat closes, it makes a puff of air that can "pop" the snare-drum mic. Place the snare mic so the air puff doesn't hit it. To prevent hi-hat leakage into the snare mic:

- Mike the snare closely.
- Bring the snare boom in under the hi-hat, and aim the snare mic away from the hi-hat.
- Use a piece of foam to block sound from the hi-hat.
- Use a de-esser on the snare.
- Overdub the hi-hat.

133

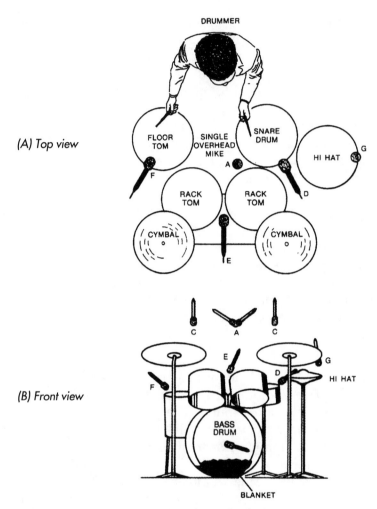

(A) Top view

(B) Front view

Figure 8.4 Typical mic placements for a drum set.

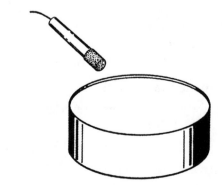

Figure 8.5
Snare drum miking.

Hi-Hat

Try a cardioid condenser mic about six inches over the cymbal edge that's farthest from the drummer (Figure 8.6). To avoid the air puff just mentioned, don't mike the hi-hat off its side; mike it from above aiming down. This also reduces snare leakage. You may not need a hi-hat microphone, especially if you use room mics.

Tom-Toms

You can mike the toms individually, or put a mic between each pair of toms. Place a cardioid dynamic about 1-inch over the drum head and 1-inch in from the rim, angled down about 45 degrees toward the head (Figure 8.7). Again, the cardioid's proximity effect gives a full sound. Another way: clip mini condenser mics to the toms, peeking over the top rim of each drum. Or you might try a bi-directional mic between two tom-toms.

If the tom mics pick up too much cymbals, aim the "dead" rear of the tom mics at the cymbals. If you use a supercardioid or hypercardioid mic, aim the null of best rejection at the cymbals.

Another way to reduce cymbal leakage: Remove the bottom heads from the toms and mike them inside a few inches from the head, off-center. This also keeps the mics out of the drummer's way. The sound picked up inside the tom-tom has less attack and more tone than the sound picked up outside.

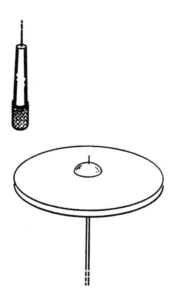

Figure 8.6
Hi-hat miking.

THE LIBRARY
GUILDFORD COLLEGE
of Further and Higher Education

Figure 8.7
Tom-tom miking.

Kick-Drum

Place a blanket or carpet pad inside the drum, pressing against the beater head to dampen the vibration and tighten the beat. The blanket shortens the decay portion of the kick-drum envelope. Tune the drum low.

A popular mic for kick drum is a large-diameter, cardioid dynamic type with an extended low-frequency response. For starters, place it inside on a boom, a few inches from where the beater hits (Figure 8.8). Mic placement close to the beater picks up a hard beater sound; off-center placement picks up more skin tone, and farther away picks up a boomier shell sound.

Other miking tips: Hang a mini omni condenser mic inside near the beater, or place an omni condenser a few inches from the beater. These mics respond to very deep frequencies, and have sharp transient response, which helps the attack.

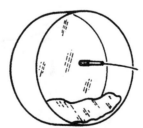

Figure 8.8
Kick-drum miking.

How should the recorded kick drum sound? Well, they don't call it KICK drum for nothing. THUNK! You should hear a powerful low-end thump plus an attack transient.

Kick drum often needs a fair amount of EQ to sound good. Typically you cut several dB around 400 to 600 Hz to remove the "papery" sound, and boost around 2.5 kHz to 10 kHz to add click or snap.

Cymbals

To capture all the crisp "ping" of the cymbals, a good mic choice is a cardioid condenser with a flat, extended high-frequency response. Place the overhead mics about 2 to 3 feet above the cymbal edges; closer miking picks up a low-frequency ring (Figure 8.9). The cymbal edges radiate the most highs. Place the cymbal mics to pick up all the cymbals equally. If your recording will be heard in mono, you might want to mount the mic grilles together and angle the mics apart (Figure 8.4, position "A"). Or use a stereo mic.

Recorded cymbals should sound crisp and smooth; not muffled or harsh.

Room Mics

Besides the close-up drum mics, you might want to use a distant pair of room mics when you record drum overdubs. Place the mics about 10 or 20 feet from the set to pick up room reverb. When mixed with the close-up mics, the room mics give an open, airy sound to the drums. Popular room mics are omni condensers or boundary mics taped to the control-room window. You might compress the room mics for special effect. If you don't have enough tracks for room microphones, try raising the overhead mics.

Boundary Mic Techniques

Boundary mics let you pick up the set in unusual ways. You can strap one on the drummer's chest to pick up the set as the drummer hears it. Tape them to hard-surfaced goboes surrounding the drummer. Put them on the floor under the toms and near the kick drum, or hang a pair over the cymbals.

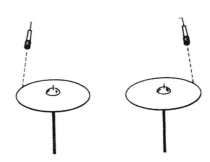

Figure 8.9
Overhead cymbal miking.

Recording With Two to Four Mics

Sometimes you can mike the set simply. Place a stereo mic (or two mics) overhead and put another mic in the kick. If necessary, add a snare-drum mic. This method works well for acoustic jazz and country music. If you want the toms to sound fuller, boost the lows in the overhead mics.

Another setup is shown in Figure 8.10. It uses only one mini omni condenser mic and one kick-drum mic. This method works well on small drum sets.

Clip a mini omni condenser mic to the snare-drum rim about four inches above the rim, in the center of the set, aiming at the hi-hat. Also mike the kick-drum.

The mini mic will pick up the snare, hi-hat, and toms all around it, and will pick up the cymbals from underneath. Move the mic closer or farther from the toms, and raise or lower the cymbals, until you hear a pleasing balance. Add a little bass and treble. You'll be surprised at the good sound and even coverage you can get with this simple setup.

Want a stereo effect? Tape another mini mic just over the floor-tom rim, on the side farthest from the drummer (Figure 8.11). Pan the mini mics left and right.

Drum Recording Tips

After you set up all the mics, ask the drummer to play. Listen for rattles and leakage by soloing each microphone. Try not to spend much time getting a

Figure 8.10 Miking a small drum set with two mics (one in the kick drum).

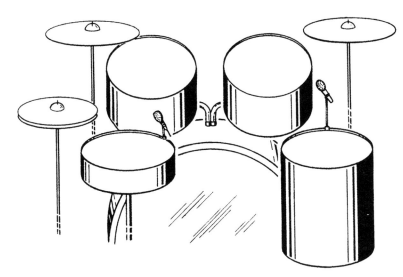

Figure 8.11 Miking a small drum set with three mics (one clipped onto the snare-drum rim, the other in the kick drum).

sound, otherwise you waste the other musicians' time and wear out the drummer. To keep the drum sound tight, turn off mics not in use in a particular tune, use a noise gate on each drum mic, or overdub the drums.

One effect for the snare drum is gated reverb. It's a short splash of bright-sounding reverberation, which is rapidly cut off by a noise gate or expander. Many effects units have a gated-reverb program.

Another trick is recording "hot." Using an analog multitrack, record the drums at a high level so they distort just a little. It's also common to compress the kick.

A drummer might use drum pads, or drum triggers, fed into a sound module. Record direct off the module. You might want to mike the cymbals anyway for best sound.

If you're recording a drum machine and it sounds too mechanical, add some real drums. The machine can play a steady background while the drummer does other things.

When miking drums on stage for P.A., you don't need a forest of unsightly mic stands and booms. Instead, you can use short mic holders that clip onto drum rims and cymbal stands, or use mini condenser mics.

In a typical rock mix, the drums are either the loudest element, or are slightly quieter than the lead vocal. The kick drum is almost as loud as the snare. If you don't want a wimpy mix, keep those drums up front!

139

Try these EQ settings to enhance the recorded sound of the drums:

- Snare and rack toms: Fat at 200 Hz, crack at 5 kHz. If the sound is too tubby, cut around 200 Hz.
- Floor toms: Fullness at 80 to 100 Hz.
- Cymbals: Sizzle at 10 kHz or higher. Roll off the lows below 500 Hz to reduce low-frequency leakage.
- Kick drum: Boost at 2.5 kHz to 10 kHz for click. Filter out highs above 9 kHz to reduce leakage from cymbals. To remove the "cardboard" sound, cut at 300 to 600 Hz.

A typical track assignment for drums might be:

1. Kick
2. Snare
3. & 4. Toms in stereo
5. & 6. Overheads in stereo

Percussion

Let's move on to percussion, such as the cowbell, triangle, tambourine, or bell tree. A good mic for metal percussion is a condenser type because it has sharp transient response. Mike at least 1 foot away so the mic doesn't distort.

You can pick up congas, bongos, and timbales with a single mic between the pair, a few inches over the top. A cardioid dynamic with a presence peak gives a full sound with a clear attack.

For xylophones and vibraphones, place two cardioid mics 1 1/2 feet above the instrument, aiming down. Cross the mics 135 degrees apart or place them about 2 feet apart. You'll get a balanced pickup of the whole instrument.

Acoustic Guitar

The acoustic guitar has a delicate timbre which you can capture through careful mic selection and placement. First prepare the acoustic guitar for recording. To reduce finger squeaks, try a commercial string lubricant, a household cleaner/waxer, talcum powder on fingers, or smooth-wound strings.

Replace old strings with new ones. Experiment with different kinds of guitars, picks, and finger picking to get a sound that's right for the song.

For acoustic guitar, a popular mic is a condenser with a smooth, extended frequency response from 80 Hz up. This kind of mic has a clear, detailed sound. You can hear each string being plucked in a strummed chord. Usually the sound picked up is as crisp as the real thing.

Now let's look at some mic positions. To record a classical guitar solo in a recital hall, mike about 3 to 6 feet away to pick up room reverb. Try a stereo pair (Figure 8.12 A), such as XY, ORTF, MS, or spaced pair (described in Chapter 7).

When you record pop, folk, or rock music, try a spot about 6 to 12 inches from where the fingerboard joins the guitar body—at about the 12th fret (Figure 8.12 B). That's a good starting point for capturing the acoustic guitar accurately. Still, you need to experiment and use your ears. Close to the bridge, the sound is woody and mellow.

In general, close miking gives more isolation, but tends to sound harsh and aggressive. Distant miking lets the instrument "breath." You hear a more gentle, open sound.

Another spot to try: Tape a mini omni mic onto the body, halfway between the sound hole and bridge, about 1/2 inch from the low E string (Figure 8.12 C).

The guitar will sound more real if you record in stereo. Try one mic near the 12th fret, and another near the bridge (Figures 8.12 D, 8.12 E). Pan left and right.

Is feedback or leakage a problem? Mike close to the sound hole (Figure 8.12 F). The tone there is very bassy, so turn down the low-frequency EQ on your mixer until the sound is natural. Also cut a few dB around 3 kHz to reduce harshness.

You get the most isolation with a contact pickup. It attaches to the guitar, usually under the bridge. The sound of a pickup is something like an

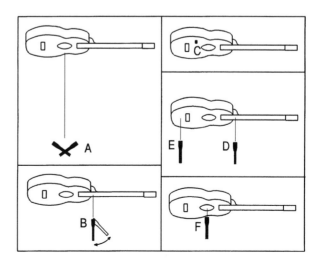

Figure 8.12
Some mic techniques for
acoustic guitar.

electric guitar. You can mix a mic with a pickup to add air and string noise to the sound of the pickup. That way, you get good isolation and good tone quality.

Singer/Guitarist

Normally you overdub the guitar and vocal separately. If you have to record both at once, try one of these methods:

- Angle the vocal mic up and angle the guitar mic down to isolate the two sources. Follow the 3:1 rule.
- Use a pickup or mini mic on the guitar.
- Use a coincident pair of figure eights crossed at 90 degrees. Aim the front of one mic at the voice; aim the front of the other mic at the guitar.
- Use a stereo mic or stereo pair about 1 foot out front; raise or lower the mics to adjust the voice/guitar balance.

Grand Piano

This magnificent instrument is a challenge to record well. First have the piano tuned, and oil the pedals to reduce squeaks. You can prevent thumps by stuffing some foam or cloth under the pedal mechanism.

For a classical music solo, record in a reverberant room such as a recital hall or concert hall. Reverb is part of the sound. Set the piano lid on the long stick. Use condenser mics with a flat response. Place a stereo mic, or a stereo pair of mics, about 7 feet away and 7 feet high, up to 9 feet away and 9 feet high (Figure 8.13). Move the mics closer to reduce reverb, farther to increase it. You might need to mix in a pair of hall mics. Try cardioids aiming away from the piano about 25 feet away.

When recording a piano concerto, give the piano a spot mic about 3 feet away. Put the mic in a shock mount.

Pop music demands close miking. Close mics pick up less room acoustics and leakage, and give a clear sound that cuts through the mix. Try not to mike the strings closer than eight inches, or else you'll emphasize the strings closest to the mics. You want equal coverage of all the notes the pianist plays.

One popular method uses two spaced mics inside the piano. Use omni or cardioid condensers, ideally in shock mounts. Put the lid on the long stick. If you can, remove the lid to reduce boominess. Center one mic over the treble strings and one over the bass strings. Typically, both mics are 8 to 12 inches over the strings and 8 inches horizontally from the hammers

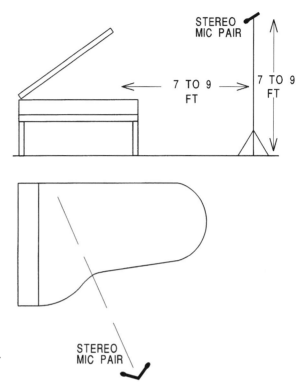

Figure 8.13
Suggested grand piano miking for classical music.

(Figure 8.14, top, bass and treble mics). Pan the mics partly left and right for stereo. As an alternative, try two ear-spaced omni condensers about 12 to 18 inches above the strings.

The spaced mics might have phase cancellations when mixed to mono, so you might want to try coincident miking (Figure 8.14, top, stereo pair). Boom-mount a stereo mic, or an XY pair of cardioids crossed at 120 degrees. Miking close to the hammers sounds percussive, toward the tail has more tone.

For more clarity and attack, boost EQ around 10 kHz or use a mic with a rising high-frequency response. Boundary mics work well too. If you want to pick up the piano in mono, tape a boundary mic to the underside of the raised lid, in the center of the strings, near the hammers. Use two for stereo over the bass and treble strings. Put the bass mic near the tail of the piano to equalize the mic distances to the hammers (Figure 8.14 bottom). If leakage is a problem, close the lid, and cut EQ around 250 Hz to reduce boominess.

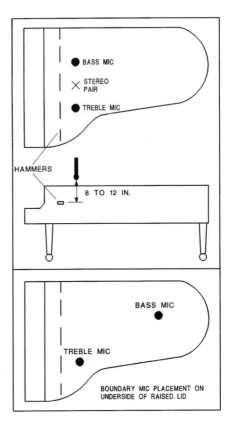

Figure 8.14
Suggested grand piano miking for popular music.

Upright Piano

Here are some ways to mike an upright piano:

Figure 8.15A: Remove the panel in front of the piano to expose the strings over the keyboard. Place one mic near the bass strings and one near the treble strings about 8 inches away. Record in stereo and pan the signals left and right for the desired piano width. If you can spare only one mic for the piano, just cover the treble strings.

Figure 8.15B: Remove the top lid and upper panel. Put a stereo pair of mics about a foot in front and a foot over the top. If the piano is against a wall, angle the piano about 17 degrees from the wall to reduce tubby resonances.

Figure 8.15C: Aim the soundboard into the room. Mike the bass and treble sides of the soundboard a few inches away. In this spot, the mics pick up less pedal thumps and other noises. Try cardioid dynamic mics with a presence peak.

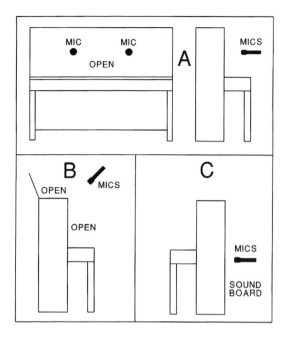

Figure 8.15
Some mic techniques for
upright piano.

Acoustic Bass

The acoustic bass (string bass, bass viol) can be recorded many ways. This instrument puts out frequencies as low as 41 Hz, so use a mic with an extended low-frequency response.

It's common to mix a pickup with a mic. For a well-defined sound, place the mic a few inches in front of the bridge, on the side toward the G string (top string). For more fullness, move the mic toward the f-hole. Another spot is a few inches from the top of the treble f-hole. Watch out for proximity effect (up-close bass boost) if you use a directional mic up close. If you need more isolation, place a cardioid dynamic mic near the treble f-hole and roll off the excess bass on your mixer.

Here are some methods which isolate the bass and let the player move around. They work well for P.A.:

- Wrap a mini omni condenser mic in foam rubber (or in a foam wind-screen) and mount it in an f-hole.
- Tape the cable of a mini omni mic to the bridge.
- Wrap a regular mic in foam padding (except the front grille) and squeeze it behind the bridge or between the tailpiece and the body.
- Try a direct feed from a pickup. This method adds clarity and edge, but might sound electric. Also wrap a mini omni condenser in foam

and stuff it in an f-hole. Mix this mic with the pickup to round out the tone. You may need to roll off the bass of the f-hole mic. Flip the polarity of the mic and use whatever setting sounds best.

Banjo

Try a flat-response mic about 1 foot away from where the fingerboard joins the body. If you need more isolation, mike closer and roll off some bass. The sound is thinner toward the edge of the head. Cloth stuffed inside will reduce feedback in P.A. situations.

For the most isolation, tape a mini omni condenser mic to the head about one inch in from the bottom edge, or on the tailpiece, or on the bridge. You can wedge a pickup between the strings below the bridge and the banjo head. Put the pickup flat against the head surface.

Mandolin, Dobro, Bouzouki, Lap Dulcimer

Mike these about 8 to 12 inches away with a condenser mic. If you need more lows and more isolation, mike close to an f-hole. You can tape a mini omni condenser mic near an f-hole and tweak EQ for the best sound.

Hammered Dulcimer

Place a flat-response condenser mic about 2 feet over the center of the soundboard (Figure 8.16 A). On stage, place a cardioid dynamic or condenser 6 to 12 inches over the middle of the top end (Figure 8.16 B). For the best gain-before-feedback in a P.A. system, mix in a mini omni condenser mic (or a cardioid with bass rolloff) very near the sound hole (Figure 8.16 C).

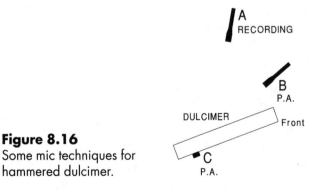

Figure 8.16
Some mic techniques for hammered dulcimer.

Fiddle (Violin)

Listen to the fiddle itself to make sure it sounds good. Correct any instrument problems before miking.

First try a flat-response condenser mic (omni or cardioid) about 2 to 3 feet over the bridge. This distant miking gives an airy, silky sound; close miking sounds nasal and scratchy. If the ceiling is low, reflections might color the sound. In that case, cover the ceiling over the fiddle with a sleeping bag or foam. Or have the fiddle player sit down. If you have to mike close—say, for a singing fiddler—aim the mic horizontally at the mouth.

Other methods: Get a mini omni mic, and clip its holder to the violin's tailpiece. Mount the mic a few inches from an f-hole or over the bridge. Or clip the mini omni to the strings between the bridge and tailpiece; aim the mic at the fiddle body. If necessary, cut a little at 3 kHz to reduce harshness and boost around 200 Hz for warmth. A good spot for a pickup is on the left side of the top (player's view), on the player's side of the bridge.

To record a classical violin solo, try a stereo mic (or a stereo pair) 12 to 20 feet away in a reverberant room.

String Section

Place the strings in a large, live room and mike them at a distance to pick up a natural acoustic sound. A common mic choice is a condenser with a flat response. First try a stereo mic or a stereo pair of mics about 4 to 20 feet behind the conductor, raised about 15 feet.

If the room is noisy or too dead, or the balance is poor, you'll need to mike close and add digital reverb. Try one mic on every two to four violins, 6 feet off the floor, aiming down. Same for the violas. Mike the cello about 2 feet from the bridge. When you mix the strings to stereo, pan them evenly between the monitor speakers. Spread them left, center, and right to make a "curtain of sound." If you can spare only one track for the strings, use a stereo-izer effect during mixdown.

String Quartet

Record a quartet in stereo using a stereo mic or a pair of mics. Place them about 6 to 10 feet away to capture the room ambience. The monitored instruments should not spread all the way between speakers. If you want to narrow the stereo stage, angle or space the mics closer together.

Bluegrass Band, Old-Time String Band

Suppose you're recording a group that has a good acoustic balance. Try a stereo mic or stereo pair of mics about 3 feet away and 6 feet high. Move the players toward or away from the mics to adjust their balance.

You'll have more control if you mike all the instruments up close and mix them. This also gives a more "commercial" sound. The production style aims for a natural timbre on all the instruments, either with no effects or with slight reverb.

Harp

Use a condenser mic with a flat response. If the harp is playing with an orchestra, mike the harp about 18 inches from the front of the soundboard, or 18 inches from the player's left hand. You can mike a harp solo about 4 feet over the top.

Tape a mini omni condenser mic to the soundboard if you need more isolation. A mic on the inside of the soundboard has more isolation; a mic on the outside sounds more natural. Also try a cardioid condenser wrapped in foam, stuck into the center hole from the rear.

Horns

"Horns" in studio parlance refers to the brass instruments: trumpets, cornets, trombones, baritones, french horns, and tubas. All the brass radiate strong highs straight out from the bell, but do not project them to the sides. A mic close to and in front of the bell picks up a bright, edgy tone. To mellow out the tone, mike the bell off axis with a flat-response mic (Figure 8.17). The sound on axis to the bell has a lot of spiky high harmonics which can overload a condenser mic, mixer input, or analog tape. That's another reason to mike off axis.

Mike the trumpet with a dynamic—a moving coil or ribbon—to take the edge off the sound. Use a condenser mic if you want a lot of sizzle. Mike

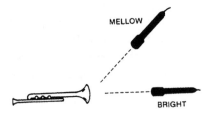

Figure 8.17
Miking for trumpet tone control.

about a foot away for a tight sound; mike several feet away for a fuller, more dramatic sound.

You can pick up two or more horns with one microphone. Several players can be grouped around a single omni mic, or around a stereo pair of mics. The musicians can play to a pair of boundary mics taped on the control-room window or on a large panel.

Record a classical brass quartet in a reverberant room. Use a stereo mic, or a stereo pair of mics, about 6 to 12 feet away.

Saxophone

A sax miked very near the bell sounds bright, breathy, and rather hard (Figure 8.18). Mike it there for best isolation. To get a warm, natural sound, mike the sax about 1 1/2 feet away, halfway down the wind column (Figure 8.18). Don't mike too close, or else the level varies when the player moves. A compromise position for a close-up mic is just above the bell, aiming at the holes. You can group a sax section around one mic.

Woodwinds

With woodwinds, most of the sound radiates not from the bell, but from the holes. So aim a flat-response mic at the holes about 1 foot away.

When miking a woodwind section within an orchestra, you need to reject nearby leakage from other instruments. To do that, try aiming a bidirectional mic down over the woodwind section. The side nulls of the mic cut down on leakage.

To pick up a flute in a pop-music group, try miking a few inches from the area between the mouthpiece and the first set of finger holes

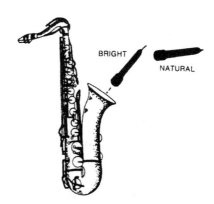

Figure 8.18
Two ways to mike a saxophone.

Figure 8.19
One effective placement for
flute miking.

(Figure 8.19). You may need a pop filter. If you want to reduce breath noise, roll off high frequencies or mike farther away. You also can attach a mini omni mic to the flute a few inches above the body, between the mouthpiece and finger holes. For classical music solos, try a stereo pair 4 to 12 feet away.

Harmonica, Accordion, Bagpipe

One way to mike a harmonica (harp) is to use a cardioid dynamic mic with a ball grille. Place the mic very close to the harmonica or have the player hold it. A condenser mic about 1 foot away gives a natural sound. To get a bluesy, dirty sound, use a "bullet" type harmonica mic or play the harmonica through a miked guitar amp.

For accordion, try a mic about 6 to 12 inches from the sound holes near the keyboard. Some accordions have sound holes on both sides, so you'll need two mics. Follow the 3:1 rule. The distance between mics should be at least 3 times the mic-to-source distance.

A bagpipe has two main sound sources: the chanter, which the musician plays with the fingers, and the drone pipes, which make a steady tone. Mike the chanter about a foot away from the side, and mike the drone pipes a foot from the end. Again, follow the 3:1 rule. You could also mike the bagpipe a few feet away with one mic.

Lead Vocal

The lead vocal is the most important part of a pop song, so it's critical to record it right. First set up a comfortable environment for the singer. Put down a rug, add some flowers or candles, dim the lights. Set up a good cue mix with effects to help the singer get into the mood of the song.

With any vocal recording, there are some problems to overcome, but we can deal with them. Among these are proximity effect, breath pops, wide dynamic range, sibilance, and sound reflections from the music stand. Let's look at these in detail.

Miking Distance

When you sing or talk close to most directional mics, the microphone boosts the bass in your voice. This is called the proximity effect. We've come to accept this bassy sound as normal in a P.A. system, but the effect just sounds boomy in a recording.

To prevent boomy bass, mike the singer at a distance, about 8 inches away (Figure 8.20). A popular mic choice is a flat-response condenser mic with a large diaphragm (1 1/4-inch diameter). As always, you can use any mic that sounds good to you. If the mic has a bass rolloff switch, set it to "flat."

Singers should maintain their distance to the mic. I ask the singer to spread the fingers, touch lips with the thumb, and touch the mic with the pinky. The hand forms a spacer for keeping a constant distance.

Some singers can't help but "eat" the mic. You can mike them at a distance, and also give them a dummy mic to hold while singing.

If you must record the singer and the band at the same time—as in a concert—you'll have to mike close to avoid picking up the instruments with the vocal mic. Try a cardioid mic with a bass rolloff and a foam pop filter. The sound will be bassy due to proximity effect, so roll off the excess lows at your mixer. For starters, try –6 dB at 100 Hz. Some mics have a bass filter switch for this purpose. Aim the mic partly toward the singer's nose to prevent a nasal or closed-nose effect. This closeup method works well if you want an intimate, breathy sound.

When recording a classical-music singer who is accompanied by an orchestra, place the mic about 1 to 2 feet away. If the singer is a soloist (maybe accompanied by piano), use a stereo pair about 8 to 15 feet away to pick up room reverb.

Figure 8.20
Typical miking technique for
a lead vocal.

Breath Pops

When you sing a word with "p" or "t" sounds, a turbulent puff of air shoots out of the mouth. The puff hits the mic and makes a thump or small explosion called a pop. To reduce it, put a foam-plastic pop filter on the mic. Some mics have a ball grille screen to cut pops, but foam works better. The pop filter should be made of special open-cell foam to pass high frequencies. For best pop rejection, allow a little air space between the foam and the front of the mic grille.

Foam pop filters reduce the highs a little. So they should be left off instrument mics, except for outdoor recording or dust protection. Pop filters do not reduce breathing sounds or lip noises. To get rid of these problems, mike farther away or roll off some highs.

The most effective pop filter is a hoop with a nylon stocking stretched over it. You can buy this type, or make one with an embroidery hoop from a fabric or craft store. Place the filter a few inches from the mic.

Another way to get rid of pops is to put the mic at forehead height, aiming at the mouth (Figure 8.20). This way the puffs of air shoot under the mic and miss it. Make sure the vocalist sings straight ahead, not up at the mic, or the mic will pop. Caution: If your studio has a low ceiling, the recorded vocal might have a colored tone quality due to phase cancellations from ceiling reflections. Try putting the mic lower and use a hoop-type pop filter, or cover the ceiling with absorbent material. You can also mike the singer from the side to prevent pops (Figure 8.21).

Wide Dynamic Range

During a song, vocalists often sing too loud or too soft. They blast the listener or get buried in the mix. That is, many singers have a wider dynamic range than their instrumental backup. To even out these extreme level variations, ask the singer to use proper mic technique. Back away from the mic

MUSIC SHEET IS
ANGLED AWAY FROM
MICROPHONE

Figure 8.21
Miking a singer from the side.

on loud notes; come in closer for soft ones. Or you can ride gain on singers: gently turn them down as they get louder, and vice versa.

The best solution is to pass the vocal signal through a compressor, which acts like an automatic volume control. Plug the compressor into the vocal channel's access jacks. A typical compressor setting is 2:1 ratio, –5 dB threshold.

If the singer moves toward and away from the mic while singing, their average level will go up and down. Try to mike the singer at least 8 inches away, so that small movements of the singer won't affect the level.

If you must mike close to prevent leakage or feedback, ask the vocalist to sing with lips touching the pop filter to keep the same distance to the mic. Turn down the excess bass using your mixer's low-frequency EQ (typically –6 dB at 100 Hz).

Sibilance

Sibilance is the emphasis of "s" or "sh" sounds, which are strongest around 5 kHz to 10 kHz. They help intelligibility. In fact, many producers like sizzly "s" sounds, which add a bright splash to the vocal reverb. But the sibilance should not be piercing or strident.

If you want to reduce sibilance, use a mic with a flat response—rather than one with a presence peak—or cut the highs a little around 8 kHz on your mixer. Better yet, use a de-esser signal processor, which cuts the highs only when the singer makes sibilant sounds.

Reflections from the Music Stand

Suppose that a lyric sheet or music stand is near the singer's mic. Some sound waves from the singer go directly into the mic. Other sound waves reflect off the lyric sheet or music stand into the mic (Figure 8.22, top). The delayed reflections will interfere with the direct sound, making a colored tone quality like mild flanging.

To prevent this, lower the music stand and tilt it almost vertically (Figure 8.22, bottom). This way, the sound reflections miss the mic. Or tape the lyric sheet behind the singer's cardioid mic (Figure 8.20). The cardioid mic will reject reflections that enter the mic from the rear.

Vocal Effects

Some popular vocal effects are stereo reverb, echo, and doubling. You can record real room reverb by miking the singer at a distance in a hard-surfaced room. Slap echo provides a 1950's Rock 'n' Roll effect.

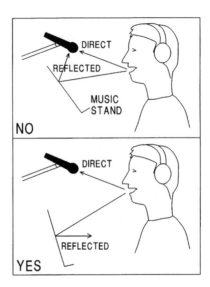

Figure 8.22
Preventing reflections from
a music stand.

Doubling a vocal gives a fuller sound than a single vocal track. Over-dub a second take of the vocal on an empty track, in sync with the original take. During mixdown, mix the second vocal take with the original, at a slightly lower level than the original. Or double the vocal by running it through a digital delay set to 15 to 35 msec.

Background Vocals

When you overdub background vocals (harmony vocals), you can group two or three singers in front of a mic. The farther they are from the mic, the more distant they will sound in the recording. Pan the singers left and right for a stereo effect. Since massed harmonies can sound bassy, roll off some lows in the background vocals.

Barbershop or gospel quartets with a good natural blend can be recorded with a stereo mic or stereo pair of mics about 2 to 4 feet away. If their balance is poor, close-mike each singer about 8 inches away, and balance them with your mixer. This also gives a more "commercial" sound. If you close-mike, spread the singers at least 2 feet apart to prevent phase cancellations.

Summary

We can sum up mic placement like this: If leakage or feedback are problems, place the mic near the loudest part of the instrument, and add EQ to get a natural sound. Otherwise, place the mic in various spots until you find a position that sounds good over your monitors. There is no single "correct" mic technique for any instrument. Just place the mic where you hear the desired tonal balance and amount of room reverb.

Try the techniques described here as a starting point, then explore your own ideas. Trust your ears! If you capture the power and excitement of electric guitars and drums, if you capture the beautiful timbre of acoustic instruments and vocals, you've made a successful recording.

9

ANALOG TAPE RECORDING

Thanks to the tape recorder, a musical performance can be captured permanently and relived again and again. In this chapter we'll look at these aspects of tape recording:

1. The analog tape recorder—parts, functions, alignment, cleaning
2. Noise reduction systems (Dolby and dbx)
3. Tape handling, storage, and editing

Although the analog tape recorder might seem outdated in these days of digital tape recording, it is still a viable format. Many top engineers prefer the sound of analog, claiming smoother and more extended highs, and a gentler, warmer sound than digital.

How does the analog tape recorder work? As it records, it changes the incoming audio signal into a magnetic signal on tape. The tape itself is a strip of plastic, usually Mylar, with a thin coating of magnetic particles: ferric oxide, ferric-cobalt, or chromium dioxide. Each particle is a tiny magnet. These particles aim in random directions. During recording, an external magnetic field aligns them into magnetic patterns. During playback, the magnetic patterns on tape generate the original audio signal.

Tape Recorder Parts and Functions

A tape recorder has three main parts: the heads, the electronics, and the transport.

- The heads are electromagnets that change electric signals to magnetic fields, and vice versa.
- The electronics amplify and equalize the signals going to and from the heads.
- The transport pulls the tape past the heads, which contact the tape.

Let's look more closely at each of the three main parts.

The Heads

Professional open-reel recorders have three heads placed left-to-right: erase, record, and playback (Figure 9.1). Semi-pro recorders, and cassette multitrackers, use a single head to record and play back.

The **erase head** makes a magnetic field that vibrates at an ultrasonic frequency. As the tape rolls past the erase head, the tape is exposed to this vibrating magnetic field, which weakens as the tape moves away from the head. This action orients the magnetic particles randomly and erases any signal on tape.

The **record head** changes the incoming electric signal into a magnetic field that varies as the signal does. As the tape passes the record head, the head magnetizes or aligns the tape particles in a pattern that corresponds to the audio signal. This pattern stays on the tape forever unless it's erased.

The pattern stored on tape is a varying magnetic field. As the tape passes the **playback head**, the head picks up this magnetic field and changes

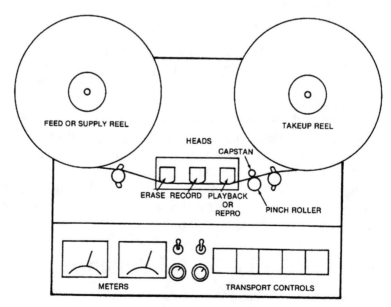

Figure 9.1 Major parts of a typical tape recorder.

it back into a similar electric signal. This signal is amplified and is sent to the recorder's output connectors.

In the center of the head's front face, where it contacts the tape, is the **gap**: a vertical line that is the break in the electromagnet. The record-head gap puts the signal on the tape as it passes the head; the playback-head gap reads the signal off the tape as it passes the head. Most heads have more than one gap: one gap per track (Figure 9.2).

There are limits to the signal level you can put on tape. When the signal is strong enough to align all the magnetic particles, higher levels can't increase the magnetic signal on tape. This is called **tape saturation.** It causes the fuzzy, grungy sound of distortion.

If the recording level is too low, though, you start to hear tape noise (hiss). **Hiss** is the random noise signals generated by non-aligned particles. You hear hiss when the recorded signal is weak compared to the tape noise.

So, you should record most signals at as high a level as possible without distortion. That way, you get a clean recording with no grunge and almost no hiss. But sometimes a little tape saturation sounds good on drums.

The Electronics

Recorder electronics have these functions:

* Amplify and equalize the incoming audio signal
* Send the audio signal to the record head
* Amplify and equalize the signal from the playback head

The equalization done by the electronics is called **record equalization** and **playback equalization.** Record equalization is a slight boost at low and high frequencies to improve the signal-to-noise ratio at those frequencies. Part the of playback equalization is a bass cut to compensate for the bass boost during recording.

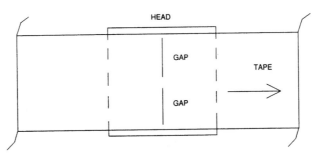

Figure 9.2 Gaps in a tape head.

During playback, the output from the playback head rises 6 dB per octave because the head's output depends on the rate of change of magnetic flux, which doubles with each higher octave. (Magnetic flux is the lines of magnetic force.) To compensate for this rise, playback equalization falls 6 dB per octave.

The magnetism on tape tends to slightly erase the high frequencies on tape. But playback equalization has a high frequency boost to compensate for this, and for high frequency losses within the head.

The frequency response of the playback EQ has been standardized in the United States to a curve called the NAB (National Association of Broadcasters) curve. Other countries may use different playback EQ.

An ultrasonic oscillator in the electronics drives the erase head. The ultrasonic signal, called **bias,** is also mixed with the audio fed to the record head. Adding bias is needed to reduce distortion. The amount of bias is adjustable. It affects the recording's level, frequency response, distortion, and **drop-outs** (short signal losses).

The bias setting is critical. Too high a setting reduces the level recorded on tape and rolls off high frequencies. Too low a setting also reduces the level on tape, results in distortion and drop-outs, and raises the highs. The bias is set at the factory for a particular type of tape. That's why it's important to use the type of tape your user manual recommends. If your cassette deck has a bias adjustment, set it according to the user manual. We'll describe how to set the bias in pro machines later in this chapter.

The Tape Transport

The job of the transport is to move the tape past the heads. During recording and playback, the transport should move the tape at a constant speed and with constant tape tension. During rewind or fast forward, the tape shuttles rapidly from one reel to the other.

Most pro open-reel decks have three motors in the transport: two for shuttling and tape tension, and a third to drive the capstan. The **capstan** is a post that rotates against a rubber **pinch roller.** The tape is pressed between the capstan and pinch roller. As the capstan rotates, it pulls the tape past the heads. Also in the transport are rollers that reduce **wow** (a slow periodic variation in tape speed) and **flutter** (rapid variation in tape speed).

The **tape counter** usually shows the elapsed time on tape. You can mark a particular point on tape—say, the beginning of a song—by resetting the tape counter to zero. On some machines, a return-to-zero button shuttles the tape to the zero point and then stops automatically. This function is useful when you want to repeatedly practice an overdub or a mix.

A pro open-reel tape deck moves tape at 7 1/2, 15, or 30 ips (inches per second). Cassette decks run at 1 7/8 ips or 3 3/4 ips. Faster speeds sound better. As tape speed increases, cymbals sound cleaner, there's less tape hiss, and the pitch is more stable. But a slower tape speed consumes less tape and allows more running time.

Tracks

A track is a path on tape containing a single channel of audio. The wider the track (that is, the more tape it covers), the greater the signal-to-noise ratio (S/N). Doubling the track width improves the S/N by 3 dB.

Some tape-recorder heads have one gap that records and plays one track over the full width of the tape. Other heads have several gaps that record and play two or more tracks. Each track covers part of the tape width. Figure 9.3 shows some track-width standards for 1/4-inch tape.

- A full-track mono head records over nearly the full width of the tape in one direction (Figure 9.3 A).
- A half-track mono head records one track in one direction and one track in the opposite direction when the tape is flipped over (Figure 9.3 B). Each track covers about one-third of the tape. The unused third between the tracks is a guard band to prevent crosstalk between tracks.
- A 2-track stereo or half-track stereo head records 2 tracks in one direction (Figure 9.3 B). This format is used for stereo master tapes. Track widths are the same as half-track mono.
- A quarter-track stereo head records 2 tracks in one direction and 2 tracks in the opposite direction when the tape has been flipped over (Figure 9.3 C). Stereo cassette decks use this format on 1/8-inch wide tape.
- A multitrack head records 4 more tracks in one direction (Figure 9.3 C). Depending on the width of the tape, the number of tracks can be 4, 6, 8, 16, 24, 32, or 48. Cassette recorder-mixers use this format (4 or 8 tracks only).

Tape Width

Magnetic recording tape comes in various widths to accommodate the track formats:

- 1/8-inch for 1/4-track stereo, and cassette (4- or 8-track)
- 1/4-inch for full-track mono, half-track mono, quarter-track stereo, 2-track stereo, 4-track, and 8-track
- 1/2-inch for 4, 8, or 16 tracks
- 1-inch for 8, 16, or 24 tracks
- 2-inch for 16, 24, 32, or 48 tracks

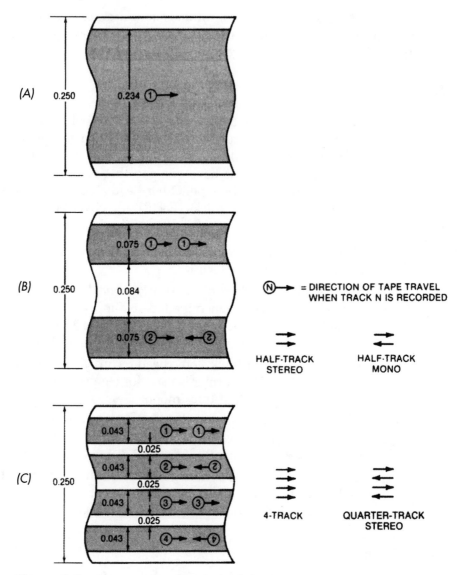

Figure 9.3 Some track-width standards for 1/4-inch tape.

Cassette Deck

The cassette deck from a home stereo is a low-cost, convenient way to make proof tapes and demos. The sound is not good enough, though, for masters that will be duplicated commercially. You can get an excellent deck for $200 and up. Some useful features are Dolby B, C, and S which reduce tape hiss, and Dolby HX which reduces high-frequency distortion. Look for a wow and flutter spec below 0.05 percent, or you may hear wobbly pitch. Also look for a front-mounted adjustable bias control, so you can fine tune the bias to the type of tape you're using.

Multitrack and Synchronous Recording

The multitrack recorder can be either a stand-alone unit or can be built into a recorder-mixer. Chapter 2 describes the features of the multitrack recorders used in recorder-mixers.

Multitrack machines are available that record 4, 6, 8, 16, 24, 32, or 48 tracks side-by-side on a single tape. It's like having several stereo tape decks in sync. Each track contains the signal of a different instrument, or a different mix of instruments. You can record the tracks all at once, one at a time, or in any combination. After you record the tracks, you combine and balance them through a mixing console. Unlike 2-track recording, multitrack recording lets you fine tune the mix after the recording session. You can practice the changes in the mix until you get them right.

Multitrack recordings can sound clearer than live-to-2-track recordings because you can overdub instruments without mic leakage rather than recording them all at once. If you record several instruments and vocals at the same time, leakage, or off-mic sound can add a muddy, loose sound to the mix. But when you overdub there is no leakage, so the final mix can be cleaner.

Multitrack recording, though, needs an extra generation, since you must record the multitrack mix onto a 2-track tape. Each generation or tape copy adds 3 dB of tape hiss. In addition, every time the number of tracks used in the mix doubles, the noise increases 3 dB. It's not a lot, but it is audible.

Another trade-off of multitrack recording is that the recording process takes much longer. With live-to-2-track recording, the recording is done when the performance is done. But with multitrack recording, you must record first and then do a mixdown. Overdubbing is optional, but is the norm, and adds more time. Plus, multitrack equipment is more complicated and expensive, and is more time-consuming to set up. Still, the ability to fine tune the mix after the session makes multitrack the preferred choice for pop-music recording.

Overdubbing

One advantage of multitrack recorders is that tape tracks can be recorded at different times. For instance, a musician can listen to recorded tracks off the playback head and overdub a new part. During playback, the new part is delayed relative to the original tracks because the playback head is located a small distance from the record head.

To remove the delay caused by overdubbing, and to sync the original tracks with the overdub, you play the original tracks through the record head temporarily. At the same time, the record head records the overdub on an open track. This process is called **simul-sync, selsync,** or **synchronous recording.** To enable it, set each track's tape-monitor switch to the SYNC position. When you play a tape through the record head, the signal loses highs but is adequate for synching.

Semi-pro tape decks and cassette multitrackers combine the record and playback functions in one head. So there's no sync problem; recorded tracks and overdubs are always in sync. The same is true of digital multitracks.

Meters and Level Setting

Meters on the tape recorder (one per track) show the record and playback levels. These meters may be VU meters, VU meters with built-in peak LEDs, or LED bargraph indicators showing peak levels.

The VU Meter

A VU meter is a voltmeter that shows approximately the relative volume or loudness of the audio signal. The meter is calibrated in VU or Volume Units. The Volume Unit is the same as a decibel only when you're measuring a steady sine-wave tone. That is, 1 VU = 1 dB only when you apply a steady tone.

A **0 VU recording level** (0 on the record level meter) is the normal operating level of a recorder. It indicates that the magnetic signal is being recorded on tape at the optimum level.

Recording levels that are too high (greater than +3 VU) saturate or overload the tape, which causes distortion. Too-low levels (say, consistently below −10 VU) result in audible tape hiss.

When you apply a rapidly changing signal to a VU meter, the meter reads less than the peak voltage of the waveform. This is because the response of a VU meter is not fast enough to accurately track rapid transients, such as drum beats.

This inaccuracy can cause problems with level setting. For example, if you record drums at 0 VU on the meter, peaks may be 8 to 14 dB higher, resulting in tape distortion. So, whenever you record instruments having sharp attacks or a high peak-to-average ratio (such as drums, piano, percussion, or horns), record at −6 to −8 VU to prevent tape distortion. Note that mild distortion on drum peaks (recording "hot") may give a desirable effect. Instruments with a low peak-to-average ratio, such as organ or flute, can be recorded around +3 VU without audible distortion.

Peak Indicators

Unlike the VU meter, the peak indicator shows peak recording levels more accurately because it responds very rapidly. If your recorder has an LED peak light in each VU meter, set the levels so the LEDs flash only occasionally. For setting recording levels, an LED flash takes precedence over the VU meter reading. If the recorder has LED bargraph peak indicators, set all tracks to peak at 0 to +6 dB, according to the user manual.

Matching Mixer and Recorder Meters

If you're using a separate mixer and recorder, it's common practice to set the mixer meters and recorder meters to match each other. That way you have to watch only the mixer meters while recording. Also, when the mixer and recorder are both peaking around 0 VU, this minimizes the noise and distortion in both units.

If your mixer meters and recorder meters have the same ballistics (the same speed), match the meters with a steady 1 kHz tone from a signal generator or a synthesizer note (C or B two octaves above middle C). Otherwise, hum a steady tone into a mic plugged into your mixer, and set the levels on the mixer and tape deck to 0.

If your tape-deck meters move faster or slower than your mixer meters, follow this procedure:

1. Play a program with a lot of loud drum hits through your mixer, peak the mixer at 0, and set your recorder level to peak at 0 on its meters. Then play a 0 VU tone through your mixer, and note the level on the recorder meters for future reference.

2. At this point you could watch just the mixer meters. But you get the most accurate reading by watching your recorder meters. You also need to watch the recorder meters if you use dbx noise reduction (explained later).

3. Once the mixer and recorder meters match each other, leave the recorder controls alone. Set levels only with the mixer faders or mixer gain-trim controls.

Judging Machine Specifications

The following specs indicate excellent performance:

- Wow and flutter 0.04 percent or less
- S/N 70 dB with Dolby C, 85 dB with Dolby S or SR, 90 dB with dbx
- Record/play response 40 Hz – 18 kHz (+/– 3 dB)

Operating Precautions

Here are some operating tips for tape recorders that may prevent some accidents:

- Don't put the machine in record mode until levels are set. If you record an extremely high-level high-frequency signal, the crosstalk within the head might erase other tracks.
- Keep tape away from recorder heads when turning the machine on or off, or you may put a click on tape.
- Keep degaussers and bulk tape erasers several feet from tapes you don't want to erase.
- Before you start recording on a track, make sure you won't be erasing something you wanted to keep. Listen to the track first or refer to your track sheet.
- Edge tracks of multitrack tapes are prone to drop-outs due to edge damage. Since drop-outs occur mostly at high frequencies, use the edge tracks only to record instruments with little high-frequency output (such as bass or kick drum).
- Repeated passes of a recording past the heads may gradually erase high frequencies. You may want to make a copy of the multitrack tape (or a quick 2-track mix) for musicians to practice overdubs with. Then go back to the original tape when the musicians are ready to record.
- Bouncing or ping-ponging tracks tends to lose high-frequency response and increase tape hiss, so try to limit bounced tracks to bass or midrange instruments.

Noise Reduction

An analog tape recorder adds undesirable tape hiss and **print-through** to the recorded signal, degrading its clarity. Print-through is the transfer of a magnetic signal from one layer of tape to the next, causing an echo. Tape hiss is especially easy to hear when you do a multitrack mixdown, because every track mixed in adds to the overall noise level. Noise increases 3 dB whenever the number of tracks in use doubles, assuming they are mixed at equal levels.

Fortunately, noise-reduction devices such as Dolby or dbx can reduce tape hiss and print-through. However, these units do not remove noise in the original signal from the mixing console. If your signal is noisy before you record it, Dolby or dbx will not remove this noise. They work only on tape noise.

Dolby is used with consumer cassette decks, Fostex cassette multitrackers and open-reel recorders. dbx is used with TASCAM cassette multitrackers and some open-reel recorders.

You need one channel of noise reduction per tape track. Noise-reduction units connect between the mixer output busses and the corresponding tape-track inputs, and also between the tape-track outputs and the mixer tape inputs (Figure 9.4). Some open-reel recorders and most cassette recorders have built-in noise reduction; it is permanently connected.

These noise-reduction devices compress the signal during recording and expand it in a complementary way during playback. The compressor part of the circuit boosts the recorded level of quiet musical passages. During playback, the expander part works in a complementary way, turning down the volume during quiet passages, reducing noise added by the tape. During loud passages (when noise is masked by the program), the gain returns to normal.

A compressed tape is called **encoded;** the expanded tape is called **decoded.** If an encoded tape is played without decoding, it sounds compressed and trebley.

The encode and decode sections must track each other. For example, a 10 dB level change at the input of the encode section should yield a 10 dB level change at the output of the decode section. Otherwise, the dynamics sound unnatural. To prevent this problem, avoid excessive recording levels and adjust the noise-reduction unit for correct tracking if it can be adjusted (see the operating manual for instructions).

dbx

With dbx noise reduction, the compression ratio is 2:1. That is, a program with a 90 dB dynamic range is compressed to 45 dB, which is easily handled by a tape recorder with a 60 dB S/N. During playback, dbx expands the dynamic range back to the original 90 dB. Use of dbx improves S/N by 30 dB and increases headroom by 10 dB. The dbx circuit also includes pre-emphasis (treble boost) of 12 dB during recording and complementary de-emphasis (treble cut) during playback to reduce modulation noise. dbx operates at all signal levels and across the entire audible spectrum.

Dolby

Dolby operates only on quiet passages—those below –10 VU. High-level passages do not need noise reduction because the program masks the noise. There are five different types of Dolby noise reduction (listed in increasing order of effectiveness):

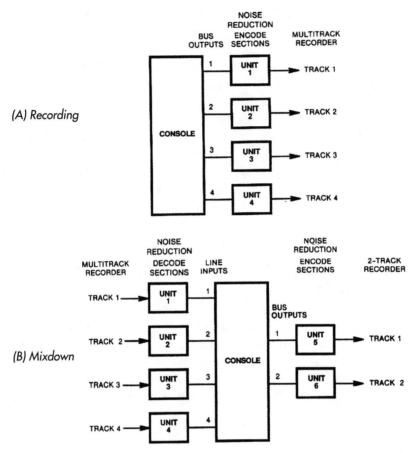

Figure 9.4 Noise reduction applied to multitrack tape and to 2-track master tape.

- Dolby B
- Dolby C
- Dolby A
- Dolby S
- Dolby SR

Dolby B is a lower-cost system for cassette decks, and operates only at high frequencies to reduce tape hiss by up to 10 dB.

Dolby C, for cassette and open-reel, works over a slightly wider range and reduces noise by up to 20 dB.

Dolby A, an early system for open-reel, divides the audible spectrum into four separate frequency bands which are compressed and expanded independently. This system reduces noise by 10 dB below 5 kHz and up to 15 dB at 15 kHz.

Dolby S for cassettes reduces tape hiss by 24 dB and low-frequency noise by 10 dB.

Dolby SR (Spectral Recording), for open-reel, is the most effective Dolby system, reducing noise by more than 25 dB over most of the audible spectrum. As a result, a recorder operated at 15 ips with Dolby SR can have a maximum S/N exceeding 105 dB. During recording, Dolby SR boosts the gain of parts of the spectrum that are low-to-medium in level. During playback, it reduces the gain of the same regions in a complementary way.

When using Dolby A, you must record a calibration signal called a **Dolby tone** on tape before the regular program. An oscillator in the Dolby unit generates this tone. During playback, the level of the recorded Dolby tone is indicated on a Dolby meter. You set the Dolby input level so that the meter indication lines up with the Dolby-level mark on the meter. Then the expander circuitry tracks the recording properly.

If the level is set improperly, the frequency response and dynamic range are slightly changed. Fortunately, there is room for some error since these changes occur in low-level signals and so are hard to hear.

Dolby vs. dbx

Both Dolby and dbx have pros and cons. Compared to Dolby, dbx gives more noise reduction. On the other hand, dbx exaggerates drop-outs more than Dolby does. Dolbyized recordings are relatively free of noise "breathing"—hiss or fuzziness that varies with the signal level, especially on bass or percussion tracks. dbx can change the dynamics at low frequencies.

Many professional engineers record at 30 ips, without noise reduction, or use Dolby SR. Dolby-encoded and dbx-encoded tapes are not compatible with each other, and cannot be played properly without decoding through the appropriate unit. So, if you plan to send your tapes to another studio, check that the studio has the same type of noise reduction that you want to use. On mass-produced cassettes, use either no noise reduction or Dolby B because most consumer stereo cassette decks have Dolby B. You can use whatever you want on your multitrack tape.

When using noise reduction, avoid saturating the tape while recording. Otherwise, the attack transients may be altered during playback through the noise-reduction unit. If you are using noise reduction, you can record at 3 VU lower than normal for 3 dB more headroom. When you copy a tape, switch in the noise reduction on the playback deck (if it has been recorded with noise reduction), and also on the recording deck. Be sure to copy the Dolby tone if the master tape has one.

Preventive Maintenance

The analog tape recorder needs periodic maintenance—cleaning, demagnetizing, alignment, and calibration—for best performance.

Cleaning the Tape Path

Over time, dust and oxide shed from the tape build up on your deck's heads. This layer of deposits separates the tape from the heads, causing high-frequency loss and drop-outs. In addition, buildup of oxide on the tape guides, capstan, and pinch roller can cause flutter. So it's very important to clean the entire tape path before every recording session.

Use the cleaning fluid recommended in your recorder manual. Denatured alcohol (from hardware stores or drugstores) and a dense-packed cotton swab are often used. Don't use rubbing alcohol or isopropyl alcohol because they can leave a film on the head, and they contain water. Use rubber cleaner on rubber parts—rather than alcohol—to prevent swelling or cracking. Allow the cleaning fluid to dry before putting in a cassette or threading on a tape.

Demagnetizing the Tape Path

Tape heads and tape guides can build up a magnetic field which can partly erase high frequencies, add tape hiss, and cause clicks at splices. You can get rid of this magnetism with a tape-head **demagnetizer** or **degausser**. It's an electromagnet with a probe tip. Generally, only the gapped types are strong enough to be effective; the pencil-shaped types may cost less but don't work as well.

The demagnetizer produces a magnetic field that vibrates at 60 Hz. You touch the probe tip to the head in order to magnetize it; then slowly pull the tip away so that the magnetism tapers off until none is left.

The technique of using a demagnetizer is critical:

1. If necessary, cover the probe tip with electrical tape or a handkerchief to avoid scratching the heads.
2. Turn off your recorder. Be sure that no tapes are near the demagnetizer.
3. With the demagnetizer at least one foot from your recorder, plug it in.
4. Bring the demagnetizer slowly to the part to be demagnetized.
5. After touching the part with the probe tip, remove the demagnetizer SLOWLY to at least one foot away. In this way, the induced magnetic field diminishes to zero gradually.

Move slowly. If you touch the demagnetizer to a head and quickly remove it, you'll magnetize the head worse than when you started!

Demagnetize each tape head and tape guide in this manner—one at a time. Finally, withdraw the degausser at least one foot away and unplug it. Demagnetize your machines after every eight hours of use and before playing an alignment tape. The same precautions about slow operation apply to a bulk tape eraser.

Alignment and Calibration for a Cassette Deck

For the best high-frequency response (brilliance and clarity), the gap in each head must be exactly at a right angle to the tape edge. This is called **azimuth alignment** (Figure 9.5). The azimuth is the left-right head angle relative to the tape edge. Cassette decks are aligned at the recorder factory, and usually stay aligned if you treat the recorder gently and avoid bumping it. In pro studios, however, engineers align the heads periodically with the aid of a **standard alignment tape** (described later).

In a home studio, if you clean and demagnetize the heads, and still think that the high-frequency response of your recorder is diminished, consider aligning the heads yourself or letting a professional technician do it. You need either a standard alignment tape or a commercially recorded cassette. You may need to remove the cassette door cover first.

Look for a small spring-loaded screw next to the record/playback head. This screw affects the azimuth. If you have an alignment tape, put it in your cassette deck and play the 15 kHz tone. Adjust the azimuth screw for maximum signal level as shown on your meters.

If you lack an alignment tape, put on a good commercially recorded cassette that has lots of cymbals or high-hat. Adjust the azimuth screw to the point where these instruments sound the most crisp and clear. You'll find a peak in the clarity at a certain screw rotation, and a duller sound on either side of that.

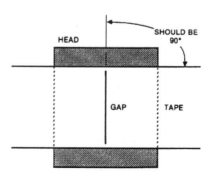

Figure 9.5
Azimuth alignment.

Before a cassette deck or recorder-mixer is shipped to your dealer, its electronics are factory calibrated—adjusted for best performance from a certain brand of tape. If your machine is calibrated correctly, the playback signal should sound just like the input signal (except for some added tape hiss, and perhaps a little loss of clarity or high end). The calibration is usually left alone, and you use the brand of tape for which the machine was adjusted.

Occasionally, these circuit adjustments drift, so the circuit may need to be recalibrated. It's a complicated procedure best left to a service technician. In fact, some home and semi-pro recorders are not designed for easy calibration. The internal parts to be adjusted may not be easily accessible.

Alignment and Calibration for Professional Tape Machines

Professional recording engineers align and calibrate their machines periodically. This ensures flat frequency response, the best S/N, and the least distortion. They also align and calibrate to correctly reproduce tapes made at other studios, using the alignment tones on those tapes.

To perform a complete alignment, you need:

- A small screwdriver
- An audio-frequency generator
- A standard playback alignment tape

Information about standard playback alignment tapes is available from various tape-recorder manufacturers, and from Standard Tape Laboratory (415-851-6600) or Magnetic Reference Laboratory (415-965-8187).

Follow your owner's manual about calibration. Clean and demagnetize the tape heads before starting. Basically, you follow these steps:

1. Using the alignment tape, play the 15 kHz tone and adjust playback-head azimuth for maximum output or for best phase match between channels (using an oscilloscope).

2. Adjust the high-frequency playback equalization (if any) to achieve the same output level at 700 Hz and 10 kHz. Or try for the flattest overall response if several tones are on the tape. Don't adjust the low-frequency equalization yet.

3. The magnetic field strength on tape (the fluxivity) is measured in nanowebers per meter (nWb/m). If you're using an alignment tape made at a standard operating level of 185 nWb/m (old Ampex standard level), set the playback level to read –3 VU or –6 VU as recommended

by the tape manufacturer. If you're using an elevated-level alignment tape that uses a standard operating level of 250 nWb/m or 320 nWb/m, set the playback level to read 0 VU (or as recommended by the tape or recorder manufacturer). Do not touch the playback level for the rest of the calibration.

4. Thread on some blank tape of the desired brand.

5. Record a 15 kHz tone and adjust record-head azimuth for maximum playback output, or for best phase match between channels. (Skip this step if your recorder combines the record and playback functions in a single head.)

6. While recording a 1 kHz tone, set the bias to achieve maximum playback level. Then go back to 10 kHz, and turn up the bias past that point (**overbias**) until the output drops 0.5 to 1 dB. Overbiasing reduces drop-outs and modulation noise. For other overbias settings, read the suggestions on the tape box.

7. While recording tones of 10 kHz, 100 Hz, and 700 Hz, adjust the high-frequency record equalization and low-frequency playback equalization (if any) to achieve the same playback output level at all frequencies. Or use many tones to achieve the flattest overall response. Record the tones at 0 VU for 15 ips, –10 VU for 7 1/2 ips, and –20 VU for cassettes. The slower tape speeds need lower recording levels to prevent tape saturation at high frequencies.

8. Feed a 1 kHz tone at 0 VU from the mixing console to the recorder. Record the tone. Set the record level so the recorder reads 0 VU on playback.

9. Set the "record cal" or "meter cal" so that the meter reads 0 VU on "input" or "source."

After calibration, your tape machine will operate as well as possible with the particular type of tape you're using. The playback signal should sound identical to the input signal (except for some added tape hiss).

Using Magnetic Tape

Now that you have been introduced to tape recording hardware, it's time to consider the tape itself—its editing, handling, and storage.

Editing and Leadering

When you edit a tape, you cut and rejoin the tape to delete unwanted material, to insert leader tape, or to rearrange material into the desired sequence.

If you're doing all your work on cassettes, or if you're editing with a digital audio workstation, you can skip this section.

Equipment and Preparation

To edit, you need these materials: demagnetized single-edge razor blades, a light-colored grease pencil, splicing tape, **leader tape,** and an **editing block.** Leader tape is plastic or paper tape without an oxide coating, which is used for a spacer between takes (i.e., silence between recorded songs). Paper leader is preferred over plastic because plastic can generate static electricity pops. An editing block holds the tape during the splicing operation. It's easier to use than a tape splicer with hold-down tabs and allows more-precise cuts.

Before editing, wash your hands to avoid getting oily spots on the tape. Cut several 1-inch pieces of splicing tape and stick them on the edge of the tape deck or table. Also cut several sections of leader at the 45-degree slot in the editing block. A typical leader length between songs is 4 seconds, which is 60 inches long for 15 ips or 30 inches long for 7 1/2 ips. While editing, try to hold the magnetic tape lightly by the edges.

Leadering

Suppose you've recorded a reel full of takes and you want to remove the out-takes, count-offs, and noises between the good takes. You also want to insert leader between each song. This process, called **leadering,** can be done like this:

1. Wind several turns of leader onto an empty take-up reel and cut the leader at the 45-degree slot.
2. Remove this take-up reel, put on an empty one, and play the tape to be edited.
3. Locate the beginning of the first song's best take. Stop the tape there.
4. Put the machine in cue or edit mode so the tape presses against the heads.
5. While monitoring the tape-recorder output, rock the tape back and forth over the heads by rotating both reels by hand—first rapidly, then more and more slowly. You'll hear the music slowed down and low in pitch.
6. Find the exact point on tape where the song starts; that is, where it passes over the playback-head gap. Align the beginning sound with the gap.
7. Using the grease pencil, mark the tape about 1/2 inch to the right of the gap (that is, at a point on tape just before the song starts).

8. Loosen or "dump" the tape by rotating the supply reel counterclockwise and the take-up reel clockwise simultaneously.

9. Remove the tape from the tape path and press it into the splicing block, oxide side down.

10. Align the mark with the 45-degree-angled slot (Figure 9.6).

11. Slice through the tape with a razor blade, drawing the blade toward you. Don't use the 90-degree slot because such an abrupt cut can cause a pop noise at the splice.

12. Remove the unwanted tape to the right of the cut and put the take-up reel aside.

13. Slide the cut end of the tape to the right of the editing-block slot (Figure 9.7).

14. Put on the take-up reel containing the turns of leader tape, and insert the end of the leader into the right half of the block.

15. Slide together the ends of the leader tape and magnetic tape so that they butt or touch together with no overlap.

16. Take a piece of splicing tape and stick a corner of it onto a hand-held razor blade.

(A) Mark for beginning of song.

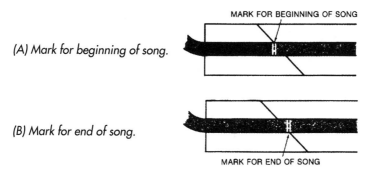

(B) Mark for end of song.

Figure 9.6 Aligning edit marks with cutting slot.

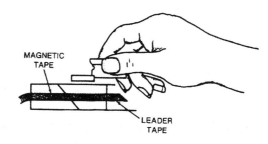

Figure 9.7
Applying splicing tape.

17. Align the splicing-tape piece parallel to the recording tape.

18. Apply the piece over the cut onto the non-oxide side, and stick it down by rubbing with your fingernail.

19. Slide the splice out of the block. Gently pop the tape out of the block by pulling up on the ends of the tape extending from both sides of the block. Twist the tape toward you while pulling.

20. Check that there is no gap or overlap at the splice.

21. Wind the tape onto the take-up reel and locate the ending of the first song.

22. As it ends, turn up the monitors and listen for the point where the reverberant "tail" of the music fades into tape hiss. Stop the tape there and mark it lightly at the playback-head gap (at the center line of the head).

23. After pressing the tape into the block, cut the tape at the mark. Remove the tape to the left of the cut.

24. Splice the end of the first song to a 4-second length of leader and again check the splice.

25. Wind the first song and the leader onto the take-up reel and remove it.

26. Put on the take-up reel containing unwanted material you set aside previously. Splice it to the rest of the master tape.

27. Locate the beginning of the next good take you want in the program. Mark it and cut the tape.

28. Put the reel containing the first song on the take-up spindle.

29. Splice the tail end of the leader onto the beginning of the second song, then wind the second song onto the take-up reel. You now have two songs joined by leader tape.

Repeat these steps until all the good takes are joined by leader. You then have a reel of tape with several songs separated by white leader, which makes it easy to find the desired selection.

Joining Different Takes

What if you want to join the verse of Take 1 to the chorus of Take 2? You have to cut into both takes at the same point in the song, then join them. It takes practice to make an inaudible splice in this manner, but it's done every day in professional studios.

The two takes must match in tempo, balance, and level for the edit to be undetectable. To mask any clicks occurring at the splice, cut the tape just before a beat—at the beginning of a drum attack, for instance. An alternative is to cut into a silent pause. If you cut into a continuous sound such as a steady chord, a cymbal ring, or reverberation, the splice is noticeable.

Follow these steps to join different takes:

1. Play Take 1 and locate the point where you want Take 1 to stop and Take 2 to start—at the beginning of the chorus, for instance. Stop the tape there.

2. Put the recorder in cue or edit mode, rock the tape, and try to identify a beat or attack transient.

3. At the point on tape where this beat just starts to cross the playback-head gap, mark the tape.

4. Cut the tape at the mark and remove the take-up reel containing the verse of Take 1.

5. Put on an empty take-up reel, thread the master tape, and fast-wind to Take 2.

6. Find the same spot in Take 2 that you marked in Take 1. Mark and cut it.

7. Using splicing tape, join Take 2 (in the supply reel) to Take 1 (in the take-up reel you just set aside). Again, check that there is no gap and no overlap at the splice.

Play the spliced area to see if the edit is detectable. If not, congratulations! It should sound like a single take. If Take 2 comes in a little late, carefully remove the splice and cut out just a little tape surrounding the cut. Resplice and listen again.

Suppose you've recorded most of a good take, but the musicians make a mistake and stop playing. Rather than repeating the entire song, the musicians can start playing a little before the point where they stopped and then finish the song. You splice the two segments into a complete and perfect take. Editing is also useful for inserting sound effects in the middle of a song, or for making tape loops. You can even record a difficult mixdown in segments and then edit the segments together.

Reducing Print-Through

Print-through is the transfer of a magnetic signal from one layer of tape to the next, causing an echo. If the echo follows the program, it is called **post-echo.** If the echo precedes the program, it is called **pre-echo.** Print-through is easy to hear in recordings that have many silent areas, such as narration. To minimize print-through:

- Demagnetize the tape path (because stray magnetic fields increase print-through).

- Use 1 1/2 mil tape (thinner tapes increase print-through). C-60 cassette tape is thicker than C-90, so C-60 is preferred.

- Use noise-reduction devices (discussed earlier in this chapter).
- Store tapes at temperatures under 80 degrees Fahrenheit, and don't leave tapes on a hot machine (heat increases print-through).
- Rewind tapes in storage at least once a year. This action allows print-through to decay by separating and re-aligning adjacent layers of tape.
- Store tapes tail out. That is, after playing or recording a tape, leave it on the take-up reel. Rewinding a tape about 15 minutes before playing helps to reduce print-through that may have occurred during storage. Also, tail-out storage results mainly in post-echo, which is less audible than the pre-echo you get in tapes stored rewound.

Tape Handling and Storage

Tape reels must be handled and stored carefully to avoid damaging the tape and the signals recorded on it. If you examine a reel of used recording tape, you may see some edges or layers of tape sticking out of the tape pack. These edges can be crushed by pressure from the reel flanges, causing drop-outs and high-frequency loss. Figure 9.8 shows how to hold reels to prevent tape damage.

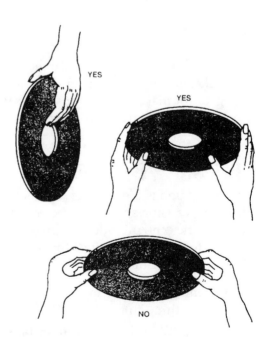

Figure 9.8
Handling tape reels.

To prevent edge damage during storage, leave tapes tail out after playing or recording to ensure a smooth tape pack. Repair or discard reels with a bent flange. Reels left out in the open can collect dust, so keep them in boxes. Store tape boxes vertically, not stacked. The preferred storage conditions are 60 to 75 degrees Fahrenheit and 35 to 50 percent relative humidity. Keep tapes away from magnetic fields such as those caused by speakers, headphones, or telephones.

10

DIGITAL RECORDING

The types of recorders described so far are analog recorders. That is, the magnetic particles on tape are oriented in patterns analogous to the audio waveform. Digital recorders convert the audio signal to a numerical code of ones and zeros.

Let's venture into the world of digital audio. We'll overview digital recording, DAT recorders, digital editing, CD-R recorders, Modular Digital Multitracks, hard disk recorders, and MiniDisc multitrack recorders.

Analog vs. Digital

Analog and digital recorders don't sound the same. Analog decks sound reasonably accurate, but they add a little warmth to the sound. It's due to slight third harmonic distortion, head bumps (bass boost), and tape compression. Analog decks also add some tape hiss, frequency response errors, wow and flutter, modulation noise, and print-through.

Digital recorders don't have these problems, so they sound very clean. While some digital recorders sound harsh compared to analog, they are improving with each generation. Both analog and digital have their colorations. Just use whatever works artistically for the particular music you're recording. Compared to analog recorders and open-reel tape, digital recorders and their tape tend to cost less, are smaller, and allow easier tape loading.

Digital Recording

Like an analog tape deck, a digital tape recorder puts audio on magnetic tape, but in a different way. Here's what happens:

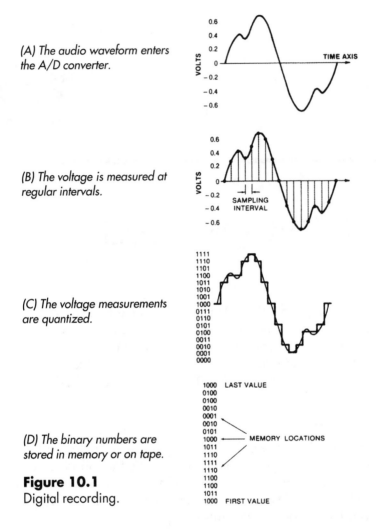

(A) *The audio waveform enters the A/D converter.*

(B) *The voltage is measured at regular intervals.*

(C) *The voltage measurements are quantized.*

(D) *The binary numbers are stored in memory or on tape.*

Figure 10.1
Digital recording.

1. The signal from your mixer (Figure 10.1 A) is run through a lowpass filter (**anti-aliasing filter**) which removes all frequencies above 20 kHz.

2. Next, the filtered signal passes through an **analog-to-digital (A/D) converter.** This converter measures (samples) the voltage of the audio waveform several thousand times a second (Figure 10.1 B).

3. Each time the waveform is measured, a binary number (made of 1's and 0's) is generated that represents the voltage of the waveform at the instant it is measured (Figure 10.1 C). This process is called **quantization.** Each 1 and 0 is called a **bit,** which stands for binary digit.

4. These binary numbers are stored magnetically on tape as a modulated square wave recorded at maximum level (Figure 10.1 D).

The playback process is the reverse:

1. The binary numbers are read from tape.

2. The **digital-to-analog (D/A) converter** translates the numbers back into an analog signal made of voltage steps.

3. An **anti-imaging filter** (lowpass filter) smoothes the steps in the analog signal, resulting in the original analog signal.

Since the digital playback head reads only 1's and 0's, it is insensitive to tape hiss and tape distortion. Numbers are read into a buffer memory and read out at a constant rate, eliminating speed variations. A process called **interpolation** restores lost data. If a bit is missing, an error-correction circuit "guesses" at its correct value based on the bytes that came before and after.

All digital recording devices employ the same A/D, D/A conversion process, but use different storage media. A DAT machine records on tape; a hard disk drive records on magnetic hard disk; an MO drive records on a magneto-optical disc; a compact disc and DVD disc record on an optical disc; and a sampler records into computer memory.

As we said, the audio signal is measured many thousand times a second to generate a string of binary numbers. The longer each binary number is (the more bits it has), the greater is the accuracy of each measurement. Short binary numbers give poor resolution of the signal voltage; long binary numbers give good resolution. A **quantization** of 16 bits is adequate (but not optimum) for hi-fi reproduction. It is the current standard for compact disc. Some digital recorders offer 18, 20, or 24 bit quantizing.

Sampling rate is the rate at which the A-to-D converter samples or measures the analog signal while recording. For example, a rate of 48 kHz is 48,000 samples per second. That is, 48,000 measurements are generated for each second of sound. The higher the sampling rate, the wider the frequency response of the recording. The upper frequency limit is slightly less than half the sampling rate. If the sampling rate is 44.1 kHz, this gives a flat response up to 20 kHz. A 96 kHz sampling rate is being considered for the DVD (Digital Versatile Disc).

The Clock

Each digital audio device has a **clock** that sets the timing of its signals. The clock is a series of pulses running at the sampling rate. When you transfer digital audio from one device to another, their clocks must be synchronized. If you send digital audio from one device, the receiver syncs to the sender's clock, which is embedded in its digital signal. If you send data from many sources at once, use a separate cable to connect one device's **word clock** output to all other devices.

Digital Audio Signal Formats

Digital audio signals are in two basic formats, and use the connectors described below:

- AES/EBU or AES3-1985: Professional type. Uses a balanced 110-ohm mic cable terminated with XLR-type connectors. The signal contains digital audio plus a word clock, or a separate word clock on another cable.
- IEC 958 or S/PDIF (Sony/Philips Digital Interface) or EIAJ CP-340 Type II: Consumer type. Uses a 75-ohm coaxial cable terminated with RCA or BNC connectors. The signal contains digital audio plus a word clock.
- Optical: S/PDIF format signal on a fiber optic cable, terminated with a Toslink connector. Optical interfaces prevent ground loops and cable losses.

In a digital transmission, the two channels of a stereo program are **multiplexed.** That is, one word from channel 1 is followed by one word from channel 2, which is followed by one word from channel 1, and so on.

AES and S/PDIF signals are similar but not necessarily compatible. You can convert one to the other using a **format converter.** Some digital audio devices do not implement AES or S/PDIF correctly, so they do not interface with some other devices.

DAT

Like analog recorders, digital recorders come in 2-track and multitrack formats. Currently, 2-track open-reel digital recorders are quite expensive. There's a low-cost alternative: a DAT or R-DAT recorder. R-DAT stands for Rotating-head Digital Audio Tape. Costing $700 and up, a DAT machine records 2 tracks of audio digitally on a small cassette. This convenient format has become the standard for mastering and audio backup. However, DAT is likely to be replaced by rewritable CD-R.

DAT sound quality is at least as good as a compact disc. What you put in, you get out, virtually without any added hiss, distortion, or wow and flutter. Sonic drawbacks might include a little phase shift at high frequencies, quantization error (imprecise measurement of the analog signal), and jitter (unstable timing of the samples). All these can make the sound slightly harsh or veiled. It's a subtle effect, though.

The DAT machine records on a small cassette about half the size of a standard analog cassette. The shell has a hinged door that flips open to expose the tape. Thanks to the DAT's fast transport, a two-hour tape winds in about 45 seconds.

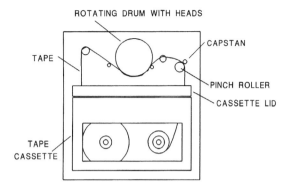

ROTATING DRUM WITH HEADS

CAPSTAN

TAPE

PINCH ROLLER

CASSETTE LID

TAPE CASSETTE

Figure 10.2
Inside a DAT recorder.

When you load the cassette, two spindles enter the two hubs from below the cassette and lock it into place. Then the tape is drawn into the machine and wraps around a rotating drum, which holds the record and playback heads (Figure 10.2).

As the tape moves past the rotating heads, they scan diagonally across the tape as they write the digital signal by pulse code modulation. The tracks laid down by the rotating heads are about one-tenth the thickness of a human hair! Although the linear tape speed is slow, the tape speed as seen by the record head is very high, so the ultrasonic digital signal can be recorded. Tape drop-outs can cause data losses, but most of these are restored by a circuit that corrects the errors.

As the DAT machine records, it writes absolute time (ABS) on tape. ABS time is the running time in hours, minutes, and seconds, where 00:00:00 is at the head of the tape. You use ABS time to locate selections and time them.

DAT Tape

A DAT cassette measures 73 x 54 x 10.5 mm. Its tape is 3.81 millimeters (1/8-inch) wide. On one end of the shell is a hinged door that flips open to expose the tape so it can be drawn around the rotating head drum. The cassette has a sliding safety tab on one end to prevent accidental erasure. If the little door is closed, you can record on the tape. If it's open, you can't.

DAT machines can record up to two hours on a 60-meter tape. Using a metal-powder oxide, DAT tapes come in various lengths. Computer-grade DAT tapes (DDS) are more error-free than audio-grade DAT tapes.

Before recording on a DAT tape, it's a good idea to exercise the tape: fast forward to the end and rewind to the top. This distributes the tape lubricants more evenly, aligns the tape with your machine's tape guides, and loosens the tape pack so it travels more freely. If audible errors occur, clean the heads with a dry cleaning cassette.

DAT Subcodes

A **subcode** is information on tape that is independent of the audio data. Subcodes tell the number of each song, where it starts, and whether to play it. You can record or erase subcodes without affecting the audio program. The three main types of subcodes are **Start IDs, Program Numbers,** and **Skip IDs.**

- A Start ID marks the beginning of each song. You can write it manually or automatically. Manual Start IDs can be placed anywhere, say, to mark cues. You can set the DAT machine to record a Start ID automatically whenever a signal appears after a silence of three or more seconds. You should record manual Start IDs after recording all your mixes onto a DAT tape. Like other subcodes, Start IDs can be recorded or erased without altering the audio program. You can enter them during recording or playback. If the cassette's safety tab is set to prevent accidental erasure, you can't record or erase subcodes.

- A Program Number is the number of each song. A DAT machine can automatically assign Program Numbers to the Start IDs in consecutive order. Then you can locate a particular song quickly by having the DAT recorder search for the song's program number.

- A Skip ID makes the machine skip the song. This ID can only be written manually. Whenever the machine senses a Skip ID during playback, it stops and fast winds to the next Start ID, and begins playing. You can turn this function on or off.

DAT Features

A DAT recorder has the usual transport controls, and some unique features, indicators, and controls. Some DAT machines include these features:

- **Search:** When you enable the search function, the machine fast winds to the selected program number.

- **Memory Rewind or Return to Zero:** When enabled, this function rewinds the tape to a preset "0" position on the tape counter.

- **Renumber:** This function automatically numbers the Start IDs consecutively from 1 upwards.

- **SMPTE time code:** This is available on a few models, digitally encoded as part of the subcode. It's used to synchronize the DAT audio program with a video tape or a MIDI sequencer.

- **Mic inputs:** These let you plug two mics directly into the DAT recorder. Generally the sound quality of the DAT mic preamps is not as good as an external mixer or mic preamp. You have a choice of pro-type balanced XLR connectors or consumer-type unbalanced phone jacks.

- **Digital inputs and outputs:** These let you make digital copies with no A/D-D/A conversion required. A copy made from digital out to digital in is a clone of the original recording, with no loss in sound quality (unlike an analog copy).

- **Confidence head playback:** Some units have an extra head so you can hear the tape playback as you're recording.

Other features: a built-in monitor speaker, a clock/calendar, phantom power for condenser mics, an input attenuator to prevent mic preamp overload, a limiter to prevent recording beyond 0 dB, and a stereo mic. Some DAT recorders are portable, and they run off batteries or AC.

DAT Indicators

The DAT recorder has an LED or LCD display window which shows various aspects of the recording:

- **Time:** Absolute time, remaining time on the tape, or elapsed time for the current selection.

- **Meters:** Peak-reading bargraph meters that show recording level. Unlike with an analog tape deck, "0" on the meter is the absolute maximum recording level. It's called 0 dBFS, or 0 dB full scale. When the signal level hits 0, all 16 bits are on.

- **Error:** If this indicator flashes, there was a loss of data. Usually the electronics can correct for this loss. Some units show the error rate.

- **Sampling rate:** The frequency at which the machine is sampling audio signals, such as 44.1 kHz. This indicator works both during recording and playback.

- **Subcodes:** Program number, Start ID, and Skip ID.

- **Search mode:** This shows whether the search mode is on.

DAT Controls

- **Sampling Frequency:** This is the rate at which the DAT's A to D converter samples or measures the analog signal while recording. On most models you can switch between 48 kHz and 44.1 kHz. Some units can also record at 32 kHz.

Which rate should you use? For CD release, use 44.1K (same as the CD). Otherwise, use 48K for better sound. For longer playing time or four channels, use 32K.

What if you have a 48K tape that you want to release on CD? The CD mastering engineer can (1) convert the 48K digital signal to 44.1K, which might degrade the sound, or (2) use the analog output of their DAT machine, and convert the analog signal to digital at 44.1K. This may or may not sound better. During playback, a DAT recorder senses the sampling rate on tape and automatically switches to that rate.

- **Analog/digital input selector:** This chooses between analog and digital input signals.

- **Emphasis:** Used in older consumer DAT recorders, this boosts high frequencies during recording and turns them down during playback. The result is lower noise. When you play a DAT tape recorded with emphasis, the DAT machine detects it and turns on de-emphasis.

    Don't use both emphasized and non-emphasized recordings in the same project. If you copy them to hard disk, they all will play with emphasis, or without it. Then some recordings will sound too bright or too dull.

- **Copy protect:** This puts a flag in the subcode that prevents the tape from being digitally copied. A DAT deck will identify a copy-inhibit flag in its subcode, and will not digitally copy that recording. Be sure copy inhibit is OFF if you want to duplicate your DAT master digitally.

    Consumer DAT machines have a copy inhibit system called SCMS (**Serial Copy Management System**). This feature lets you make a digital-to-digital copy, but prevents regenerations from that copy. That is, you can't make digital copies of the copy. Most compact-disc duplicators can disable or ignore the SCMS copy-code bit.

    For more on DAT, check out the DAT-head's mailing list on the Internet. Subscribe to DAT-heads-request@fedney.near.net.

    When you edit a DAT recording, splicing the tape doesn't work because the tracks won't align perfectly after the splice, and you may remove some data. Instead, you can copy from one DAT machine to another, changing the order of selections during the copying process. Tight edits are impossible unless you use a digital audio workstation (DAW), covered next.

The Digital Audio Workstation

The digital audio workstation (DAW) allows you to record, edit, and mix audio programs entirely in digital form, providing the highest sound quality. The DAW can store up to several hours of digital audio or MIDI data. You can edit this data with great precision on a computer-monitor screen. What's more, you can add digital effects and perform automated mixdowns.

DAWs are available in two general forms:

1. A stand-alone system or computer plus controller interface
2. A personal computer system, plus a special plug-in card, high-capacity hard drive, and editing software (Figure 10.3)

The stand-alone system uses real controls—faders and pushbuttons—while the personal computer system uses **virtual controls** simulated on your monitor screen. Stand-alone systems cost more than computer systems, but are faster and easier to use. Both record digital audio on a hard disk. The stand-alone system is a hard-disk recorder-mixer (covered later) or a computer and a controller interface.

The personal computer workstation includes a sound card you plug into a computer, and editing software. The sound card converts audio into computer data which is recorded on hard disk. The card also converts hard-disk data back into audio. Some cards have an A/D converter built in; others accept a digital signal directly from your DAT or digital mixer.

Once the audio data is on hard disk, it can be read by the read head in the hard drive. Since the head can be controlled to jump to any particular location on disk, it allows **random access,** and nearly instant access to different parts of the audio program. You can quickly locate or edit any portion of the program. To rearrange or repeat song sections, you simply tell the computer which order you want them played.

Regarding the hard drive, you need about 11 megabytes per minute of stereo program, plus more for temporary audio files and regular programs. A 1.2 gigabyte hard drive is big enough to record a full-length CD program. The drive should be an A/V (audio/video) type with embedded servo. It

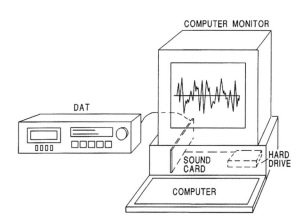

Figure 10.3
A personal computer digital
audio workstation.

postpones thermal recalibration until the drive is idle. The drive should be capable of high throughput (at least 1000 kilobytes/sec for 2 tracks, and 2000 to 4000 kilobytes/sec for 4 stereo tracks).

One function of a DAW is for compact disc pre-mastering. You start with your rough DAT mixdown master containing music, studio noises, and long spaces between songs. After copying it to hard disk, edit this program with the workstation. That is, define the start and stop points of each song. Finally you dump the edited program onto DAT, and that is your CD premaster. You can even record your own CDs off the workstation with a compact disc recorder (CD-R writer).

Below is a general editing procedure for DAWs:

1. Play the DAT tape containing your program, and record it onto the hard disk. This recording might be just one song, several sound effects, or an entire album.

2. The waveform of the audio program appears on your monitor screen (Figure 10.4). You can **zoom out** to see the entire program, or **zoom in** to see individual samples.

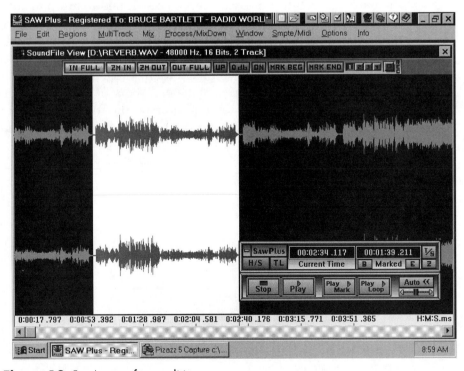

Figure 10.4 A waveform editing screen.

3. Using a mouse, mark the beginning and end of **regions** in the audio waveform. These are sections of the program that you want to keep. In Figure 10.4, a region is highlighted. Examples of regions are an entire song, the chorus of a song, a drum riff, or a vocal lick. The computer keeps track of these start/stop points—called **pointers**—which are addresses on your hard disk. Also define some regions of silence—4 seconds, for example—to insert between songs.

4. Once you have defined the regions, put them into a stereo track, or into a playlist. A **playlist** (**Edit Decision List**, **EDL**, or **Sequence View**) is a sequence of regions in the order you want them to play (Figure 10.5).

5. While editing, you can delete regions, move them around, copy them, insert silences, and re-sequence the playlist so that the songs progress in any desired order. Some edits may be **destructive:** they change the data on disk. Most edits are **non-destructive:** they change only the pointers, so the data on disk is not changed or destroyed. Non-destructive edits are not permanent. If you don't like an edit, you can undo it and do it over.

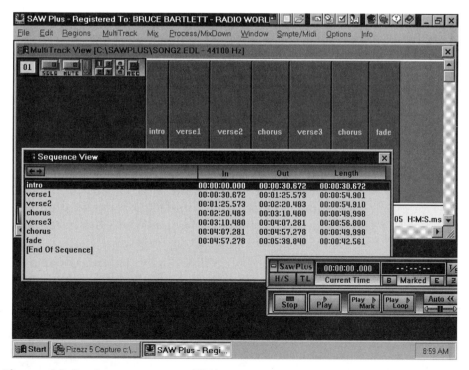

Figure 10.5 A sequence view (EDL) screen.

6. Once the program is edited the way you want it, copy it in realtime onto a DAT tape. During this process, the software plays the program by running down the playlist of pointers. The pointers tell the hard-disk head where on the disk to read the data from (Figure 10.5). That is, the playlist tells the head to jump around from point to point on the disk to play selected portions in the desired order.

DAW Functions

Here are some of the functions found in most digital audio workstations:

MULTITRACK DIGITAL RECORDING: From 2 tracks on up.

CUT AND PASTE: Remove a section of a song and put it somewhere else in the song.

COPY: Duplicate a section of a song (say, a chorus) and put it somewhere else in the song.

CROSSFADE: Fade out of one song while fading into another.

FADE-IN, FADE-OUT: Do an automated fade.

SLIP: Move a section of a song in time, either within or across tracks. Or adjust the start times of sound effects by moving them on-screen with a mouse. You can slip one track forward or backward in time, independent of other tracks.

SCRUB: Play the program slowly forward and backward (like rocking tape reels to move the tape back and forth across the head). This lets you find edit points accurately.

TRIM: Remove or truncate unwanted portions of the waveform or program.

DSP (Digital Signal Processing): Digital control and effects such as mixing, volume adjustments, panning, equalization, reverberation, echo, chorus, limiting, looping, compression, expansion, noise removal, and noise gating. Third-party manufacturers make **plug-ins,** which are DSP effects in the form of software. When you install the software, it becomes part of the DAW editing program.

TIME COMPRESSION/EXPANSION: This makes an audio program shorter or longer without changing its pitch. It's useful for making radio spots fit their allotted time.

AUTOMATED MIXING: The computer remembers your mix moves and sets the mixer controls accordingly during the mixdown. For more information, see the section on Automated Mixing in Chapter 16.

SPECTRUM ANALYSIS: A display of level vs. frequency of the audio program as it progresses in time.

MIDI SEQUENCER: This lets you import or record several tracks of MIDI performance data.

SYNC: Synchronization to SMPTE time code or MIDI time code.

PQ EDITING: This allows you to add song start-and-stop times and index points to create a pre-master tape for compact discs.

CD Recordable

Another form of digital recording is the compact disc. Right on your own desktop, you can cut a compact disc by using a CD-R recorder. The disc will play in any CD-ROM drive or audio CD player. It's exciting to hear one of these CD's playing your music with the purity of digital sound.

CD-R stands for Compact Disc Recordable. This optical medium is a **write-once** (non-erasable) format, but rewritable CD-RW discs are available. Sound quality meets or exceeds compact disc standards, and the expected lifetime is about 70 years.

How can you use the CD-R format? You could make demos that sound much cleaner than cassettes. Or make a one-off copy of your stereo mixes for clients. Use the CD-R as a premaster (PMCD) to send to a CD replicator. Another function is to compile sound libraries of production music, samples, and sound effects. The CD format is a more dependable storage medium than DAT, so it's a great way to archive your recordings.

CD-R Equipment Needs

Want to try CD-R? First you'll be faced with two basic choices of CD-R recorder:

1. A **stand-alone CD-R recorder,** sometimes called a consumer CD recorder.
2. A **computer peripheral CD-R recorder,** also called a CD-ROM recorder. You plug it into your computer system.

Both types produce discs that sound equally good. They are not compatible with each other: a consumer disc will not play in a CD-ROM recorder, and vice-versa. But both types of disc will play on any audio CD player.

The stand-alone CD recorder has everything you need in one chassis. Inside is a CD transport, laser, and microprocessor. On the back are analog and digital ins and outs. On the front are the level meters, record-level knob, display, and keypad. Because the stand-alone unit needs no external computer, it's user-friendly. Just connect your audio source containing an edited program, either on DAT or analog tape. Set the recording level and start recording.

The stand-alone unit can write audio but not data. The 63-minute blank discs it uses are "CD for Consumer" format, and sell for around $20. (An exception is the Marantz CDR610, which can use cheaper blanks.) Prices for stand-alone CD-R writers start around $1700.

A peripheral CD recorder costs less: about $250 and up. The unit plugs into a computer expansion port or SCSI connector. You can write audio or computer data. Its blank discs, in the "CD for Professionals" format, cost $7 to $10. Disc length is 18, 21, 63, or 74 minutes. Maximum program length is 71.5 minutes or about 650 MB.

The peripheral CD-R recorder also requires a CD recording program, usually packaged with the recorder. You don't have to use that program; other ones are available that you might prefer. Be sure the program is compatible with your recorder. Most software requires at least a 486DX2-66 computer with 16 MB RAM. You'll also need a sound card (44.1K, 16-bit), and some software to record audio onto hard disk.

Regarding the hard disk drive, use an A/V (audio/video) type of at least 1 GB capacity and high throughput. If you want to burn more than one CD at once, you can daisy-chain CD-R recorders together.

CD-R Technology

While conventional CD players follow the Sony-Philips Red Book standard, CD-R's conform to the Orange Book part II standard. Once recorded, a CD-R disc meets the Red Book standard.

A recordable CD is the same size as a standard compact disc, but is more colorful. On top is a layer of gold; on the bottom is a recording layer made of blue cyanine dye. Actually, the blue layer appears green because of the gold layer behind it. Some CD-R discs use a yellow (gold) dye of phthalocyanine. Yellow dye lasts a little longer in accelerated aging tests, and it may work better with high speed CD-R drives. Still, most CD-R writers are optimized for cyanine dye, which can handle a wider range of laser power.

A blank CD-R is made of four layers:

1. Clear plastic (protects the gold layer)
2. Gold (reflects laser light)
3. Dye (for the recording)
4. Clear plastic (protects the dye layer)

The dye fills a spiral groove which is etched in the bottom clear-plastic layer. This groove guides the laser.

To record data on disc, the laser melts holes in the dye layer. The plastic layer flows into the holes to form pits. During playback, the same laser reads the disc at lower power. At each pit, laser light reflects off the gold layer. The reflected light enters the laser reader, which detects the varying reflectance as the pits go by.

In contrast with a standard CD, a CD-R disc has two more data areas:

- The **Program Calibration Area** (PCA). The CD recorder uses this area to make a test recording, which determines the right amount of laser power to burn the disc (4 to 8 milliwatts).

- The **Program Memory Area** (PMA). This area stores a temporary table of contents (TOC) as the CD-R tracks are being assembled. The TOC is a list of the tracks, their start times, and the total program time. The recorder uses the Program Memory Area for this information until it writes the final TOC.

CD-R Configurations

Before we look at the differences among CD-R recorders, we need to understand the concept of a **session**. A session on disc is a lead-in, program area, and lead-out. Each session has its own TOC. Each lead-in and lead-out consume 13 MB of disc space.

With the **multisession** feature, you can write several sessions on a disc at different times. This feature comes in handy when you need to add information to a disc a little at a time. Only the first session on disc will play on an audio CD player, so the discs are just for your own use—not for distribution.

Some CD-R recorders permit **Disc-at-Once** recording, in which the entire disc must be recorded nonstop. You can't add new material once you write to the disc. With the right software, Disc-at-Once lets you set the length of silence between tracks (down to 0 seconds), and lets you control how the tracks are laid out on disc. Disc-at-Once is the pro audio format.

Most CD-R recorders allow **Track-at-Once** recording. They can record one track (or a few tracks) at a time—up to 99 tracks. You can play a partly recorded disc on a CD-R recorder. But the disc will not play on a regular CD player until the final TOC is written. Track-at-Once is not recommended for audio because it puts 2-second spaces and clicks between audio tracks.

If you want no pauses between tracks (as on a live album), get a CD-R writer with Disc-at-Once, and some software that can adjust the pause length down to zero. Note that a self-contained CD-R writer will copy your edited program as it is, with or without pauses.

A good CD-R writer has a buffer of at least 1 MB. Some writers include SCMS copy code. Be sure that you can return the CD-R writer if it proves to be unreliable.

How to Use a Stand-Alone CD-R Writer

Let's say that you've compiled a DAT of song mixes. You want to copy them onto a CD by using a self-contained CD-R writer. The first step is to edit your mixes into a finished program. To do this you'll need a sound card and a 2-track digital editing program. Here are the steps:

1. Connect your DAT recorder to the sound card. Transfer your mixes from DAT to hard disk.
2. While viewing the soundfile waveform, mark each song as a region (a selected segment of audio). Using a mouse, fine-tune the start and stop points of each region so that there are no noises between songs.
3. Assemble the songs in order into an Edit Decision List (a playlist or cue sheet). The total playing time should be under 60 minutes to fit on the CD-R disc.
4. Between songs, put silent spaces or crossfades. Three to four seconds of space is typical.
5. Have the computer change the level of songs that need it.
6. Start recording on your DAT. Play the cue list off the hard drive and copy it to DAT.
7. Once your edited program is on DAT, write a start ID 1/2 second before each song. Renumber the program numbers.

Now you're ready to play the edited DAT into the CD-R writer. Connect the DAT's output to the CD-R writer's input. Set your levels and begin recording. Your program will copy to disc in realtime.

Depending on the CD-R writer, the DAT start ID's may or may not convert to CD track numbers. If not, you can use a converter box, or add the track numbers manually while you record.

How to Use a Peripheral CD-R Writer

The procedure for this type of CD-R is different. First make the right connections. If your mixes are on DAT, plug the DAT's digital output into your sound card's digital input. If your sound card has no digital input, plug your DAT's analog output to your sound card's analog input. Other sources can be analog tape, LPs, or cassettes. Just connect the analog source signal to the sound card's analog input.

You'll need two types of software: sound recording/editing software and CD-R recording software. The first type creates WAV files on hard disk. The second is used to select and sequence the WAV files, and transfer them to CD. Some software performs both functions.

Follow these steps to record and edit:

1. Start with some recording/editing software of your choice. Record each song to hard disk as a separate WAV file. (Note that some programs let you record all your mixes as one long WAV file, then break it into separate files or start-ID pointers.)

2. If your software permits, trim the start and stop points of each WAV file. That lets you remove noises before and after each song.

3. Next, boot up the CD-R software that came with the recorder, or use other compatible CD-R software. Choose which WAV files you want to put on CD, and arrange them in order in your playlist.

The total playing time must be less than the CD-R length. CD-R discs cannot be erased and used over again, so try to make everything right before you burn a disc. With some software, you can create start ID's manually and fine-tune their timing; other software adds the start ID's automatically.

Track-at-Once inserts a 2-second gap between each song whether you want it or not. To make the gaps longer, record a 1-second silent WAV file and put it into your playlist where needed or edit silence onto each song's end.

PQ Subcode

At this point, you have completed the playlist. But you're not quite ready to burn the CD yet. Do you plan to have your CD-R replicated by a CD mastering house? If so, you need to understand **PQ subcode.** Basically the code provides track timing information. In all standard CDs, PQ subcode is inserted in the audio data stream.

The P channel part of the subcode indicates the currently read part of the disc: the lead-in, track beginning, data, or lead-out. The Q channel designates the track numbers, track timing, index numbers, absolute time, 2- or 4-channel format, pre-emphasis, copy prohibit, error detection, and the catalog number for the disc. Sometimes PQ subcode is called **Q subcode.**

When a CD mastering engineer cuts a commercial CD, he or she sends the PQ subcode in the form of a **PQ burst** to the CD mastering machine. Some CD-R software can put PQ code on disc, but not all. If your CD-R disc does NOT have PQ code, the mastering engineer will add it while transferring your program to a Sony PCM 1630 tape or Exabyte DDP tape. But if your CD-R disc DOES have PQ code, it can be used as is, without transferring to 1630 tape. It's a CD premaster.

Some PC software that does PQ code is CDRWIN by Goldenhawk Technology, CD Architect by Sonic Foundry, and Red Roaster by Microboards of America. On the Mac side, there's Toast CD-DA by Astarte, Sadie, Sonic Solutions, and Digidesign's Masterlist CD.

Cutting the CD

Now that your program is in final form, it's time to burn the CD. Connect your CD recorder, turn it on, and insert a blank disc.

At this point, you can either simulate a recording or actually burn a disc. The simulation (**CD image** or **physical image**) lets you check for errors in advance by writing the program to hard disk. You'll need up to 650 MB of free disk space for this simulation. As the image is written, the recorder light might flash, but the laser is running at low power to avoid burning pits.

With a **physical image,** the computer re-writes the sound files to a space on your hard disk and puts them in order. That's the physical image. Then you copy the physical image from hard disk to CD-R during a burn. With a **virtual image,** the computer grabs the sound files from random locations on hard disk and puts them in order as the CD-R does a burn.

A physical image takes up disk space, but it increases throughput so you can use a slower computer. A virtual image takes no extra disk space, but it requires a fast Pentium.

You might have a choice of transfer speed—up to 2, 4, or 6 times normal speed, depending on the throughput of your hard drive. High speeds do not degrade sound quality. Normal speed is 172 kilobytes per second, double speed is 344 kBps, etc.

At high speeds, a physical image is more likely to copy without errors than a virtual image. Also at high speeds, a yellow blank is more likely to write without errors than a green blank.

When you're ready, begin the transfer to the CD-R disc. The CD writer will burn the tracks in the order you chose, and will put in spaces and track numbers. You can check how the transfer is going by looking at the display. As soon as the recording is done, the display will indicate that the TOC is being written. Eventually the system will beep and eject the disc.

If you wish, you can burn more discs of the same program, or start over with a new one. Some software can even print liner notes, including the track numbers, titles, and timing.

Now you have a finished CD! To prevent error-causing fingerprints, be sure to handle the disc only by the edges. Pop the disc in an audio CD player, press Play, and check that all the tracks play correctly. Mission accomplished.

MDM—Modular Digital Multitrack

Another digital recording format is the MDM (Figure 10.6). It records 8 digital tracks on a videocassette, using a rotating drum like a DAT recorder. Two popular models are the Alesis ADAT-XT which records on S-VHS tape, and the TASCAM DA-88 or DA-38 which record on Hi-8mm tape. ADAT records up to 40 minutes on a single tape; DA-88 records up to 1 hour 48 minutes. It's great for recording concerts nonstop.

With both types, you can sync several 8-track units by a cable to add more tracks. Unlike SMPTE time code, MDM sync does not use up any tracks. MDM options include remote controls, remote editors, circuit boards with enhanced converters, and circuit boards that allow sync to SMPTE and MIDI.

To prevent data errors, be sure to format the MDM tape correctly. Fast forward the tape to the end, rewind to the top, clean the heads with a dry cleaning cassette, then format the tape.

Random-Access Multitracks

While MDMs record on tape, random-access units record on hard disk (HD), magneto-optical (MO) disk, or optical disc. Multitrack random-access units come in four formats:

- Computer DAW
- HD recorder or MO recorder
- HD recorder-mixer (Figure 10.7)
- MiniDisc recorder-mixer

All these units allow editing and can undo edits. Listed below are the characteristics of each type of random-access recorder:
Computer DAW:

- Records 2 to 16 tracks at a time per sound card
- Mixes to stereo in the computer

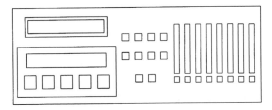

Figure 10.6
Modular Digital Multitrack
(MDM).

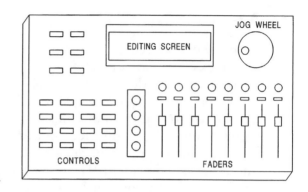

Figure 10.7
A hard-disk recorder-mixer.

- Uses on-screen faders that you control with a mouse
- Has a large screen for graphical editing of the waveform

HD recorder:

- Records and plays up to 24 tracks at once on a built-in hard drive
- Requires an external mixer
- Has a small LCD screen for editing
- Compared to a DAW, an HD recorder is less likely to crash, is cheaper and more portable, and tends to have better converters

HD recorder-mixer (stand-alone digital workstation):

- Records and plays up to 8 tracks at once on a built-in hard drive (some units can be daisy-chained for more tracks)
- Has a built-in mixer with real faders and knobs
- Has a small LCD screen for editing (some units have a monitor screen output)
- Compared to a DAW, an HD recorder-mixer is less likely to crash, is cheaper and more portable, and tends to have better converters

MiniDisc recorder-mixer

- Records and plays up to 8 tracks at once on a built-in MiniDisc
- Has a built-in mixer with real faders and knobs
- Has a small LED screen for non-graphical editing

Most random-access recorders let you associate several virtual tracks, or takes, with a single channel. Up to 8 tracks play at once, but you can choose which virtual track (take) plays on each track.

How do random-access units compare to MDMs? Random-access offers instant locating of song parts, instant computer editing of tracks, and fewer errors than tape. But tape costs much less than a disk of the same recording time. MDMs make it easy to remove a project and start a new one: just change the tapes. With a DAW or HD recorder, you must back up the project on DAT tape, delete the project, and start over. Or you could use a removable hard disk drive to change projects.

It's common to record music on an MDM, dump the MDM tracks to hard disk for editing, then dump the edited tracks back to MDM. Most HD recorders offer MDM interfaces for this purpose.

HD Recorder-Mixer Features

Listed below are some features to look for in HD recorder-mixers:

- Expansion ports for SMPTE and MTC sync, MDM interface, and extra ins and outs
- SCSI port to backup data to a removable hard drive
- Cue and locate points
- Jog/shuttle wheel to "scrub" audio—play it slowly forward and backward to locate an edit point
- Remote control
- Automated mixing
- Digital effects
- Monitor output to plug in a computer monitor screen

MiniDisc Recorder-Mixer

Another random-access device is the MiniDisc (MD). It is like a miniature compact disc inside a 2.5-inch square housing. It is a rewritable, magneto-optical storage medium that is read by a laser. A write-protect tab prevents accidental erasure. The disc type used for multitrack recording is called MD Data. One 140 MB disc can record up to 37 minutes of 44.1 kHz, 16-bit audio on 4 tracks.

To fit all this data on a small disc, MD recorders use a data compression scheme called ATRAC: **Adaptive Transform Acoustic Coding.** It reduces by 5:1 the storage needed for digital audio. ATRAC is a perceptual coding method, which omits data deemed inaudible due to masking.

For example, if an audio signal has two sounds that are about the same frequency, and one sound is louder than the other, the quieter sound will be inaudible due to masking. So ATRAC removes the quieter sound, which would be inaudible anyway.

The sound quality of ATRAC is nearly as good as a compact disc. However, there is a slight generation loss when tracks are copied or bounced. The signal is ATRAC-processed with each copy. Cumulatively, the sound begins to take on a mid-to-low rumble and a high-frequency squeak.

A MiniDisc recorder-mixer is a mixer and multitrack MD recorder in a single, affordable package. It uses an MD Data disc to record audio. Compared to a cassette recorder-mixer, a MiniDisc recorder-mixer has many advantages:

- Cleaner sound (no tape hiss, wow and flutter, or crosstalk)
- Much less distortion
- Instant access to any point in the recording (no waiting for fast-forward and rewind)
- No open track needed for bouncing (you can bounce 4 tracks down to 2 tracks on the same disc)
- Ability to write disc and track titles
- No need to clean heads
- Less generation loss
- More accurate recording timer
- More durable medium
- Tighter punch-ins
- Jog/shuttle wheel
- Markers (pointers) can be added
- Cue list playback and Program Play List (explained later)
- Editing: Track and song copy/erase, song divide, song combine

Let's go over editing in more detail. **Song Copy** duplicates a song at a new location, while **Track Copy** copies data from one track to another. **Song Divide** lets you divide a song in two at the current counter location. With this function, you can create sections—verse, chorus, etc.—that you can play or repeat in any order. You can also remove noises before and after songs. **Song Combine** joins divided parts from the same song. Some units let you move individual tracks in time.

The **Cue List** feature is a list of the song sections in the order you want them played. Sections can be looped or repeated. The recorder will play down the cue list, assembling the song from its sections. A **Program Play List** is a list of the songs in the order you want them played. All units offer an output for MIDI Time Code or MIDI Clock, in order to sync the MD to an external MIDI sequencer.

As we've seen, there are many digital formats to choose from. Read all you can on digital technology, then select the systems that meet your needs.

11

EFFECTS AND SIGNAL PROCESSORS

With effects, your mix sounds more like a real "production" and less like a bland home recording. You might simulate a concert hall with reverb. Put a guitar in space with stereo chorus. Make a kick drum punchy by adding compression.

Used on all pop-music records, effects can enhance the basic sounds on tape by adding spaciousness and excitement. They are essential if you want to produce a commercial sound. But many jazz, folk, and classical groups sound fine without any effects.

To add an effect to a track, you feed its signal from your mixer's aux send to an effects device, or **signal processor** (Figure 11.1). It modifies the signal in a controlled way. Then the modified signal returns to your mixer.

This chapter describes the most popular signal processors and suggests how to use them.

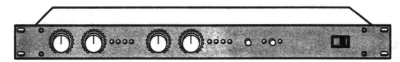

Figure 11.1 A signal processor.

Equalizer

Recall from Chapter 2 that an equalizer (usually in the mixer) is a sophisticated tone control, something like the bass and treble controls in a stereo system. Equalization (EQ) lets you improve on reality: add crispness to dull cymbals; add bite to a wimpy electric guitar. Or EQ can make a track sound more natural: for instance, remove tubbiness from a close-miked vocal.

To understand how EQ works, we need to know the meaning of a spectrum. Each instrument or voice produces a wide range of frequencies called its **spectrum**—the fundamentals and harmonics. The spectrum gives each instrument its distinctive tone quality or timbre.

If you boost or cut certain frequencies in the spectrum, you change the tone quality of the instrument. EQ adjusts the bass, treble, and midrange of a sound by turning up or down certain frequency ranges. That is, it alters the frequency response. For example, a boost (a level increase) in the range centered at 10 kHz makes percussion sound bright and crisp. A cut at the same frequency dulls the sound.

Types of EQ

Equalizers range from simple to complex. The most basic type is a **bass and treble control** (labeled LF EQ and HF EQ). Figure 11.2 shows its effect on frequency response. Typically, this type has up to 15 dB of boost or cut at 100 Hz (for the low-frequency EQ knob) and at 10 kHz (for the high-frequency EQ knob).

With a three-band EQ you can boost or cut the lows, mids, and highs at fixed frequencies (Figure 11.3). Sweepable EQ is more flexible because you can "tune in" the exact frequency range needing adjustment (Figure 11.4). If your mixer has sweepable EQ, one knob sets the center frequency while another sets the amount of boost or cut.

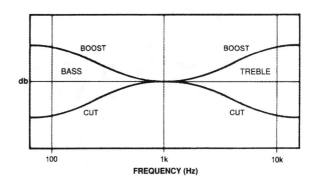

Figure 11.2
The effect of the bass and treble control.

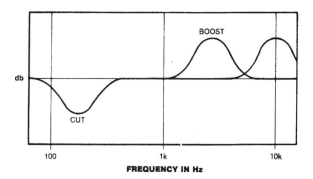

Figure 11.3
The effect of multiple-frequency equalization.

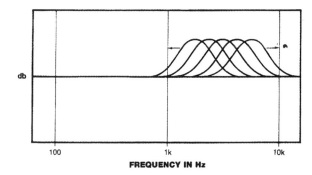

Figure 11.4
The effect of sweepable equalization.

Parametric EQ lets you set the frequency, amount of boost/cut, and bandwidth—the range of frequencies affected. Figure 11.5 shows how a parametric equalizer varies the bandwidth of the boosted part of the spectrum.

A graphic equalizer (Figure 11.6) is usually outside the mixing console. This type has a row of slide pots that work on 5 to 31 frequency bands. When the controls are adjusted, their positions graphically show the resulting frequency response. Usually, a graphic equalizer is used for monitor-speaker EQ, or is patched into a channel for sophisticated tonal tweaking.

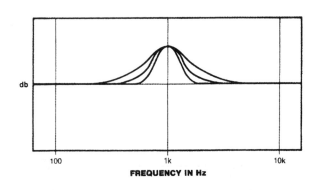

Figure 11.5
Curves that illustrate varying the bandwidth of a parametric equalizer.

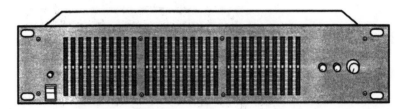

Figure 11.6   A graphic equalizer.

Equalizers can also be classified by the shape of their frequency response. Peaking EQ shapes the response like a hill or peak when set for a boost (Figure 11.7). With shelving EQ, the shape of the frequency response resembles a shelf (Figure 11.8).

A filter causes a rolloff at the frequency extremes. It sharply rejects (attenuates) frequencies above or below a certain frequency. Figure 11.9 shows three types of filters: lowpass, highpass, and bandpass.

For example, a 10 kHz lowpass filter (high-cut filter) removes frequencies above 10 kHz. Its response is down 3 dB at 10 kHz and more above that. This reduces hiss-type noise without affecting tone quality as much as

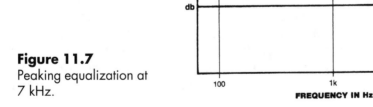

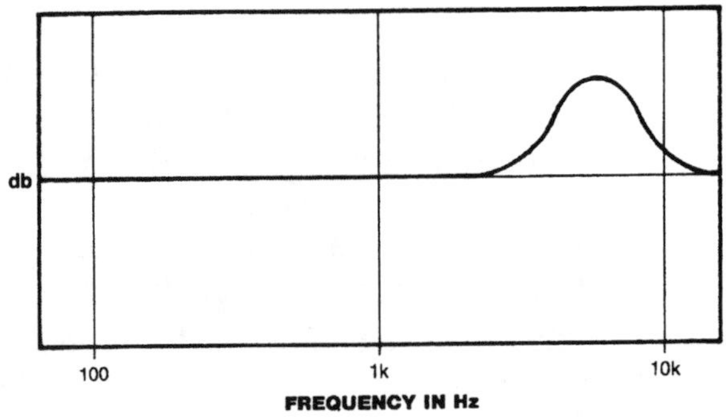

Figure 11.7
Peaking equalization at
7 kHz.

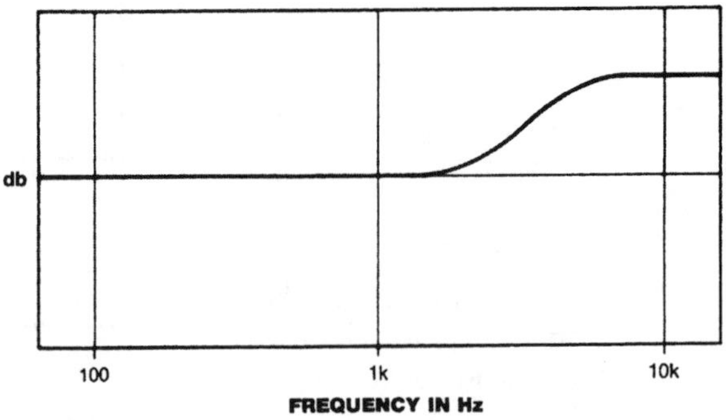

Figure 11.8
Shelving equalization at
7 kHz.

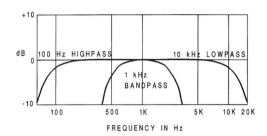

Figure 11.9
Lowpass, highpass, and
bandpass filters.

a gradual treble rolloff would. A 100 Hz highpass filter (low-cut filter) attenuates frequencies below 100 Hz. Its response is down 3 dB at 100 Hz and more below that. This removes low-pitched noises such as air-handler rumble or breath pops. A 1 kHz bandpass filter cuts frequencies above and below a frequency band centered at 1 kHz.

The crossover filter in most monitor speakers consists of lowpass, highpass, and bandpass filters. They send the lows to the woofer, mids to the midrange, and highs to the tweeter. A filter is named for the steepness of its rolloff: 6 dB/octave (first-order), 12 dB/octave (second-order), 18 dB/octave (third-order) and so on.

How to Use EQ

If your mixer has bass and treble controls, their frequencies are preset (usually at 100 Hz and 10 kHz). Set the EQ knob at 0 to have no effect ("flat" setting). Turn it clockwise for a boost; turn it counter-clockwise for a cut. If your mixer has sweepable EQ, one knob sets the frequency range and another sets the amount of boost or cut.

Table 11.1 shows the fundamentals and harmonics of musical instruments and voices. The harmonics given represent an approximate range. For each instrument, turn up the lower end of the fundamentals to get warmth and fullness. Turn down the fundamentals if the tone is too bassy or tubby. Turn up the harmonics for presence and definition; turn down the harmonics if the tone is too harsh or sizzly.

Here are some suggested frequencies to adjust for specific instruments. If you want the effects described below, apply boost. If you don't, apply cut. Try these suggestions and accept only the sounds you like:

- Bass: Full and deep at 60 Hz, growl at 600 Hz, presence at 2.5 kHz, string noise at 3 kHz and up.
- Electric guitar: Thumpy at 60 Hz, full at 100 Hz, puffy at 500 Hz, presence or bite at 2 to 3 kHz, sizzly and raspy above 6 kHz.

Table 11.1 Frequency Ranges of Musical Instruments and Voices

Instrument	Fundamentals	Harmonics
Flute	261–2349 Hz	3–8 kHz
Oboe	261–1568 Hz	2–12 kHz
Clarinet	165–1568 Hz	2–10 kHz
Bassoon	62–587 Hz	1–7 kHz
Trumpet	165–988 Hz	1–7.5 kHz
French horn	87–880 Hz	1–6 kHz
Trombone	73–587 Hz	1–7.5 kHz
Tuba	49–587 Hz	1–4 kHz
Snare drum	100–200 Hz	1–20 kHz
Kick drum	30–147 Hz	1–6 kHz
Cymbals	300–587 Hz	1–15 kHz
Violin	196–3136 Hz	4–15 kHz
Viola	131–1175 Hz	2–8.5 kHz
Cello	65–698 Hz	1–6.5 kHz
Acoustic Bass	41–294 Hz	1–5 kHz
Electric Bass	41–300 Hz	1–7 kHz
Acoustic Guitar	82–988 Hz	1–15 kHz
Electric Guitar	82–1319 Hz	1–3.5 kHz (through amp)
Electric Guitar	82–1319 Hz	1–15 kHz (direct)
Piano	28–4196 Hz	5–8 kHz
Bass (voice)	87–392 Hz	1–12 kHz
Tenor (voice)	131–494 Hz	1–12 kHz
Alto (voice)	175–698 Hz	2–12 kHz
Soprano (voice)	247–1175 Hz	2–12 kHz

- Drums: Full at 100 Hz, wooly at 250 to 600 Hz, trashy at 1 to 3 kHz, attack at 5 kHz, sizzly and crisp at 10 kHz.

- Kick drum: Full and powerful below 60 Hz, papery at 300 to 800 Hz (cut at 400 to 600 Hz for better tone), click or attack at 2 to 6 kHz.

- Sax: Warm at 500 Hz, harsh at 3 kHz, key noise above 10 kHz.

- Acoustic guitar: Full or thumpy at 80 Hz, presence at 5 kHz, pick noise above 10 kHz.

- Voice: Full at 100 to 150 Hz (males), full at 200 to 250 Hz (females), honky or nasal at 500 Hz to 1 kHz, presence at 5 kHz, sibilance ("s" sounds) above 6 kHz.

Example: A vocal track sounds too full or bassy. Reach for the LF EQ knob (say, 100 Hz) and turn it down until the voice sounds natural.

Set EQ to the approximate frequency range you need to work on. Then apply full boost or cut so the effect is easily audible. Finally, fine-tune the frequency and amount of boost or cut until the tonal balance is the way you like it.

What if an instrument sounds honky, tubby, or harsh, and you don't know what frequency to tweak? Set a sweepable equalizer for extreme boost. Then sweep the frequencies until you find the frequency range matching the coloration. Cut that range by the amount that sounds right. For example, a piano miked with the lid closed might have a tubby coloration—maybe too much output around 300 Hz. Set your low-frequency EQ for boost, and vary the center frequency until the tubbiness is exaggerated. Then cut at that frequency until the piano sounds natural.

In general, avoid excessive boost because it can distort the signal. Try cutting the lows instead of boosting the highs.

When to Use EQ

Before using EQ, try to get the desired tone quality by changing the mic or its placement. This gives a more natural effect than EQ. Many purists shun the use of EQ, complaining of excessive phase shift or ringing caused by the equalizer—a "strained" sound. Instead, they use high quality microphones, carefully placed, to get a natural tonal balance without EQ.

Suppose you still need some EQ. Should you EQ while recording or mixing? If you mix more than one instrument to the same track, you can't EQ them independently during mixdown unless their frequency ranges are far apart. To explain, suppose a recorded track contains lead guitar and vocals. If you add a midrange boost to the guitar, you'll hear it on the vocals too. The only solution is to EQ the lead guitar by itself when you record it.

If you assign each instrument to its own track, the usual practice is to record flat (without EQ) and then equalize the track during mixdown. Sometimes the instruments need a lot of EQ to sound good. If so, you might want to record with EQ so that the playback for the musicians will sound good. When you play the multitrack recording through your monitor mixer, the recording may not sound right unless the tracks are already EQ'd. (That's assuming the monitor mixer in your board has no EQ.)

When you do a bass cut or treble boost, you'll get a better S/N by applying this EQ during recording, instead of during mixdown. But if you're doing a treble cut, apply it during mixdown to reduce tape hiss.

Uses of EQ

Here are some applications for EQ:

- Improve tone quality. This is the main use of EQ: to make an instrument sound better tonally. For example, you might use a high-frequency rolloff on a singer to reduce sibilance, or on a direct-recorded electric guitar to take the "edge" off the sound. You could boost 100 Hz on a floor tom to get a fuller sound, or cut around 250 Hz on a bass guitar for clarity. Cut around 100 Hz to reduce bass buildup on massed harmony vocals. The frequency response and placement of each mic affect tone quality as well.

- Create an effect. Extreme EQ reduces fidelity, but it also can make interesting sound effects. Sharply rolling off the lows and highs on a voice, for instance, gives it a "telephone" sound. A 1 kHz bandpass filter does the same thing.

- Reduce noise and leakage. You can reduce low-frequency noises—bass leakage, air-conditioner rumble, mic-stand thumps—by turning down the lows below the range of the instrument you're recording. For example, a fiddle's lowest frequency is about 200 Hz, so you'd set the equalizer's frequency range to 40 or 60 Hz and apply cut. This rolloff won't change the fiddle's tone quality because the rolloff is below the range of frequencies that the fiddle produces. Similarly, a kick drum has little or no output above 9 kHz, so you can filter out highs above 9 kHz on the kick drum to reduce cymbal leakage. If you do this filtering during mixdown, it will also reduce tape hiss. Filtering out frequencies below 100 Hz on most instruments reduces air-conditioning rumble and breath pops. Try rolling off the lows on audience mics to prevent muddy bass.

- Compensate for the Fletcher-Munson effect. As discovered by Fletcher and Munson, the ear is less sensitive to bass and treble at low volumes than at high volumes. So, when you record a very loud instrument and play it back at a lower level, it might lack bass and treble. To restore these, you may need to boost the lows (around 100 Hz) and the highs (around 4 kHz) when recording loud rock groups. The louder the group, the more boost you need. It also helps to use cardioid mics with proximity effect (for bass boost) and a presence peak (for treble boost).

- **Make a pleasing blend.** If you mix two instruments that sound alike, such as lead guitar and rhythm guitar, they tend to mush together—it's hard to tell what each is playing. You can make them more distinct by equalizing them differently. For example, make the lead guitar edgy by boosting 3 kHz, and make the rhythm guitar mellow by cutting 3 kHz. Then you'll hear a more pleasing blend and a clearer mix. The same philosophy applies to bass guitar and kick drum. Since they occupy about the same low-frequency range, they tend to mask or cover each other. To make them distinct, either fatten the bass and thin out the kick a little, or vice versa. The idea is to give each instrument its own space in the frequency spectrum. For example: the bass fills in the lows, synth chords emphasize midbass, lead guitar adds edge in the upper mids, and cymbals add sparkle in the highs.

- **Compensate for mic placement.** Sometimes you are forced to mike very close to reject background sounds and leakage. But a close mic emphasizes the part of the instrument that the mic is near. This gives a colored tone quality, but EQ can partly compensate for it. Suppose you had to record an acoustic guitar with a mic near the sound hole. The guitar track will sound bassy because the sound hole radiates strong low frequencies. But you can turn down the lows on your mixer to restore a natural tonal balance. This use of EQ can save the day by fixing poorly recorded tracks in live concert recordings. During a concert, the stage monitors might be blaring into your recording/P.A. microphones, so you're forced to mike close in order to reject monitor leakage and feedback. This close placement, or the monitor leakage itself, can give the recording an unnatural tone quality. In this case, EQ is the only way to get usable tracks.

- **"Re-mix" a single track.** If you have a track that contains two different instruments, sometimes you can change the mix within that track by using EQ. Imagine a track that has both bass and synth. By using LF EQ, you can bring the bass up or down without affecting the synth very much. Mixing with EQ is more effective when the two instruments are far apart in their frequency ranges.

Whenever you record, the ideal situation is to use the right mic in the right position, and in a good-sounding room. Then you don't need or want equalization. Otherwise, though, your recordings will sound better with EQ than without it.

Compressor

A compressor acts like an automatic volume control, turning down the volume when the signal gets too loud. Here's why it's necessary: Suppose you're recording a female vocalist. Sometimes she sings too softly and gets buried in the mix; other times she hits loud notes, blasting the listener, and saturating the tape. Or she may move toward and away from the mic while singing, so that her average recording level changes.

To control this problem, you can **ride gain**—turn her down when she gets too loud; turn her up when she gets too quiet. But it's hard to anticipate these changes. You might prefer to use a compressor, which does the same thing automatically. It reduces the gain (amplification) when the input signal exceeds a preset level (called the threshold). The greater the input level, the less the gain. As a result, loud notes are made softer, so the dynamic range is reduced (Figure 11.10).

Compression keeps the level of vocals or instruments more constant, so they are easier to hear throughout the mix. And it prevents loud notes that may saturate the tape. Also, it can be used for special effect—say, to make drums sound fatter, or to increase the sustain on a bass guitar. In pro studios, compression is used almost always on vocals, often on bass guitar, kick drum, and acoustic guitar; and sometimes on other instruments.

Doesn't compression rob the music of its expressive dynamics? Yes, if overdone. But a vocal that gets too loud and soft is annoying. You need to tame it with a compressor. Even then, you can tell when the vocalist is singing loudly by the tone of the voice. It also helps to compress the bass and kick drum to ensure a uniform, driving beat.

You can avoid vocal compression if the singer uses proper mic technique. He or she should back away from the mic on loud notes, and come in

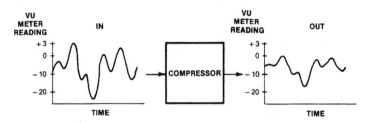

Figure 11.10 Compression.

close on soft notes. To tell whether you need a compressor, listen to your finished mix. If you can understand all the words, and no notes are too loud, omit the compressor.

Using a Compressor

Normally, you compress individual tracks or instruments, not the entire mix. You want to compress only the stuff that needs it. Compressing the overall mix is wise when your mix will be duplicated on cassette, because cassettes have a small dynamic range. To compress a stereo mix, you need a 2-channel compressor with a stereo link, which keeps the left-right balance from changing.

Should you compress while tracking or mixing? If you compress while tracking, that improves the S/N of the tape track, but you must decide on compressor settings during the recording session. If you compress tracks during mixdown, you can change the settings at will.

Let's describe the controls on the compressor. Some compressors have few controls; most of their settings are pre-set at the factory.

Compression Ratio or Slope

This is the ratio of the change in input level to the change in output level. For example, a 2:1 ratio means that for every 2 dB change in input level, the output changes 1 dB. A 20 dB change in input level results in a 10 dB change in the output, and so on. Typical ratio settings are 2:1 to 4:1.

A "soft knee" or "over easy" characteristic is a low compression ratio for low-level signals and a high ratio for high-level signals. Some manufacturers say that this characteristic sounds more natural than a fixed compression ratio.

Threshold

This is the input level above which compression occurs. Set the threshold high (about –5 VU) to compress only the loudest notes; set it low (–10 or –20 VU) to compress a broader range of notes. A setting of –10 to –5 is typical. If the compressor has a fixed threshold, adjust the amount of compression with the input level control.

Gain Reduction

This is the number of dB that the gain is reduced by the compressor. It varies with the input level. You set the ratio and threshold controls so that the gain is reduced on loud notes by an amount that sounds right. The amount of gain reduction shows up on a meter.

213

Attack Time

This is how fast the compressor reduces the gain when it's hit by a musical attack. Typical attack times range from 0.25 to 10 milliseconds (msec). Some compressors adjust the attack time automatically to suit the music; others have a factory-set attack time. The longer the attack time, the larger the peaks that are passed before gain reduction occurs. So, a long attack time sounds punchy; a short attack time reduces punch by softening the attack.

Release Time

This is how fast the gain returns to normal after a loud passage ends. It's the time the compressor takes to reach 63 percent of its normal gain. You can set the release time from about 50 msec to several seconds. For bass instruments, the release time must be longer than about 0.4 second to prevent harmonic distortion.

Short release times make the compressor follow rapid volume changes in the music, and keep the average level higher. But since the noise rises along with the gain, short release times can give a pumping or breathing sound. Long release times sound more natural. If the release time is too long, though, a loud passage will reduce the gain during a subsequent quiet passage. In some units, the release time varies automatically, or is factory-set to a useful value.

Output-Level Control

This sets the compressor's output signal to the level your mixer wants to see. Some compressors keep the output level constant when other controls are varied.

Spend some free time playing with all the settings so you learn how they affect the sound. Play various instruments and vocals through a compressor, vary the settings, and take notes on what you hear.

Some compressors have a **side chain**. This is a pair of in/out jacks for connecting an equalizer. To compress only the sibilant sounds on a vocal track, boost the side-chain EQ around 10 kHz. To compress only the breath pops on a vocal track, boost the side-chain EQ around 20 Hz.

Connecting a Compressor

Connect a compressor in series with the signal you want to compress, in one of the following ways:

- To compress one instrument or voice while recording: On your mixer, locate the input module of the instrument you want to compress.

Connect the compressor between the access (insert) in-and-out jacks (Chapter 12 explains these terms). Or, take a signal from the input module's direct out. Feed that into the compressor, and feed the compressor output to the tape track.

- To compress a group of instruments while recording: Locate the bus output of the instruments you want to compress. Go from bus out to compressor in, and go from compressor out to tape-track in. If the bus has access jacks, you could connect to them instead.

- To compress one tape track during mixdown: Go from track out to compressor in, and go from compressor out to mixer tape in. Or just patch into the access jacks for that track's mixer module.

Limiter

A limiter keeps signal peaks from exceeding a preset level. Limiters are used to prevent DAT overload during field recording, or to prevent P.A. power amps from clipping.

The compression ratio in a limiter is very high—10:1 or greater—and the threshold is set high, say at 0 VU. For input levels up to 0 VU, the output level matches the input. For input levels above 0 VU, the output level stays at 0 VU. This prevents overload in the device following the limiter.

While a compressor reduces the overall dynamic range of the music, a limiter affects only the highest peaks (Figure 11.11). To act on these rapid peaks, limiters have a very fast attack time—1 microsecond to 1 millisecond.

A compressor/limiter does both functions in its name. It compresses the average signal levels over a wide range, and limits peaks to prevent overload. It has two thresholds—one low for the compressor and one high for the limiter.

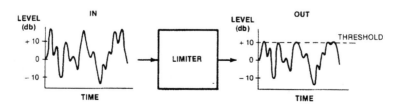

Figure 11.11 Limiting.

Noise Gate

A noise gate (expander) acts like an on-off switch that removes noises during pauses in an audio signal. It reduces the gain when the input level falls below a preset threshold. That is, when an instrument stops playing for a moment, the noise gate drops the volume—removing any noise and leakage during the pause (Figure 11.12).

NOTE: The gate does NOT remove noise while the instrument is playing.

Where is it used? The noise gate helps to clean up drum tracks by removing leakage between beats. It can shorten the decay time of the drums, giving a very tight sound. During a mixdown, you might gate each track of an analog recorder to reduce tape hiss. If you're recording a noisy guitar amp, try a gate to cut out the buzz and hiss between phrases.

How do you use a noise gate? Patch it between a tape-track output and a mixer tape input. Play the tape. Set the gate's threshold so that hiss and leakage go away during pauses. If the gate chops off each note, gradually turn down the threshold. Set the release time short for drums and longer for instruments that have a long sustain.

Excellent tapes can be made without gating. But if you want a tighter sound, gates come in handy. Some signal processors have compression, limiting, and noise gating in a single package.

Some gates have a sidechain input or key input. It's an input for an external signal that controls the gating action. The control signal triggers the output of the gate's main audio path. For example, you could feed a bass guitar through the noise gate, and gate the bass with a kick drum signal fed into the side chain. Then the bass will follow the kick drum's envelope.

Delay

A digital delay (or a delay program for a computer) takes an input signal, holds it in memory chips, then plays it back after a short delay—about 1 msec to 1 second (Figure 11.13). Delay is the time interval between the input signal and its repetition at the output of the delay device.

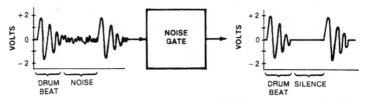

Figure 11.12 Noise gating.

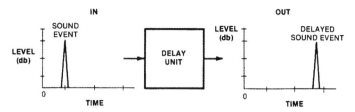

Figure 11.13 Delaying the signal.

If you listen to the delayed signal by itself, it sounds the same as the undelayed ("dry") signal. But if you combine the delayed and dry signals, you may hear two distinct sounds: the signal and its repetition. By delaying a signal, a processor can create several effects such as echo, repeating echo, doubling, chorus, and flanging.

Echo

If the delay is about 50 msec to 1 second, the delayed repetition of a sound is called an **echo**. This is shown in Figure 11.14 by the two pulses. Echoes occur naturally when sound waves travel to a distant room surface, bounce off, and return later to the listener—repeating the original sound. A delay unit can mimic this effect.

In setting up a mix with echo, you want to hear both the dry sound and its echo. You do this by creating an effects loop: from the mixer, to the effects box, back to the mixer. Here's how:

1. On the delay unit, set the dry/wet mix control all the way to "wet." Then the output of the delay unit will be only the delayed signal.

2. Suppose you want to use aux1 as the echo control. Connect aux1 send to delay unit IN. Connect delay unit OUT to bus 1 and 2 IN (or to the effects-return jacks).

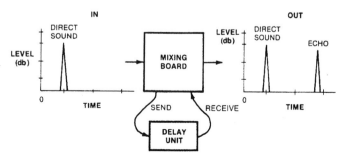

Figure 11.14 Echo.

217

3. Find the mixer module for the instrument you want to add echo to.

4. Assign the instrument to busses 1 and 2. Monitor busses 1 and 2.

5. Find the knobs labeled Bus 1 IN and Bus 2 IN. They might be called "Aux Return" or "Effects Return." Turn them up to 0, about 1/2 to 3/4 up.

6. Turn up the aux1 send knob, and there's your echo.

The delayed sound mixes with the dry sound in busses 1 and 2. You hear both sounds, which together make an echo. Each aux knob controls the amount of echo on each track, while the effect-return knobs control the overall amount of echo.

Slap Echo

A delay from 50 to 200 msec is called a slap echo or slapback echo. It was often used in 1950's rock 'n' roll tunes, and still is used today.

Repeating Echo

Most delay units can be made to feed the output signal back into the input, internally. Then the signal is re-delayed many times. This creates a repeating echo—several echoes that are evenly spaced in time (Figure 11.15). The regeneration control sets the number of repeats.

Repeating echo is most musical if you set the delay time to create an echo rhythm that fits the tempo of the song. A slow repeating echo—0.5 second between repeats, for example—gives an outer-space or haunted-house effect.

Doubling

If you set the delay around 15 to 35 msec, the effect is called doubling or automatic double tracking (ADT). It gives an instrument or voice a fuller sound, especially if the dry and delayed signals are panned to opposite sides. The short delays used in doubling sound like early sound reflections in a studio, so they add some "air" or ambience.

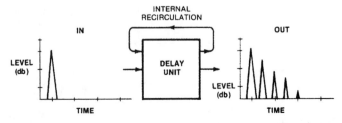

Figure 11.15   Repeating echo.

Doubling a vocal can be done without a delay unit. Record a vocal part, then overdub another performance of the same vocal part. Mix the parts, pan them both to center, or pan them left and right.

Chorus

This is a wavy or shimmering effect. The delay is 15 to 35 msec, and the delay varies at a slow rate. Sweeping the delay time causes the delayed signal to bend up and down in pitch, or to detune. When you combine the detuned signal with the original signal, you get chorusing.

Stereo Chorus

This is a beautiful effect. In one channel, the delayed signal is combined with the dry signal in the same polarity. In the other channel, the delayed signal is inverted in polarity, then combined with the dry signal. Thus, the right channel has a series of peaks in the frequency response where the left channel has dips, and vice versa. The delay is slowly varied or modulated.

Bass Chorus

This is chorus with a highpass filter so that low frequencies are not chorused, but higher harmonics are. It gives an ethereal quality to the bass guitar.

Flanging

If you set the delay around 0 to 20 msec, you usually can't resolve the direct and delayed signals into two separate sounds. Instead, you hear a single sound with a strange frequency response. The direct and delayed signals combine and have phase interference, which puts a series of peaks and dips in the frequency response. This is called a **comb-filter effect** (Figure 11.16). It gives a very colored, filtered tone quality. The shorter the delay, the farther apart the peaks and dips are spaced in frequency.

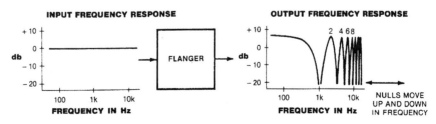

Figure 11.16 Flanging (or positive flanging).

The flanging effect varies or sweeps the delay between about 0 and 20 msec. This makes the comb-filter nulls sweep up and down the spectrum. As a result, the sound is hollow, swishing, and ethereal, as if the music were playing through a pipe. Flanging is easiest to hear with broad-band signals such as cymbals but can be used on any instrument, even voices. Some examples of flanging are on many Jimi Hendrix records, on the oldies "Itchycoo Park" by the Small Faces, and "Listen to the Music" by the Doobie Brothers.

Positive flanging refers to flanging in which the delayed signal is the same polarity as the direct signal (Figure 11.16). With **negative flanging**, the delayed signal is opposite in polarity to the direct signal, which makes a stronger effect. The low frequencies are canceled (the bass rolls off), and the "knee" of the bass rolloff moves up and down the spectrum as the delay is varied. The high frequencies are still comb-filtered (Figure 11.17). Negative flanging makes the music sound like it's turning inside out.

When the flanger feeds some of the output signal back into the input, the peaks and dips get bigger. It's a powerful "science fiction" effect called **resonant flanging**.

Reverberation

This effect adds a sense of room acoustics, ambience, or space to instruments and voices. To know how it works, we need to understand how reverb happens in a real room.

Natural reverberation in a room is a series of multiple sound reflections which makes the original sound persist and gradually die away or decay. These reflections tell the ear that you're listening in a large or hard-surfaced room. For example, reverberation is the sound you hear just after you shout in an empty gymnasium. A reverb effect simulates the sound of a room—a club, auditorium, or concert hall—by generating random multiple echoes that are too numerous and rapid for the ear to resolve (Figure 11.18).

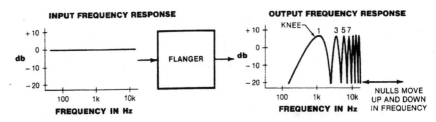

Figure 11.17 Negative flanging.

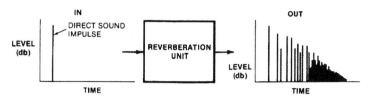

Figure 11.18   Reverberation.

Reverb and echo are not the same thing. Echo is a repetition of a sound (HELLO hello hello); reverb is a smooth decay of sound (HELLO-OO-oo-oo).

Digital reverb is available either in a dedicated reverb unit, or as part of a multi-effects processor. A digital reverb has a lot of reverb patterns stored as programs or algorithms. You can get the sound of a small room, a big room, or a warm room (with more lows and less highs). Reverb algorithms with high density of delayed signals sound smoother when they decay than low-density algorithms.

A **plate reverb** setting duplicates the bright sound of a metal-foil plate, which used to be the most popular type of reverb in pro studios. Some units let you adjust tone quality, decay time, and other parameters. Unnatural effects are available, such as non-linear decay, **reverse reverb** that builds up before decaying, or **gated reverb**. With gated reverb, the reverb cuts off suddenly shortly after a note is hit. It's often used on a snare drum. A good example is the oldie "You Can Call Me Al" on Paul Simon's album, "Graceland."

Another feature in some units is **pre-delay** (pre-reverb delay). This is a short delay (30 to 100 msec) before the reverb to simulate the delay that happens in real rooms before reverb starts. The longer the pre-delay, the greater the sense of room size. If your reverb unit does not have pre-delay built in, you can create it by connecting a delay unit between your mixer's aux send and the reverb input.

To connect a reverb unit to your mixer, connect a cable from the mixer aux send to the reverb input. Connect a cable (two for stereo) from the reverb outputs to the mixer aux returns (effects returns or bus inputs). Set the mix control on the reverb unit all the way to "wet" or "reverb." Turn up the mixer's aux-return or bus-in knobs (if any) about 1/2 to 2/3 up, then adjust the amount of reverb on each track with the aux-send knobs. Try to get an overall reverb-send level near 0 VU, then fine-tune the aux-return level for the desired amount of reverb.

"Wet" lead vocals (with reverb) are the norm. But some recordings are made with totally dry vocals.

Enhancer

If a track or a mix sounds dull and muffled, you can run it through an enhancer to add brilliance and clarity. An enhancer works either by adding slight distortion (as in the Aphex Aural Exciter) or by boosting the treble when the signal has high-frequency content (as in the Alesis Micro Enhancer and the Barcus Berry 402 Sonic Maximizer).

This last device also divides the frequency range into three bands. The lows are delayed about 1.5 msec; the midrange is delayed about 0.5 msec, and the highs are delayed only a few microseconds. In this way, the Maximizer aligns the harmonics and fundamentals in time for added clarity.

Octave Divider

This unit takes a signal from a bass guitar and provides deep, growling bass notes one or two octaves below the pitch of the bass guitar. It does this by dividing the incoming frequency by 2 or 4: If you put 82 Hz in, you get 41 Hz out. Some MIDI sound modules have bass patches with extra-deep sound, and some bass guitars have an extra string tuned especially low.

Harmonizer

Basically a delay unit with time modulation, a harmonizer makes a variety of pitch-shifting effects. It can create harmonies, change pitch without changing the duration of the program, change duration without changing pitch, and many other oddities. You've heard harmonizers on radio station breaks when the announcer's voice sounds like a munchkin or Darth Vader. The latest vocal pitch-shifters maintain the voice formant structure when they shift pitch; this prevents the "chipmunk" effect.

Preverb

With this effect, you can make reverb precede the attack of each note. This can turn a snare hit into a whip—ssSSHHKK! Here's how to add preverb to a drum track recorded on an analog multitrack:

1. Reverse the reels so the tape plays backwards.
2. Add reverb to the drum track.
3. Record the reverb output on an open track.
4. Again reverse the reels and play the tape.

What if you're using an MDM recorder, which won't play backwards? You'll need a multitrack digital audio workstation (DAW) to create preverb.

1. Get a signal processor with a **reverse reverb** effect: The reverb builds up before it fades out.
2. Add reverse reverb to a drum hit, and record the reverb output on an open MDM track.
3. Using the DAW, copy the tape tracks to hard disk.
4. On your computer screen, slide the reverse-reverb track earlier in time (to the left) until its timing sounds correct.

Tube Processor

This device uses a vacuum tube, or a transistor simulation of one, to add a warm sound, or to reduce digital harshness. Tubes have euphonic even-order harmonic distortion, which adds richness. There are tube mics, tube mic preamps, tube compressors, and stand-alone tube processors.

Analog Tape Simulator

Analog tape saturation is mainly third-harmonic distortion and slight compression. An analog tape simulator adds these effects to digital recordings in an attempt to smear or warm-up the sound in a pleasant way.

Spatial Processor

Spatial processors enhance the stereo imaging or spatial aspects of the mix. Some units have joystick-type pan pots, which move the image of each track anywhere around the listener. Other units make the stereo stage wider, so that images can be placed to the left of the left speaker, and to the right of the right speaker. The listener might hear images toward the sides of the listening room.

Multi-Effects Processor

This device provides several effects in a single chassis. Some units let you combine up to four effects in any order. Others have several channels, so you can put a different effect on different instruments. With most processors, you can edit the sounds and save them in memory as new programs.

A multi-effects processor uses a digital signal processing (DSP) chip and RAM memory. The amount of memory is limited, so the more memory that one effect uses, the less is available for other effects. For example, suppose you're combining reverb and echo. If you use a reverb with a long decay time (which takes a lot of memory), you may have to settle for an echo with a short delay.

Most units have a frequency response up to 20 kHz and at least 16-bit resolution. They offer 100 or more programmable presets with MIDI control over any parameter. For example, with some units you can place an instrument in a simulated room, and use a MIDI controller to continuously change the size of that room.

Many signal processors can be controlled by MIDI program-change commands. You can quickly change the type of effect, or effect parameters, by entering certain program changes into a sequencer.

Suppose you want each tom-tom hit in a drum fill to have a different size room added to it. For example, put the high rack tom in a small room; put the low rack tom in a concert hall, and put the floor tom in a cave. To do this, first assign a different program number (patch or preset number) to each effects parameter. You do this with the effects device. Then, using the sequencer, punch in the appropriate program number for each note.

A MIDI program-change footswitch lets guitarists call up different effects on MIDI signal processors. By tapping a footswitch, they can get fuzz, flanging, wah-wah, spring reverb, and so on.

A MIDI mapper lets you control some effects parameters with any controller. For example, vary reverb decay time with a pitch wheel, or vary a filter with key velocity.

Software (Plug-In) Effects

Most digital-editing software includes effects that you control on your computer screen. The software runs algorithms in the sound card's DSP to make effects. An Effects Plug-In is third-party software which you load into your digital-editing program. All sorts of effects are available as plug-ins.

Looking Back

We've come a long way with effects. Looking back over the last few decades, each era had its own "sound" related to the effects used at the time. The 50's had tube distortion and slap echo; the 60's used fuzz, wah-wah, and flanging. Much of the early 70's sounded dry, and the early 90's emphasized synth, drum machines, and gated reverb. Now vacuum tubes and acoustic instruments are back. Whatever effects you choose, they can enhance your music if used with taste.

Sound-Quality Glossary

The sound of effects and EQ can be hard to translate into engineering terms. For example, what EQ should you use to get a "fat" sound or a "thin" sound? The glossary below may help. It's based on conversations with producers, musicians, and reviewers over many years. Not everyone agrees on these definitions, but they are common. This glossary is limited to sound quality; the glossary at the back of the book covers all audio terms.

AIRY—Spacious. The instruments sound like they are surrounded by a large reflective space full of air. A pleasant amount of reverb. High-frequency response that extends to 15 or 20 kHz.

BALLSY or BASSY—Emphasized low frequencies below about 200 Hz.

BLOATED—Excessive mid-bass around 250 Hz. Poorly damped low frequencies, low-frequency resonances.

BLOOM—Early reflections or a sense of "air" around each instrument in an orchestra.

BOOMY—Excessive bass around 125 Hz. Poorly damped low frequencies or low-frequency resonances.

BOXY—Having resonances as if the music were enclosed in a box. Speaker cabinet diffraction or vibration. Sometimes an emphasis around 250 Hz to 500 Hz.

BREATHY—Audible breath sounds in vocals, flute, or sax. Good high-frequency response.

BRIGHT—High-frequency emphasis. Harmonics are strong relative to fundamentals.

BRITTLE—High-frequency peaks, or weak fundamentals. Opposite of round or mellow. *See* Thin. Objects that are physically thin and brittle emphasize highs over lows when you crack them.

CHESTY—The vocalist sounds like their chest is too big. A bump in the low-frequency response around 125 to 250 Hz.

CLEAN—Free of noise, distortion, and leakage.

CLEAR—*See* Transparent.

CLINICAL—Too clean or analytical. Emphasized high-frequency response, sharp transient response. Not warm.

COLORED—Having timbres that are not true to life. Non-flat response, peaks, or dips.

CONSTRICTED—Poor reproduction of dynamics. Dynamic compression. Distortion at high levels. *See also* Pinched.

CRISP—Extended high-frequency response. Like a crispy potato chip, or crisp bacon frying. Often referring to cymbals.

CRUNCH—Pleasant guitar-amp distortion.

DARK—Opposite of bright. Weak high frequencies.

DELICATE—High frequencies extending to 15 or 20 kHz without peaks. A sweet, airy, open sound with strings or acoustic guitar.

DEPTH—A sense of closeness and farness of instruments, caused by miking them at different distances. Good transient response that reveals the direct/reflected sound ratio in the recording.

DETAILED—Easy to hear tiny details in the music; articulate. Adequate high-frequency response, sharp transient response.

DRY—Without effects. Not spacious. Reverb tends towards mono instead of spreading out. Overdamped transient response.

DULL—*See* Dark.

EDGY—Too much high frequencies. Trebley. Harmonics are too strong relative to the fundamentals. When you view the waveform on an oscilloscope, it even looks edgy or jagged, due to excessive high frequencies. Distorted, having unwanted harmonics that add an edge or raspiness to the sound.

EFFORTLESS—Low distortion, usually coupled with flat response.

ETCHED—Clear but verging on edgy. Emphasis around 10 kHz or higher.

FAT—*See* Full and Warm. Also, a diffuse spatial effect.

FOCUSED—Referring to the image of a musical instrument which is easy to localize, pinpointed, having a small spatial spread.

FORWARD—Sounding close to the listener, projected. Emphasis around 2 kHz to 5 kHz.

FULL—Opposite of thin. Strong fundamentals relative to harmonics. Good low-frequency response, not necessary extended, but with adequate level around 100 to 300 Hz.

GENTLE—Opposite of edgy. The harmonics—highs and upper mids—are not exaggerated, or may even be weak.

GLARE, GLASSY—A little less extreme than edgy. A little too bright or trebley.

GRAINY—The music sounds like it's segmented into little grains, rather than flowing in one continuous piece. Not liquid or fluid. Suffering from harmonic or I.M. distortion. Some early A/D converters sounded grainy, as do current ones of inferior design. "Powdery" is finer than "grainy!"

GRUNGY—Lots of harmonic or I.M. distortion.

HARD—Too much upper midrange, usually around 3 kHz. Or, good transient response, as if the sound is hitting you hard.

HARSH—Too much upper midrange. Peaks in the frequency response from 2 kHz to 6 kHz. Or, excessive phase shift.

HEAVY—Good low-frequency response below about 50 Hz. Suggesting an object of great weight or power, like a diesel locomotive or thunder.

HOLLOW—*See* Honky. Or, too much reverberation. Or, a mid-frequency dip.

HONKY—The music sounds the way your voice sounds when you cup your hands around your mouth. A bump in the response around 500 to 700 Hz.

LIQUID—Opposite of grainy. A sense of seamless flowing of the music. Flat response and low distortion. High frequencies are flat or reduced relative to mids and lows.

MELLOW—Reduced high frequencies, not edgy.

MUDDY—Not clear. Weak harmonics, smeared time response, I.M. distortion. Too much reverb at low frequencies.

MUFFLED—The music sounds covered up. Weak highs or weak upper mids.

MUSICAL—Flat response, low distortion, no edginess. Conveying emotion.

NASAL—The vocalist sounds like he or she is singing with the nose closed. Also applies to strings. Bump in the response around 300 to 1000 Hz. *See* Honky.

NEUTRAL—Accurate tonal reproduction. No obvious colorations. No serious peaks or dips in the frequency response.

PAPERY—Referring to a kick drum that has too much output around 400 to 600 Hz.

PINCHED—Narrowband. Midrange or upper-midrange peak in the frequency response. Pinched dynamics are overly compressed.

PIERCING—Strident, hard on the ears, screechy. Having sharp, narrow peaks in the response around 3 kHz to 10 kHz.

PRESENT, PRESENCE—Adequate or emphasized response around 5 kHz for most instruments, or around 2 to 5 kHz for kick drum and bass. Having some edge, punch, detail, closeness, and clarity.

PUFFY—Bump in the response around 500 to 700 Hz.

PUNCHY—Good reproduction of dynamics. Good transient response. Sometimes a bump around 5 kHz or 200 Hz.

RASPY—Harsh, like a rasp. Peaks in the response around 6 kHz which make vocals sound too sibilant or piercing.

RICH—*See* Full. Also, having euphonic distortion made of even-order harmonics.

ROUND—High-frequency rolloff or dip. Not edgy.

SHARP—*See* Strident and Tight.

SIBILANT, ESSY—Exaggerated s and sh sounds in singing, caused by a rise in the response around 4 to 7 kHz.

SIZZLY—*See* Sibilant. Also, too much highs on cymbals.

SMEARED—Lacking detail. Poor transient response. This may be a desirable effect in large-diameter mics. Also, poorly focused images.

SMOOTH—Easy on the ears, not harsh. Flat frequency response, especially in the midrange. Lack of peaks and dips in the response.

SPACIOUS—Conveying a sense of space, ambience, or room around the instruments. To get this effect, mike farther back, mix in an ambience mic, add reverb, or record in stereo. Components that have out-of-phase crosstalk between channels may add false spaciousness.

SQUASHED—Overly compressed.

STEELY—Emphasized upper mids around 3 to 6 kHz. Peaky, nonflat high-frequency response. *See* Glassy, Harsh, Edgy.

STRAINED—The component sounds like it's working too hard. Distorted. Inadequate headroom or insufficient power. Opposite of effortless.

STRIDENT—*See* Harsh and Edgy.

SWEET—Not strident or piercing. Flat high-frequency response, low distortion. Lack of peaks in the response. Highs are extended to 15 or 20 kHz, but they are not bumped up. Often used when referring to cymbals, percussion, strings, and sibilant sounds.

THIN—Fundamentals are weak relative to harmonics. Note that the fundamental frequencies of many instruments are not very low. For example, violin fundamentals are around 200 to 1000 Hz. So if the 300 Hz area is weak, the violin may sound thin—even if the mic's response goes down to 40 Hz.

TIGHT—Good low-frequency transient response. Absence of ringing or resonance when reproducing the kick drum or bass. Good low-frequency detail. Absence of leakage.

TINNY, TELEPHONE-LIKE—Narrowband, weak lows, peaky mids. The music sounds like it's coming through a telephone or tin can.

TRANSPARENT—Easy to hear into the music, detailed, clear, not muddy. Wide flat frequency response, sharp time response, very low distortion and noise.

TUBBY—*See* bloated. Having low-frequency resonances as if you're singing in a bathtub.

VEILED—The music sounds like you put a silk veil over the speakers. Slight noise or distortion, or slightly weak high frequencies.

WARM—Good bass, adequate low frequencies, adequate fundamentals relative to harmonics. Not thin. Or, excessive bass or midbass. Or, pleasantly spacious, with adequate reverberation at low frequencies. Or, gentle highs, like from a tube amplifier. *See* Rich.

WOOLY or BLANKETED—The music sounds like there's a wool blanket over the speakers. Weak high frequencies or boomy low frequencies. Sometimes, an emphasis around 250 to 600 Hz.

12

RECORDER-MIXERS AND MIXING CONSOLES

In a recording studio, you do most of the work with a mixer and a multi-track tape machine. The mixer accepts microphone or line-level signals and amplifies them up to the level needed by the tape machine. The multitrack tape machine records several tracks on tape. One track might be a lead vocal, another track might be a saxophone, and so on.

A recorder-mixer combines a mixer and a multitrack cassette recorder in a single portable chassis. This convenient unit is also called a **ministudio** or **portable studio**. Most recorder-mixers record 4 tracks, but high-end units record 8 or more tracks.

A mixing console (also called **board** or **desk**) is an elaborate mixer with extra controls. This chapter covers both mixers and mixing consoles. First we'll look at typical features of both, then features found only in mixing consoles.

Stages of Recording

This chapter will refer to the three stages in making a multitrack recording: recording, overdubbing, and mixdown.

- Recording (**tracking**): Record one or more instruments onto one or more tracks.

- Overdubbing: While listening to prerecorded tracks over headphones, the musician records new parts on open (unused) tracks.

- Mixdown: After all the tracks are recorded, mix or combine them into 2-track stereo. Add effects and any sequenced tracks of MIDI instruments.

Record the stereo mix onto a 2-track recorder, such as a cassette deck, open-reel deck, or DAT machine. This recording is the master tape, which can be duplicated on cassettes, vinyl records, or compact discs.

Mixer Functions

Although the knobs and meters on a mixer may appear intimidating, you can understand them if you read the manual and practice with the equipment. A mixer is rather complicated because it lets you control many aspects of sound:

- The loudness of each instrument (to control the balance among instruments in the mix)

- The tone quality of each instrument (bass, treble, midrange)

- The room that the instruments are in (reverberation)

- The left-to-right position of each instrument (panning)

- Special effects (flanging, echo, chorus, etc.)

- Track assignments (sometimes more than one instrument on each track)

- Recording level (to prevent distortion and noise)

- Monitor selection (what you want to listen to)

Inputs and Outputs

A mixer can be specified by the number of inputs and outputs it has. For example, an 8-in, 2-out mixer (8x2 mixer) has 8 signal inputs, which can be mixed into 2 output channels (buses) for stereo recording. Similarly, a 16-in, 8-out (16x8) mixing board has 16 signal inputs and 8 output channels for multitrack recording. A 16x4x2 mixing board has 16 inputs, 4 submixes (explained later), and 2 main outputs. There also are connectors for external equipment, such as effects devices and a monitor amplifier.

Most mixing boards are organized in three sections:

- Input modules
- Output section
- Monitor section

Input Modules

A mixer is made of groups of controls called **modules**. An **input module** (Figure 12.1) affects a single input signal—from a microphone, for instance. The module is usually a narrow vertical strip, one per input. Several modules are lined up side-by-side. Each input module is the same, so if you know one, you know them all.

The more inputs your mixer has, the more instruments you can record at the same time. If you're recording only yourself, you may need only two inputs. Each input module typically has these parts:

INPUT CONNECTORS: On the back of the module are connectors that accept the signal from a microphone, a direct box, an electric musical instrument, or a multitrack tape recorder.

INPUT SELECTOR: Lets you choose which input signal you want to work with. Some consoles omit this control.

PREAMP: Inside the module, the preamp amplifies your input signal.

GAIN or TRIM CONTROL: Adjusts the level of the input signal for the best compromise between noise and distortion. Also used to set recording level.

FADER: Adjusts the balance among tracks during mixdown.

EQUALIZER: Adjusts the tone quality (bass, treble, and midrange).

AUX SEND (EFFECTS SEND): Controls how much effects (e.g., reverb, chorus, etc.) are added to each input signal. Also might be used as a monitor-mix volume control.

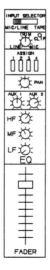

Figure 12.1
A typical input module.

ASSIGN SWITCH: Before recording, selects which tape track to send the signal to. This is usually used in combination with the pan pot.

PAN POT: Places various sounds where desired between your stereo speakers. In some mixers, the pan pot is also used with the assign switch to send signals to the desired tracks.

Figure 12.2 shows the signal flow from input to output through a typical input module.

Input Connectors

Input connectors are for microphones, direct boxes, electric musical instruments, or an external multitrack tape recorder. The following are labels for each connector:

- MIC (for mic-level signals: microphones and direct boxes)
- LINE (for line-level signals: a synthesizer, drum machine, or electric guitar)
- TAPE (for line-level signals from an external multitrack tape recorder)

In some recorder-mixers, a single 1/4-inch phone jack is used both for mic-level and line-level signals. A mic-level signal is typically about 1 to 2 millivolts. A line-level signal is about 0.3 to 1.23 volts. The two levels are handled either by a MIC/LINE switch or a TRIM control.

Some units have separate jacks for mic and line inputs. The mic input is either an unbalanced 1/4-inch phone jack (a 1/4-inch hole) or an XLR-type connector (with three small holes). The line input is either a 1/4-inch phone jack, an RCA (phono) jack (such as that on a stereo system), or an XLR-type connector. You can plug a synthesizer directly into a phone-jack input without using a direct box if the cable is under 10 feet; a longer cable may pick up hum.

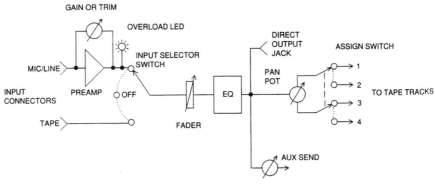

Figure 12.2 Signal flow through a typical input module.

Some newer recorder-mixers also have a **sync input**. It goes into track 4, and is used for recording a special tape-sync tone from a computer running a sequencer program. The sync tone synchronizes tape tracks with MIDI sequencer tracks. It also lets you overdub two or more sequencer tracks onto tape by keeping them synchronized. On an 8-track cassette or open-reel machine, use track 8 as the sync track.

Phantom Power (P48, +48)

The phantom power switch is found only in elaborate units, and turns on phantom power for condenser microphones. This is 12 to 48 volts DC voltage applied through resistors equally to pins 2 and 3 of XLR-type mic-input connectors. The microphone receives phantom power and sends audio along the same two conductors. In some mixing consoles, each input module has its own phantom-power switch.

Input Selector

This switch selects the input you want to process: either mic, line, or tape. The switch might be labeled in one of these ways:

Mic/line/tape
Mic/line/remix
Mic-line/off/tape
Input/off/tape
Input/mute/track
Input/off/line

The following list explains how each switch position works:

MIC: The mic signal enters the mixer.

LINE: The line or tape signal enters the mixer.

MIC-LINE or INPUT: Either the mic signal or the line signal enters the mixer, depending on what is plugged into that input.

TAPE, TRACK, or REMIX: The tape-track signal enters the mixer (for overdubbing or mixdown).

OFF or MUTE: No signal is processed. During mixdown, it's a good idea to mute tracks that have nothing playing at the moment to reduce tape hiss.

Using the input selector is simple. If you plugged in a microphone or direct box to record its signal, set the input selector to "mic" or "input." If you plugged in a synth, drum machine, or electric guitar, set the selector to "line" or "input."

Some mixers have no input selector. The mixer processes whatever signal is plugged in.

Mic Preamp

After entering the mic connector and input selector switch, a microphone signal goes into a mic preamplifier inside the mixer. The preamp boosts or amplifies the weak microphone signal up to a higher voltage, making it a line-level signal.

Trim (Gain)

If the microphone is picking up a loud instrument or vocal, the mic signal is very strong. This signal can overload the mic preamp, causing distortion—a gritty sound.

To prevent distortion, set the TRIM control as follows: First set the **master fader(s)** within the shaded portion of its travel (sometimes at 0, about 3/4 up). Do the same for the **input fader** (discussed next). Assign the signal to a tape track. Have the musician play the loudest part of the song, and adjust the TRIM control so that the meter peaks around 0.

In some mixers, in each input module is a tiny light (LED) labeled "clip" or "peak." It flashes when the mic preamp is distorting. If this light flashes when you're picking up an instrument or vocal, turn down the TRIM control just to the point where the light stays off, then turn it down another 10 dB for extra headroom.

Inexpensive recorder-mixers do not have an overload LED or a trim control. The input fader serves this function.

Input Fader (Channel Fader)

After the preamp amplifies the mic signal, it goes to the input fader. This is a sliding volume control for each input signal. During recording, you either set the input fader to **design center** (at 0, about 3/4 up) or set the fader to adjust the instrument's level in the monitor mix (depending on how your board is set up). During mixdown, you use the faders to set the loudness balance among instruments.

Direct Out

Found only in fancier units, the direct out is an output connector following each input fader and equalizer. The signal at the direct-out jack is an amplified

236

version of the input signal. The fader controls the level at the direct-output jack. Use the direct-out jack when you want to record one instrument per track (with EQ) on an external multitrack recorder. Connect the direct-out jack to a tape-track input. Since the direct output bypasses the mixing circuits farther down the chain, the result is a cleaner signal. If a mixer has eight inputs with a direct-out jack for each input, you can use the mixer with an 8-track recorder—even if the mixer has only 2 or 4 outputs.

EQ (Equalization)

The signal from the input fader goes to an equalizer, which is a tone control. With EQ you can make an instrument sound more or less bassy, and more or less trebley, by boosting or cutting certain frequencies.

Equalizers cover 2 to 4 bands of the audible spectrum. Inexpensive units have a simple two-knob bass and treble control; you can boost or cut the treble or bass. Fancier models have a sweepable EQ (sometimes called semiparametric EQ) that lets you "tune in" the exact frequency range to work on. This feature adds cost and complexity, but gives you more control over the tone quality.

In some low-cost recorder-mixers, the EQ works on two tracks at a time during recording, and on the stereo mix during mixdown. This is less flexible than a unit with EQ on each input.

Assign Switch

The equalized signal goes to the **assign switch** (sometimes called **track selector switch**). It lets you send the signal of each instrument to the tape track you want to record that instrument on.

Some units have an assign switch labeled 1, 2, 3, or 4. If you want to record bass on track 1, for instance, find the assign switch for the input module the bass is plugged into, and push switch 1. If you want to record four drum mics on track 2, push assign switch 2 for all those input modules. Other recorder-mixers assign tracks with two controls: a selector switch and a pan pot (described next).

Pan Pot

A pan pot is a device that sends a signal to two channels in adjustable amounts. By rotating the pan-pot knob, you control how much signal goes to each channel. Set the knob all the way left and the signal goes to one channel. Set it all the way right and the signal goes to the other channel. Set it in the middle and the signal goes to both channels.

Here's how you might use a pan pot to assign an instrument to a track. The track-selector switch (assign switch) might have two positions labeled

1-2 and 3-4. If you turn the pan pot left, the signal goes to odd-numbered tracks (either 1 or 3, depending on how you set the assign switch). If you turn the pan pot right, the signal goes to even-numbered tracks (2 or 4).

Suppose you want to assign the bass to track 1. Set the assign switch to 1-2, and turn the pan pot far left to choose the odd-numbered track (track 1).

During mixdown, the pan pot has a different function: it places images between your speakers. An **image** is an apparent source of sound, a point between your speakers where you hear each instrument or vocal. Set the pan pot to locate each instrument at the left speaker, right speaker, or anywhere in between. If you set the pan pot to center, the signal goes equally to both channels, and you hear an image in the center.

Aux (Effects or FX)

The aux or aux-send feature (Figure 12.3) can be used in two ways:

- To set up a monitor mix—a balanced blend of input signals you hear over speakers or headphones

- To set the amount of effects (reverb, echo) heard on each instrument in a mix.

Some recorder-mixers have no aux sends; some have one aux-send control per module; some have two (labeled aux 1, aux 2, or cue send). The more aux sends you have, the more you can play with effects, but the greater the cost and complexity. The aux number (1 or 2) is not necessarily assigned a specific function; you decide what you want aux 1 and aux 2 to do.

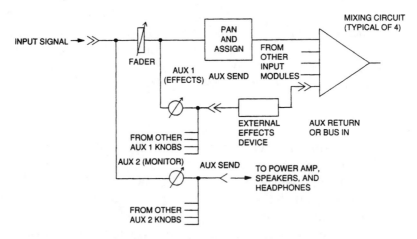

Figure 12.3 Auxiliary and output sections.

During recording and overdubbing, the aux knobs or cue-send knobs of all the input modules can be used to create a **monitor mix**. The monitor mix that you create with the aux knobs is independent of the levels going on tape. You use the gain-trims during recording to set recording levels, and you use the aux knobs to create an independent mix that is heard over your monitor system.

In Figure 12.3, the aux 1 send control is just before the fader. In this mixer, the signals from all the aux 1 knobs in the mixer combine at a connector jack labeled "aux 1 send" or "monitor." You can connect that jack to your power amplifier, which drives monitor speakers and headphones.

In this mixer, the aux 2 send control is just after the fader. During mixdown, each aux 2 knob controls how much effects (reverb, echo) you hear on each track. The aux knob controls the level of each input signal sent to an external effects device, which adds echo, reverb, or some other effect. The effects signal returns to the mixer's **aux return** or **bus in** jacks. Inside the mixer, the effects signal blends with the original signal, adding spaciousness or ambience to an otherwise "dry" track. A few recorder-mixers have an **aux return** control (also called **bus in**) that sets the overall effects level returning to the mixer.

Follow these steps to use the aux controls to adjust the amount of effects heard on each track:

1. Patch an effects unit between your mixer's aux send and aux return (bus-in) jacks.
2. On the effects unit, set the dry/wet mix control all the way to "wet" or "effect."
3. If your mixer has aux return (bus-in) knobs, turn them 1/2 to 3/4 up and pan their signals hard left and right.
4. Turn up the aux send knob for each input, according to how much effect you want to hear on that input signal.

Suppose you're using reverb as an effect. You might turn it up by different amounts for the vocals, drums, and lead guitar, and leave it turned down for the bass and kick drum. As you're setting the aux levels, check the overload indicator on the reverb unit. If it's flashing, turn down the input level on the reverb unit just to the point where the overload light stops flashing. Then turn up the output level on the reverb unit (or turn up the aux return on the mixer) to achieve the same amount of reverb you heard before.

There might be a **pre/post switch** next to the aux send knob. When an aux knob is set to pre (prefader), its level is not affected by the fader setting. You use the **pre** setting for a headphone mix during recording or overdubbing because you don't want the fader settings to affect the monitor mix.

The **post** setting (postfader) is used for effects during mixdown. In this case, the aux level follows the setting of the fader. The higher you set the track volume with the fader, the higher the effects level is. But the dry/wet mix stays the same.

Access Jacks (Insert Jacks)

These connectors on the mixer let you plug a compressor in series with an input module's signal for automatic volume control. Inexpensive units omit this feature. Some units have access jacks on only two inputs. Access jacks are usually prefader.

The access jacks also can be used to insert any other signal processor (reverb/delay, for instance) into the signal path of one track. This way, if all your aux sends are tied up, you can add another signal processor. On the reverb/delay unit, set the dry/wet mix control for the desired amount of effect.

Some insert jacks use a single connector for send and return. The tip connection is the send; the ring is the return, and the sleeve is the common ground or shield. Insert jacks can send prefader signals to a multitrack.

Output Section

The output section is the final part of the signal path; the section that feeds mixed signals to the tape tracks. It includes mixing circuits, submaster faders (sometimes), master faders, and meters (Figure 12.4).

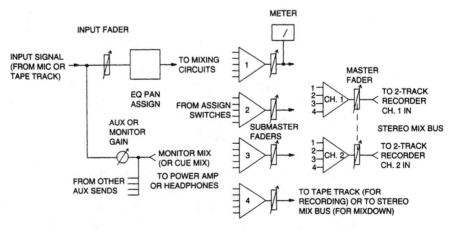

Figure 12.4 Output section of a mixer.

Mixing Circuits (Active Combining Networks)

The mixing circuits are near the right side of Figure 12.4. Recall that you use the assign switches to send each input signal to the desired channel or bus, and each bus feeds a different tape track. A bus is a channel in a mixer containing an independent mix of signals. The bus 1 mixing circuit accepts the signals from all the inputs you assigned to bus 1 and mixes them together to feed track 1 of the tape recorder. The bus 2 mixing circuit mixes all the bus 2 assignments, and so on.

Mixing circuits also accept the effects-return signals, such as the reverberated signal from an external digital reverb unit. A four-bus mixer provides four independent output channels or buses; each bus carries a signal which may contain the sounds from one or more musical instruments. The four buses feed a 4-track cassette recorder. A mixer with only two output buses can be used with a 4-track recorder by recording two tracks at a time.

Master and Submaster Faders

Located toward the right side of your mixer, the **stereo master faders** are one or two sliding volume controls—or a knob—that affects the overall level of the stereo output channels or buses. Usually, you set the master fader(s) within **design center**, the shaded area about 3/4 up on the scale. This setting minimizes mixer noise and distortion. You can fade out the end of a mix by turning down the master faders gradually.

Elaborate mixers also have faders that control the level of each bus independently, called **submaster**, **bus master**, or **group** faders. You might create a drum submix, or keyboard submix, and control the overall level of each submix with a submaster fader.

The submasters can serve double duty as master faders if you use only two of them. Then you can take your stereo mix from the bus 1 and 2 outputs. The noise is less there than after the master faders.

You feed the multitrack either from group outputs, direct outs, or insert sends. If you're mixing several instruments to track 5, for example, assign those instruments' signals to Group 5. Connect the Group 5 output to tape track 5 in. If you're recording **one** instrument on track 5, however, connect that instrument's direct out or insert send to track 5 in. The signal is cleaner at the direct out or insert send than at the group output.

Meters

Meters are an important part of the output section. They measure the voltage or level of various signals. Usually, each output bus has a meter to measure its signal level. If these buses feed the tape tracks, you use the meters to set

THE LIBRARY
GUILDFORD COLLEGE
of Further and Higher Education

the recording level for each track. Your recorder-mixer will have one of three types of meters:

VU: A voltmeter that shows approximately the relative loudness of various audio signals. Set the record level so that the meter needle reaches +3 VU maximum for most signals, and about –6 VU maximum for drums, percussion, and piano. That's necessary because the VU meter responds too slowly to show the true level of percussive sounds.

LED BARGRAPH METER: A column of lights (LEDs) that shows peak recording level. For all instruments, set the level to peak at 0 to +6 dB maximum, according to the manufacturer's recommendations.

LED PEAK INDICATOR: A light mounted in a VU meter. It flashes when peak recording levels are excessive. If you have this type of meter, set the level so the LED flashes only occasionally.

Tape-Out Jacks

Tape-out jacks are sometimes included in the output section of recorder-mixers. These are fed from the tape-track outputs, and are used for copying your four-track cassette recordings onto a multitrack studio recorder for further overdubs and processing.

Monitor Section

The monitor section controls what you're listening to. It lets you select what you want to hear, and lets you create a mix over headphones or speakers to approximate the final product. The monitor mix has no effect on the levels going on tape.

During recording, you want to monitor a mix of the input signals. During playback or mixdown, you want to hear a mix of the recorded tracks. During overdubs, you want to hear a mix of the recorded tracks and the instrument that you're overdubbing. The monitor section lets you do this.

Your mixer might have a small group of knobs called a **monitor mixer** (**monmix** or **tape cue**). Or the monitor mixer might be a row of monitor knobs, aux knobs, or output knobs, one in each input module. In either case, these knobs control the mix you hear.

The monitor mixer is made of several **monitor gain** controls, plus mixing circuits that feed headphones or an external stereo amplifier and speakers. The gain knob controls how loud each live instrument or track is in the monitor mix. Some monitor mixers also have **monitor pan** and **monitor effects** controls. The pan knob controls the left-right position of the instrument or track between your stereo speakers or headphones.

The monitor mixer also is used to blend recorded tape tracks and live microphone signals into a **cue mix** that is sent to headphones. You overdub new parts while listening to the cue mix. In most recorder-mixers, the monitor mix and cue mix are identical.

Some recorder-mixers have several extra inputs for MIDI instruments, along with aux or monitor knobs for these instruments so you can add them to the monitor mix. Because of this feature, you can use the recorder-mixer to mix tape tracks and sequencer tracks; you don't need to purchase an extra mixer to do this. Remember that the auxiliary (aux) sends in the mixer can serve multiple duty as controls for a monitor mix, headphone mix, or effects sends.

Monitor Select Buttons

Another feature of the monitor section is the monitor select buttons. They let you choose what signal you want to monitor or listen to. Since the configuration of these buttons varies widely among different recorder-mixers, they are not shown in Figure 12.4. The following are some of the monitor select buttons you may find:

- MONITOR TRACK 1, 2, 3, 4: You select which track or combination of tracks you want to hear.
- TAPE/BUS 1, 2, 3, 4: Signals from tape tracks 1, 2, 3, or 4 or from buses 1, 2, 3, or 4. Select "tape" to hear a playback, or to hear recorded tracks during an overdub. Select "bus" to hear the live signal that you're recording.
- STEREO or MIX: The 2-channel stereo mix at your mixer output.
- AUX: The aux-send signal (effects or headphone mix).
- CUE: The headphone cue mix.
- 2-TRACK or TAPE: The playback from your 2-track recorder.

Some units have no monitor-select switches. Instead, you always monitor the 2-channel stereo monitor mix. If you use insert jacks to send and receive signals to the multitrack, you can do a stereo monitor mix with the mixer faders.

Mixing-Console Styles: Split vs. In-Line

In a **split console (side-by-side console)**, a separate monitor mixer is built into the control surface. This monitor mixer has level and pan controls for each track, and perhaps more functions. In an **In-Line console (I/O console)**, the monitor mixer is a row of monitor knobs or aux knobs, one in each input module. This is the most common type of console. In either type, the monitor-mixer knobs are fed from the multitrack outputs.

An effective way to use an in-line console is to connect your multi-track recorder ins and outs to the insert jacks (Figure 12.5). Connect the insert-jack 1 tip (send) to track 1 in; connect track 1 out to the insert-jack ring (return). Make similar connections for the other tracks.

With this setup, use the trim controls to set recording levels. Use the faders to set up a monitor mix, cue mix, or mixdown with EQ and effects. Chapter 13 tells how to do this.

Additional Features Found in Mixing Consoles

Mixing consoles have more features than recorder-mixers do. If you're working only with a recorder-mixer, you might want to skip this section.

PRE/POST SWITCH: When used with an effects-send pot, the pre/post switch selects whether the effects send is derived prefader (before the fader) or postfader (following the fader). A prefader effects send is not affected by the fader level: If you turn down the input fader, the effect remains. A postfader effects send follows the fader action: If you turn down the fader, the effects-send level goes down. Normally you use prefader for monitor sends and postfader for effects sends.

FOLDBACK (FB): Another name for cue, or headphone mix.

SOLO: The SOLO button in an input module lets you monitor only that input without affecting other console functions. More than one input can be soloed at one time. On British consoles, the SOLO function is called "PFL" or "AFL," which stand for Pre-Fader Listen and After-Fader Listen (postfader), respectively. You listen to or monitor the signal before or after the fader.

Figure 12.5
Using insert jacks to send each input signal to a tape track. The track signal returns to the mixer, where you adjust level, panning, EQ, and effects.

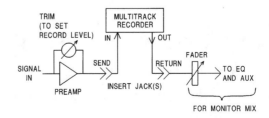

In consoles that have both PFL and SOLO, PFL is prefader and is used mainly to listen for distortion during recording. SOLO is post-fader and is used for listening to one track during mixdown.

Suppose you hear a buzz in the audio and suspect it may be in the bass guitar signal. If you push the SOLO button in the bass guitar's input module, you'll monitor only the bass guitar. Then you can easily hear whether the buzz is in that input.

PHASE (POLARITY INVERT): Used only with balanced lines, this switch inverts the polarity of the input signal. That is, it switches pins 2 and 3 to flip the phase 180 degrees. You might use it to correct a mis-wired mic cable whose polarity is reversed. If you mic a snare drum top and bottom, you need to invert the polarity of the bottom mic.

AUTOMATED MIXING CONTROLS: These controls (e.g., Read, Write, Update, etc.) are beyond the scope of this book. Basically they control the functions of a computer-assisted mixing system. The computer memory remembers and updates console settings so that a mix can be performed and refined in several stages.

EFFECTS PANNING: This feature places the images of the effects signals wherever desired between the monitor speakers. Some consoles let you pan effects in the monitor mix as well as in the final program mix.

FOLDBACK CONNECTORS: These connectors are in parallel with the TAPE IN jacks. The output of each tape track is connected to a TAPE IN jack, so the FOLDBACK connector parallels, or "mults," the tape-machine output. The FOLDBACK jacks can be used to send the tape outputs to an outboard device in addition to the console, such as an external cue mixer.

BUS TRIM: This rotary pot lets you fine-tune the bus level, along with the bus master (submaster or group) fader.

EFFECTS RETURN TO CUE: This is an effects-return level control that adjusts the amount of effects heard in the studio headphone mix. These monitored effects are independent of any effects being recorded on tape.

EFFECTS RETURN TO MONITOR: This effects-return control adjusts the amount of effects heard in the monitor mix. These monitored effects are independent of any effects being recorded on tape.

BUS/MONITOR/CUE switch for effects return: A switch that feeds the effects-return signal to your choice of three destinations: program bus (for mixdown), monitor mix, cue mix, or any combination of the three.

METER SWITCHES: In many consoles, the meters can measure signal levels other than console output levels. Switches near the meters can be set so that the meters indicate bus level, aux-send level, aux-return level, monitor-mix level, etc.

Those readings help you set optimum levels for the outboard devices receiving those signals. Too low a level results in noise; too high a level causes distortion in the outboard unit. For example, if the aux-return signal sounds garbled or distorted, the cause may be an excessive aux-send level. Verify that condition by checking the meters switched to read the aux or effects bus.

DIM: A switch that reduces the monitor level by a preset amount (as in "Dim the lights").

TALKBACK: This function lets the people in the control room talk to the musicians in the studio. A small microphone often is built into the console for this purpose.

SLATE: This function routes the control-room microphone signal to all the buses for announcing on tape the name of the tune and take number. In some consoles, a low-frequency tone is recorded on tape during slating; then the beginning of the take can be quickly located by listening for tape tones during fast-forward or rewind.

OSCILLATOR OR TONE GENERATOR: This is used to put alignment tones on tape, and to reference the tape recorder's meters to those on the console. You also can use it to check signal path, levels, and channel balance.

DIGITAL CIRCUITRY: A digital mixing console has an A/D converter on each input and a D/A converter on each output. There are also digital ins and outs. The signal stays in the digital domain for all mixer processing. Level changes, EQ, and so on are done by digital signal processing (computer calculations) rather than by analog circuits.

In some digital consoles, one knob can have several functions. Function parameters—such as EQ frequency and dB boost/cut—show up on an LCD screen. Digital consoles have built-in automated mixing (explained in Chapter 16).

Now that you understand the typical features of mixers and mixing consoles, you are ready to learn how to use them.

13

OPERATING THE MULTITRACK RECORDER AND MIXER

Get your hands on those knobs. You're going to operate a mixer as part of a recording session. This will be a basic run-through—detailed session procedures are described in Chapter 14.

First recall the stages in making a recording:

1. Session preparation
2. Recording
3. Punching in
4. Overdubbing
5. Bouncing tracks
6. Mixdown

This chapter considers each stage in turn.

Session Preparation

If you're using an analog multitrack recorder, be sure to clean the tape path before you start recording. Dust or tape oxide can accumulate on the tape heads, making the sound dull. So clean the tape heads, tape guides, and rubber pinch roller with a cotton swab and the cleaner recommended by the

247

manufacturer. A typical cleaner is denatured alcohol, available at hardware stores. Freon-based cleaners are also available. Don't use rubbing alcohol or isopropyl because they can leave a film on the heads, and they contain water.

When the heads are dry, thread on some blank recording tape or insert a cassette. Set the tape counter to zero. Use high quality blank tape. Chrome or metal cassette tape has better high-frequency response and better S/N ratio than ferric. C-60 or C-90 lengths record bass better than longer lengths. Use high-output, low-noise open-reel tape.

If you plan to use a tape-sync signal to synchronize your recorder with MIDI equipment, record it for the entire length of the song before recording any other tape tracks. Usually the sync tone is recorded on track 4 of a 4-track recorder or track 8 of an 8-track recorder. This procedure is detailed in Chapter 16.

If you're recording on a modular digital multitrack, fast-forward the tape to the end and rewind to the top. This loosens the tape pack, distributes the tape lubricants more evenly, and aligns the tape with the tape guides. Then format the tape from beginning to end. For efficiency, you might want to format an entire box of videocassettes at once.

Plan your track assignments. If you assign multiple instruments to the same track, you can't separate their images in the stereo stage. That is, you can't pan them to different positions; all the instruments on one track sound as if they're occupying the same point in space. If you're recording 4-track, you may want to do a stereo mix of the rhythm section on tracks 1 and 2; then overdub vocals and solos on tracks 3 and 4.

Recording

To start the process, first zero or neutralize the mixer by setting all the controls to "off," "flat," or "zero." That's to establish a point of reference and avoid surprises later on. Set all faders down.

If you have a separate mixer and multitrack recorder, you need to make their meter readings match. To do this, play a steady tone into the mixer to get a 0 reading on the meters for all tracks. Then set the multitrack recorder's record level (if any) to get 0 readings on all the tracks.

Suppose you're ready to record a vocal or an acoustic instrument. Place the microphone and plug it into a mic input. If you want to record a synth or drum machine on tape, connect a cable between the instrument's output and a line input on the mixer.

Attach a strip of masking tape or white removable tape along the bottom of the faders, and label each fader according to what instrument you plugged into that input.

Set the input selector (if any) to "input," "mic," or "line" depending on what is plugged into each input. If there is no input selector, turn down the TRIM control for line-level signals.

Plug in headphones to hear what you're recording. Turn up the headphone volume control. Or, if you're in a control room and the musicians are in a studio, turn up the monitor level to listen over the monitor loudspeakers. Set the MONITOR SELECT switch to hear the signal you're recording, and turn up the musicians' cue mix.

Set the master faders about 3/4 up, at 0, or within the shaded portion of fader travel. This is called **design center**. Do the same for the input fader(s) in use (Figure 13.1). These settings give the best compromise between noise and distortion.

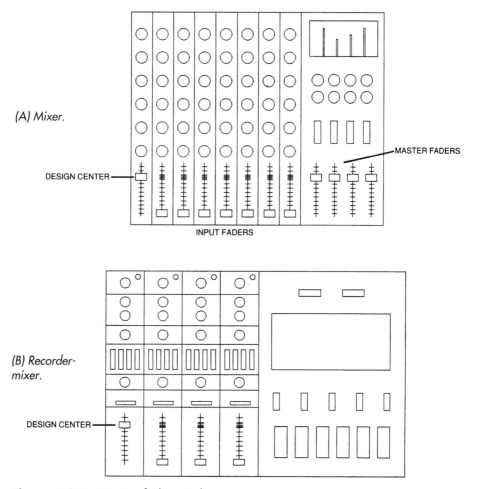

(A) Mixer.

DESIGN CENTER

MASTER FADERS

INPUT FADERS

(B) Recorder-mixer.

DESIGN CENTER

Figure 13.1 Setting faders at design center.

Assign Signals to Channels

Assign each input signal to the desired output channel (bus) as specified on your track sheet. Each bus is connected to the corresponding numbered tape track on the multitrack recorder.

If only one instrument is assigned to a track, you can eliminate the noise of the console combining amplifier by patching the instrument's signal directly to the tape track. To do this, locate the direct output jack of the input module for that instrument and patch it to the desired track. Some mixing boards also require pressing a direct button on the input module. You could also use the insert jack.

Set Recording Levels

Now you're ready to "get a level." Have each instrument play the loudest part of the music, one at a time or all at once. Set the TRIM control so the recording level is as high as possible without causing distortion.

Below are suggested recording levels for analog multitrack:

- Peak-reading LED bargraph meter: 0 to +6 dB.
- VU meter, drums and percussion: –6 to –8 VU.
- VU meter, other instruments: +3 VU maximum.
- Peak LED light in a VU meter: Occasional flashes.

While setting levels for a digital multitrack, peak each track around –5 dB. This allows some headroom for surprises. You don't want to exceed 0 dBFS (0 dB Full Scale).

If you have a separate mixer and multitrack recorder, which meters should you watch? Usually you watch the mixer meters to set levels, because you previously matched the recorder meters to the console meters. You have to watch the recorder meters for tracks that are patched direct or if you're using dbx noise reduction.

If you are mixing several instruments to one or two groups (as in a drum submix), follow this procedure: 1. Monitor the group(s). 2. Set its submaster fader to design center. 3. Set the submix balances, panning and recording level with the input faders. 4. Fine-tune each submix level with the submaster fader.

Set EQ

You may want to apply equalization at this point to each instrument heard individually. Filter out frequencies above and below the range of the instrument. However, don't spend too much time on EQ until all the instruments are mixed together. The EQ that sounds right on a soloed instrument may not sound right when all the instruments are heard together. In creating the desired tonal balance, use EQ as a last resort after trying different mics and mic placements. You also can apply EQ during playback or mixdown. This may be preferable because EQ applied when recording cannot always be undone if you're unhappy with it.

Record and Play Back

Next, set the track you want to record to "record ready" mode. Now start recording. Write down the tape-counter time for this take. "Slate" the tape: announce on tape the name of the tune and the take number. Then record a two-measure count-off. This is done to set the tempo for overdubs. For example, if the time signature is 4/4, you say, "1, 2, 3, 4, 1, 2, (rest) (rest)." The rests are silent beats. You need some silence before the song starts to make editing easier later on. If your recorder-mixer is part of a MIDI tape sync system, the count-off is unnecessary because your MIDI sequencer sets the song tempo.

After recording the track, rewind the tape to the beginning of the song using the return-to-zero or locate function. If necessary, set the monitor selectors to "tape," "track," or "remix," and play back the recording to check the performance and sound quality. You can set a rough mix with the monitor mix (aux) knobs. If your multitrack is patched to the insert jacks, use the faders, pan pots, EQ, and aux knobs to set a rough mix.

Punching In

You might want to correct musical errors in a track by punching in (doing an insert). Follow these steps:

1. Set the tape counter to "000" (or set a memory point) a few seconds before the part that needs correction.
2. Play the tape track to the musician over headphones.
3. During a rest (a pause in the track) just before the part needing correction, punch in the record button (or use a footswitch). Have the musician record the corrected musical part.
4. Immediately after the corrected part is played, punch out of record mode (or use the footswitch) to avoid erasing the rest of the track.
5. Using the return-to-zero or locate function, rewind the tape to a few seconds before the punch, and play it back. If necessary, you can rerecord the punch.

Some multitrack recorders have an **autopunch** function. You set the punch-in and punch-out points into the machine's memory. As the tape plays, the recorder automatically goes into and out of record mode at those points.

Often a musician can't get all the way through a long difficult solo or musical line without making a mistake. In this case, you can punch in and out to record the part in successive segments.

With care, you can punch in additional instruments on a completely full tape by recording them in the pauses on previously recorded tracks. For example, suppose all the tracks are full but you want to add a cymbal roll at the beginning of the chorus. Find a track that has a pause at that moment, and punch in the cymbal roll there.

Overdubbing

After your first track is recorded, you might want to add more musical parts. This procedure is called overdubbing. When you overdub, you listen to tracks you've already recorded, and record a new part on an unused track.

First, locate the monitor-mixer section. It is a row or group of knobs called "monmix," "monitor," "submix," "aux," or "tape cue." Find the monitor-mixer input selectors for the recorded tracks you want to listen to. Set them to "tape" or "track" so you can monitor them. Also, set the MONITOR SELECT switch to aux1, aux2, or whatever bus the monitor controls are affecting. If your multitrack is connected to the insert jacks, simply monitor the stereo mix.

Plug in a pair of headphones. Play the tape and set up a blend of the tracks using the monitor-level and PAN controls in the monitor mixer. If your multitrack is wired to the insert jacks, use the channel faders to set up a monitor mix.

Next, plug in the mic or instrument you want to record and assign it to an open (unused) track. Find the input-selector switch (if any) for the input signal you'll be recording. Set it to "mic," "line," or "input" as required. Do the same in the monitor mixer. Turn up the monitor-level control for the live signal so you can hear it. Turn up its fader to design center.

Set up your multitrack so it monitors the source (input signal) for the track to be overdubbed, and monitors the tape (output signal) for the recorded tracks. Have the musician play the instrument or sing into the mic. Adjust the TRIM (INPUT ATTEN or GAIN) to set the recording level.

Create a monitor mix of the recorded tracks and the live signal. Play the tape and play or sing along with it. Both the recorded tracks and the live signal can be heard in the musician's headphones. Adjust the monitor-level controls to get a good mix of the recorded tracks and the live signal.

When you're ready to record the new part, rewind or locate to the beginning of the song. Set the recorded tracks to "safe" and set the track you're recording on to "record ready." Start recording and have the musician play along with the tracks being monitored. If the musician makes a mistake, you can rerecord or punch-in the new part without affecting the other tracks.

If an overdub occurs only in the middle of a song, you don't need to start at the beginning. Set the tape counter to 000 (or set a memory point) a few seconds before the insert point. Use the return-to-zero (or locate) feature to practice the insert. Some multitrack recorders let you loop (repeat play) between two memory points before and after the insert.

If there are other parts to add, overdub them too. But you might want to leave one or two tracks open for bouncing.

Bouncing Tracks

If your recorder has too few tracks for all the parts you want to overdub, you can bounce tracks—mix several tracks to one or two open tracks, and record the mix on that track. Then you can erase the original tracks, freeing them for more overdubs.

Suppose you want to bounce tracks 1, 2, and 3 to track 4. Follow these steps:

1. Monitor only track 4.
2. Assign input modules 1, 2, and 3 to track 4.
3. Set all recorder tracks to "play" or "safe" mode.
4. Set the input selector switches for input modules 1, 2, and 3 to "tape."
5. Play the tape.
6. Using input faders 1, 2, and 3, mix the tracks as desired and set the recording level to peak around 0 on track 4.
7. When you're happy with the mix, rewind to the beginning of the song.
8. Set only recorder track 4 to "record ready" mode.
9. Start recording. The mix of tracks 1, 2, and 3 record on track 4.

To add live mic or instrument signals while you're bouncing, plug them into input 4 and assign input 4 to track 4. In this way, you can record up to ten tracks on a 4-track machine while bouncing tracks. Table 13.1 shows this bouncing procedure. Each track with an asterisk (*) is a live instrument or mic signal. Tracks without an asterisk are already recorded on tape.

Step 1 Track	Step 2 Track	Step 3 Track	Step 4 Track	Step 5 Track	Step 6 Track	Step 7 Track
1 A*	1	1 E*	1	1 H*	1	1 J*
2 B*	2	2 F*	2	2	2 HI*	2 HI
3 C*	3	3	3 EFG*	3 EFG	3 EFG	3 EFG
4	4 ABCD*	4 ABCD	4 ABCD	4 ABCD	4 ABCD	4 ABCD

Table 13.1   Bouncing procedure for recording 10 tracks with a 4-track recorder.

Follow these steps to execute this bouncing procedure:

1. Record three instruments (A, B, C) on tracks 1, 2, and 3.
2. Mix these three tracks with a live instrument signal and record the result on track 4.
3. Record two more instruments on tracks 1 and 2.
4. Bounce tracks 1 and 2 onto track 3 while mixing in another live instrument.
5. Record one more instrument on track 1.
6. Bounce track 1 to track 2 while mixing in another live instrument.
7. Record one more instrument on track 1.

You can continue this process to add even more instruments, but every rerecording adds noise, distortion, and frequency-response errors. This signal degradation is called **generation loss**. MiniDisc, hard disk, and digital tape have less generation loss than analog tape. Try a similar procedure for other track-bouncing combinations, setting up your recorder-mixer for the desired track assignments.

Mixdown

After all your tracks are recorded (maybe with some bouncing), it's time to mix or combine them to 2-track stereo. Use the mixer faders to control the relative volumes of the instruments, use panning to set their stereo position, use EQ to adjust their tone quality, and use the aux knobs to control effects.

If you're mixing only tape tracks (no sequenced-synth tracks), use the mixing console or recorder-mixer. But if you're synching a sequencer to your recorder, mix the tape tracks and synth tracks with an external mixer so you have enough inputs. Some elaborate recorder-mixers have extra line inputs for MIDI instruments so that you can do the entire mix with the recorder-mixer.

Set Up the Mixer and Recorders

To prepare for a mixdown, first locate the mixer jacks for output channels 1 and 2 (they might be called "Bus 1 and 2" or "Stereo mix bus"). Plug these outputs into the line or aux inputs of your 2-track recorder (Figure 13.2).

Clean and demagnetize any analog tape machines. Set all the mixer controls to "off," "zero," or "flat." You should start from ground zero in building a mix. Tape a strip of paper leader along the front of the mixer to write which instrument(s) each fader affects. Keep this strip with the multitrack tape for use each time you play it.

If necessary, set the input-selector switches on the mixer to "tape" (or "track" or "remix") because you'll be mixing down the multitrack tape. Monitor the 2-track stereo mix bus.

For starters, put the master faders at design center (about 3/4 up, at the shaded portion of fader travel). This sets the mixer gain structure for the best compromise between noise and headroom.

Erase Unwanted Material

Mixing will be a lot easier if you first erase noises before and after each song, and within each track. Play the multitrack tape and listen to each track alone. Erase unwanted sounds, outtakes, and entire segments that don't add to the song. To avoid mistakes, it's best to do this while the musicians are around.

What if a noise occurs just before the musician starts playing? When you erase the noise, you might erase the beginning of the performance. You can prevent this if you're using analog tape. Turn the cassette tape upside down or reverse the reels, then find the track of the desired instrument playing backwards. Play the tape section that came just after the noise. You hear it playing in reverse. Just after the reverse part ends, punch that track into record mode to erase the noise. It might be safer to mute the track during mixdown and

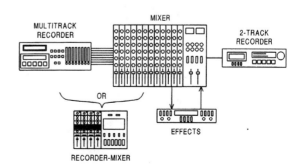

Figure 13.2
Connections for mixdown.

then unmute it just before the musician plays. If your multitrack is an MDM, set up an automatic punch-in/out at the correct times to erase the noise.

Panning

You need to pan the tracks before doing the mix, because the loudness of a track depends on where it's panned. Assign each track to busses 1 and 2, and use the pan pots to place each track where desired between your stereo speakers. Typically the bass, snare, kick drum, and vocals go to center; keyboards and guitars can be panned left and right.

Pan tracks to many points between the monitors: left, half-left, center, half-right, right. Try to achieve a stereo stage that is well balanced either side of center. For clarity, pan to opposite sides any instruments that cover the same frequency range.

You may want some tracks to be unlocalized. Harmony singers and strings should be spread out rather than appearing as point sources. Stereo keyboard sounds can wander between speakers. You could fatten a lead-guitar solo by panning it left, and panning the solo delayed to the right. (This delay might come from a distant room mic you used while recording.)

Consider creating some front-to-back depth. Leave some instruments dry so they sound close; add reverb to others so they sound farther away.

If you want the stereo imaging to be realistic (say, for a jazz combo), then pan the instruments to simulate a band as viewed from the audience. If you're sitting in an audience listening to a jazz quartet, you might hear drums on the left, piano on the right, bass in the middle, and sax slightly right. The drums and piano are not point sources, but are somewhat spread out. If spatial realism is the goal, you should hear the same ensemble layout between your speakers. In most rock recordings, the piano and drums are spread all the way between speakers—interesting but unrealistic.

Pan-potted mono tracks often sound artificial; each instrument sounds isolated in its own little space. It helps to add some stereo reverb. It surrounds the instruments and "glues" them together.

When you monitor the mix in mono, you'll likely hear **center channel buildup**. Instruments in the center of the stereo stage will sound louder in mono than they did in stereo, so the mix balance will change in mono. To prevent this, note which tracks are panned hard left or right, and bring them a little toward the center.

Compression

Sometimes the lead vocal track might be too loud or too quiet relative to the instruments because vocals have a wider dynamic range than instruments.

You can control this by running the vocal track through a compressor. It will keep the loudness of the vocal more constant, making it easier to hear throughout the mix. Patch the compressor into the access jacks for the vocal input module. Set it for the desired amount of compression (typically 3:1 ratio, –5 to –10 dB threshold). It's also common to compress the kick drum and bass. (For more information, see Chapter 11.)

Set a Balance

Now comes the fun part. The mixdown is one of the most creative parts of recording. Here are some tips to help your mixes sound terrific.

Before doing a mix, tune up your ears. Play over your monitors some CDs whose sound you admire. This helps you get used to a commercial balance of the highs, mids, and lows.

Choose a CD with tunes like those you're recording. Check out the production. How is the balance set? How about EQ, effects, sonic surprises? Try to figure out what techniques were used to create those sounds, and duplicate them. Of course, you might prefer to break new ground.

Using the input faders, adjust the volume of each track for a pleasing balance among instruments and vocals. You should be able to hear each instrument clearly. In some units, you use the MONMIX controls for this function. Some mixing consoles have trim knobs that set the gain of the tape tracks. In that case, set all faders in use to design center, and adjust the trims to get a rough mix.

Here's one way to build the mix. Make all the instruments and vocals equally loud. Then turn up the most important tracks and turn down background instruments. Or, bring up one track at a time and blend it with the other tracks. For example, first bring up the kick drum to about –10 VU, then add bass and balance the two together. Next add drums and set a balance. Then add guitars, keyboards, and finally vocals.

In a ballad, the lead vocal is usually on top. You might set the lead-vocal level to peak at –5 VU. Bring up the monitor level so that the vocal is as loud as you like to hear it, then leave the monitor level alone. Bring in the other tracks one at a time and mix them relative to the vocal track.

When the mix is right, everything can be heard clearly, yet nothing sticks out too much. The most important instruments or voices are loudest; less-important parts are in the background. In a typical rock mix, the snare is loudest, and the kick is nearly as loud. The lead vocal is next in level. Note that there's a wide latitude for musical interpretation and personal taste in making a mix.

Sometimes you don't want everything to be clearly heard. Once in a while, you might mix in certain tracks very subtly for a subconscious effect.

To reduce tape and mixer noise, mute all tracks that have nothing playing at the moment. That is, if there is a long silence during a track, mute that silent portion. Unmute these tracks just before their instruments start playing. Mute unrecorded tracks as well.

Level changes during the mix should be subtle, or else instruments will "jump out" for a solo and "fall back in" afterwards. Set faders to preset positions during pauses in the music. Nothing sounds more amateurish than a solo that starts too quietly then comes up as it plays. You can hear the engineer working the fader.

If you need to change fader levels during the mix, you might mark these levels next to each fader. Make a **cue sheet** that notes the mixer changes at various tape-counter times. For example:

0:15   Unmute vocal

1:10   Lead solo –5

1:49   Lead –10

2:42   Synth EQ +6 at 12K

3:05   Start fade, out by 3:15.

It's a good idea to monitor around 85 dB SPL. If you monitor louder, the bass and treble will be weak when the mix is played softly.

To test your mix, occasionally play the monitors very quietly, and see if you can hear everything. Switch from large monitors to small, and make sure nothing is missing.

Set EQ

Next, set EQ for the tonal balance you want on each track. If a track sounds too dull, turn up the highs or add an enhancer. If a track sounds too bassy, turn down the lows, and so on. Cymbals should sound crisp and distinct, but not sizzly or harsh; kick drum and bass should sound deep, but not overwhelming or muddy. Be sure the bass is recorded with enough edge or harmonics to be audible on small speakers.

You'll need to readjust the mix balances after adding EQ. The EQ that sounds right on a soloed track seldom sounds right when all the tracks are mixed together. So make EQ decisions when you have the complete mix happening.

In pop-music recordings, the tone quality or timbre of instruments does not have to be natural. Still, many listeners want to hear a realistic timbre from acoustic instruments, such as the guitar, flute, sax, or piano.

The overall tonal balance of the mix should be neither bassy nor trebly. That is, the perceived spectrum should not emphasize lows or highs. You should hear the low bass, mid-bass, midrange, upper midrange, and highs roughly in equal proportions. Too-loud frequency bands can tire your ears.

When your mix is almost done, switch between your mix and a commercial CD to see whether you're competitive. Play them through the same monitor speakers. If the tonal balance of your mix matches a commercial CD, you know your mix will translate to the real world. This works regardless of what monitors you use.

Add Effects

With the balances and EQ roughed in, it's time to add effects. You might want to plug in an external reverb or delay unit to add spaciousness to the sound (see Chapter 11). This device connects between your mixer's aux send and aux receive jacks (or aux send and bus-in jacks).

Find the AUX RETURN or BUS IN controls, set them 1/2 to 3/4 up, and pan them hard left and right. Using the AUX knobs on the mixer, adjust the amount of delay or reverb for each track as desired. AUX might be labeled EFFECTS or FX on your mixer.

Too much effects and reverb can muddy the mix. You might turn up the reverb only on a few instruments or vocals. Once you have the reverb set, try turning it down gradually and see how little you can get by with.

The producer of a recording is the musical director and decides how the mix should sound. Ask to hear recordings having the kind of sounds the producer desires. Try to figure out what techniques were used to create those sounds.

Also try to translate the producer's sound-quality descriptions into control settings. If the producer asks for a "warmer" sound on a particular instrument, turn up the low frequencies. If the lead guitar needs to be "fatter," try a stereo chorus on the guitar track. If the producer wants the vocal to be more "spacious," try adding reverb, and so on.

Set Levels

Set the overall recording level as you're mixing. To maintain the correct gain staging, keep the master faders at design center. Then adjust all the input faders by the same amount so your levels peak around 0. You can touch up the master faders a few dB if necessary. Don't exceed 0 dB if you're recording to DAT.

Judging the Mix

When you mix, your attention scans the inputs. Listen briefly to each instrument in turn and to the mix as a whole. If you hear something you don't like, fix it. Is the vocal too tubby? Roll off the bass on the vocal track. Is the kick drum too quiet? Turn it up. Is the lead-guitar solo too dead? Turn up its effects send.

The mix must be appropriate for the style of music. For example, a mix that's right for rock music usually won't work for folk music or acoustic jazz. Rock mixes typically have lots of production EQ, compression, and effects; and the drums are way up front. In contrast, folk or acoustic jazz is usually mixed with no effects other than slight reverb, and the instruments and vocals sound natural. A rock guitar typically sounds bright and distorted; a straight-ahead jazz guitar usually sounds mellow and clean.

If you want a realistic, natural sound, adjust the controls to create the illusion of live musicians playing in front of you. To do this you must be familiar with the sound of real instruments. Manipulate the recorded tracks until they sound like your memory of the real thing.

Try to keep the mix clean and clear. A clean mix is uncluttered; not too many parts play at once. It helps to arrange the music so that similar parts don't overlap. Usually, the fewer the instruments, the clearer the sound. Mix selectively, so that not too many instruments are heard at the same time.

In a clear-sounding recording, instruments do not "crowd" or mask each other's sound. They are separate and distinct. Clarity arises when instruments occupy different areas of the frequency spectrum. For example, the bass provides lows; keyboards might emphasize midbass; lead guitar may provide upper mids, and cymbals fill in the highs.

Often the rhythm guitar occupies the same frequency range as the piano, so they tend to mask each other's sound. You can aid clarity by equalizing them differently. Boost the guitar at, say, 3 kHz, and boost the piano around 10 kHz. Or pan them to opposite sides.

Calibration Tones

Before you record the mix, you should record calibration tones if your master tape is analog. Record the following tones on both channels simultaneously with noise reduction (20 seconds each):

- 1 kHz at 0 VU
- Optionally, 15 kHz for azimuth alignment. Record at 0 VU for 15 or 30 ips; –10 VU for 7 1/2 ips to prevent tape saturation.
- 1 kHz, 10 kHz, and 100 Hz (record at 0 VU for 15 or 30 ips; –10 VU for 7 1/2 ips)
- If Dolby-A is used, record an encoded Dolby tone at 0 VU, followed by an encoded 1 kHz tone at 0 VU. (Dolby-A tones should be generated by each track's encoder.)

- If dbx Type 1 is used, record an encoded 1 kHz tone at 0 VU. If a zero offset is used, note the offset level (for example, 0 VU program = –3 VU on tape).

The duplicating engineer or CD-mastering engineer uses the 15 kHz tone to align the repro head, the 1 kHz 0 VU tone to set overall level and channel balance, and the other tones to set playback EQ. Then the engineer's tape machine plays back the same tonal balance and stereo balance that you recorded during mixdown.

If you don't have access to a multifrequency generator, just put on a 1 kHz tone or a sine-wave synthesizer note (2 octaves above middle C) at 0 VU, both channels. DAT mixdown tapes require no tones.

If you will record the mixes on DAT, put in a blank DAT tape and exercise it: Fast-forward to the end and rewind to the top. You might want to clean the tape path with a cleaning cassette.

Record the Mix

When you're happy with the mix and recording levels, slate the tape and record the mix on your 2-track machine. Keep a tape log noting the 2-track start and stop times for each song. You'll use these times when you're ready to edit.

If the mix is very difficult, you can record it a section at a time, and then edit the sections together. To fade out the end of the tune, pull down the master faders slowly and smoothly. Try to have the music faded out by the end of a musical phrase. The slower the song, the slower the fade should be.

The mixdown is complete. Leave the tape running for a few seconds—make a blank space so you won't accidentally erase the end of the mix you just did. Play back the mix to listen for drop-outs and errors.

Repeat these mixdown procedures for the rest of the good takes, leaving about 20 seconds of silence between each recording. You might record several mixes of the same tune and choose the best one. Give your ears a rest every few hours! Otherwise, your hearing loses highs and you can't make correct judgments.

After a few days, listen to the mix on a variety of systems—car speakers, a boom box, a home system. The time lapse between mixdown and listening will allow you to hear with fresh ears. Do you want to change anything? If so, make it right. You'll end up with a mix to be proud of.

Summary

The following are summaries of the procedures for recording, overdubbing, and mixdown. Use these steps for easy reference.

Recording

1. Turn up the headphone or monitor volume control. Monitor the aux bus you're using for the monitor mix. If your multitrack is wired to the insert jacks, monitor the stereo bus instead.
2. Assign instruments to tracks. To record one instrument per track, connect its direct-out (or insert send) to a track input.
3. Turn up the channel, submaster, and master faders to design center (the shaded portion of fader travel, about 3/4 up).
4. Adjust the input attenuators (trim) to set submixes and recording levels.
5. Set the monitor/cue mix.
6. Record onto the multitrack recorder.

Overdubbing

1. Assign the instruments or vocals to be recorded to open tracks. An open track is blank or has already been bounced.
2. Turn up the monitor/cue system.
3. Turn up the submasters and master to design center.
4. Play the multitrack tape (in sync mode, if necessary) and set up a cue mix of already-recorded tracks.
5. While a musician is playing, adjust the input attenuation and recording level.
6. Set the monitor/cue mix to include the sound of the instrument or vocal being added.
7. Record the new parts on open tracks.
8. Punch in and bounce as needed.

Mixdown

1. If necessary, set the input selectors to accept the multitrack tape signals.
2. Monitor buses 1 and 2 (stereo mix bus).
3. Assign tape tracks to buses 1 and 2.
4. Turn up the master fader to design center. In some mixers, the submasters should also be up.
5. Set preliminary panning.

6. Set a rough mix with the input faders.

7. Set equalization and effects.

8. Perfect the mix and set recording levels.

9. Record onto the 2-track tape.

Automated Mixing

A multitrack mixdown is often a complicated procedure. It can be difficult to change the mixer settings correctly at all the right times. So you might want to use **automated mixing**—have a computer memory remember and set the changes for you.

As you mix a song, you might adjust the mixer controls several times. For example, raise the piano's volume during a solo, then drop it back down. Mute a track to reduce noise during pauses in the performance. An automated mixing system can remember your mix moves, and later recall and reset them.

You can even overdub mix moves, for example, do the vocal-fader moves on the first pass, drum moves on the second pass, and so on. You also can punch-in fader moves to correct them. Effects changes can be automated as well in some units.

To understand automation, you need to know several terms related to MIDI. So more information on automation is at the end of Chapter 16 on MIDI. Now that you know how to run the mixer and recorders, you're ready to do an actual session—covered next.

14

SESSION PROCEDURES

"We're rolling. Take One." These words begin the recording session. It can be an exhilarating or an exasperating experience, depending on how smoothly you run it.

The musicians need an engineer who works quickly yet carefully. Otherwise, they may lose their creative inspiration while waiting for the engineer to get it together. And the client, paying by the hour, wastes money unless the engineer has prepared for the session in advance.

This chapter describes how to conduct a multitrack recording session. These procedures should help you keep track of things and run the session efficiently.

There are some spontaneous sessions—especially in home studios—that just "grow organically" without advance planning. The instrumentation is not known until the song is done! You just try out different musical ideas and instruments until you find a pleasing combination.

In this way, a band that has its own recording gear can afford to take the time to find out what works musically before going into a professional studio. In addition, if the band is recording itself where it practices, the microphone setup and some of the console settings can be more-or-less permanent. This chapter, however, describes procedures usually followed at professional studios, where time is money.

Preproduction

Long before the session starts, you're involved in **preproduction**—planning what you're going to do at the session, in terms of overdubbing, track assignments, instrument layout, and mic selection.

Instrumentation

The first step is to find out from the producer or the band what the instrumentation will be and how many tracks will be needed. Make a list of the instruments and vocals that will be used in each song. Include such details as the number of tom toms, whether acoustic or electric guitars will be used, and so on.

Recording Order

Next, decide which of these instruments will be recorded at the same time and which will be overdubbed one at a time. It's common to record the instruments in the following order, but there are always exceptions:

1. Loud rhythm instruments—bass, drums, electric guitar, electric keyboards.
2. Quiet rhythm instruments—acoustic guitar, piano.
3. Lead vocal and doubled lead vocal (if desired).
4. Backup vocals (in stereo).
5. Overdubs—solos, percussion, synthesizer, sound effects.
6. Sweetening—horns, strings.

The lead vocalist usually sings a reference vocal or **scratch vocal** along with the rhythm section so that the musicians can get a feel for the tune and keep track of where they are in the song. The vocalist's performance in this case is recorded but probably is redone later.

In a MIDI studio, a typical order might be:

1. Drum machine (playing programmed patterns)
2. Synthesizer bass sound
3. Synthesizer chords
4. Synth melody
5. Synth solos, extra parts
6. Vocals and solos (recorded on tape or hard disk)

Track Assignments

Now you can plan your track assignments. Decide what instruments will go on which tracks of the multitrack recorder. The producer may have a fixed plan already. The outer tracks of analog multitracks are most prone to dropouts at high frequencies, so they are usually reserved for bass, kick drum, or SMPTE time code.

What if you have more instruments than tracks? Decide what groups of instruments to put on each track. In a 4-track recording, for example, you might record a stereo mix of the rhythm section on tracks 1 and 2 then over-dub vocals and solos on tracks 3 and 4. Or you might put guitars on track 1, bass and drums on track 2, vocals on track 3, and keyboards on track 4.

Remember that when several instruments are assigned to the same track, you can't separate their images in the stereo stage. That is, you can't pan them to different positions; all the instruments on one track sound as if they're occupying the same point in space. For this reason, you may want to do a stereo mix of the rhythm section on tracks 1 and 2, for instance, and then overdub vocals and solos on tracks 3 and 4.

It's possible to overdub more than four parts on a 4-track recorder. To do this, bounce or ping-pong several tracks onto one (see Chapter 13).

If you have many tracks available, leave several tracks open for exper-imentation. For example, record several takes of a vocal part using a sepa-rate track for each take, so that no take is lost. Then combine the best parts of each take into a single final performance on one track. It's also a good idea to record the monitor mix on one or two unused tracks or a cassette. The recorded monitor mix can be used to make a work-print tape for the client to take home and evaluate, or for a cue mix for overdubs.

Session Sheet

Once you know what you're going to record and when, you can fill out a session sheet (Figure 14.1). This simple document is adequate for home stu-dios. "OD" indicates an overdub. Note the tape counter time for each take, and circle the best take.

Production Schedule I

In a professional recording studio, the planned sequence of recording basic tracks and overdubs is listed on a **production schedule** (Figure 14.2).

Figure 14.1
A session sheet for a home studio.

```
SONG: Escape to Air Island

TRACK   INSTRUMENT          MICROPHONE
  1     Kick                AKG D-112
  2     Drums               Crown GLM-100
  3     Lead Voc   OD       Neumann U-87
  4     Harm Voc   OD       Neumann U-87
  5     Lead Guit  OD       Shure SM-57
  6     Keys L              Direct
  7     Keys R              Direct
  8     Bass                Direct

TAKES:   1, 2, ③
```

```
Tape Speed:  15 ips              Artist:  Muffin
8 Track                          Producer:  B. Brauning
Noise Reduction:  dbx

1.  Song:  "Mr. Potato Head."
    Instrumentation:  Bass, drums, electric rhythm guitar,
    electric lead guitar, acoustic piano, sax, lead vocal.
    Comments:  Record rhythm section together with reference
    vocal. Overdub sax, acoustic, piano, and lead vocal later.

2.  Song:  "Sambatina."
    Instrumentation:  Bass, drums, acoustic guitar, percussion,
    synthesizer.
    Comments:  Record rhythm section with scratch acoustic
    guitar. Overdub acoustic guitar, percussion, and
    synthesizer.

3.  Song:  "Mr. Potato Head."
    Overdubs:  (1) acoustic piano, (2) lead vocal, (3) sax.

4.  Song:  "Sambatina."
    Overdubs:  (1) acoustic guitar, (2) synthesizer,
    (3) percussion.

5.  Mix:  "Mr. Potato Head."
    Comments:  Add 80-msec delay to toms.
               Double lead guitar in stereo.
               Increase reverb on sax during solo.

6.  Mix:  "Sambatina."
    Comments:  Add flanger to bass on intro only.
               Manually flange percussion.
```

Figure 14.2 A production schedule.

Track Sheet

Another document used in a pro studio is the track sheet or multitrack tape log (Figure 14.3). Write down which instrument or vocal goes on which track. The track sheet also has blanks for other information.

Microphone Input List

Make up a microphone input list similar to the following:

Input	Instrument	Microphone
1	Bass	Direct
2	Kick	EV RE-20
3	Snare/Hi Hat	AKG C451
4	Drums Overhead	Shure SM81

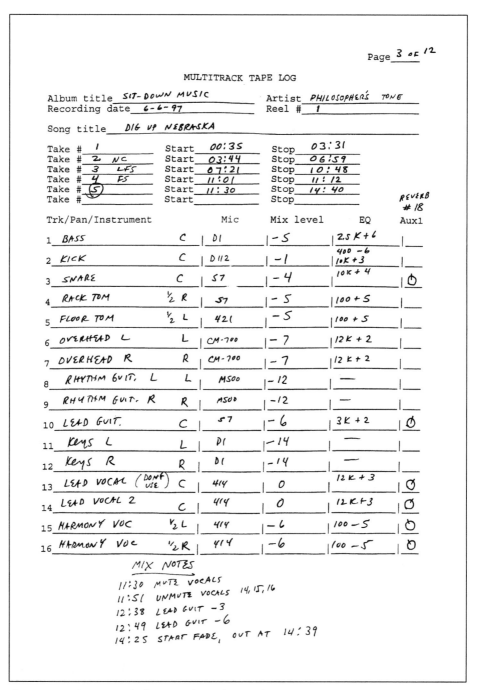

MULTITRACK TAPE LOG

Album title SIT-DOWN MUSIC   Artist PHILOSOPHER'S TONE
Recording date 6-6-97   Reel # 1

Song title DIG UP NEBRASKA

Take # 1   Start 00:35   Stop 03:31
Take # 2 NC   Start 03:44   Stop 06:59
Take # 3 LFS   Start 07:21   Stop 10:48
Take # 4 FS   Start 11:01   Stop 11:12
Take # (5)   Start 11:30   Stop 14:40
Take #   Start   Stop   REVERB #18

Trk/Pan/Instrument		Mic	Mix level	EQ	Aux1
1 BASS	C	DI	-5	2.5K+6	
2 KICK	C	D112	-1	400-6 10K+3	
3 SNARE	C	57	-4	10K+4	⏀
4 RACK TOM	½ R	57	-5	100+5	
5 FLOOR TOM	½ L	421	-5	100+5	
6 OVERHEAD L	L	CM-700	-7	12K+2	
7 OVERHEAD R	R	CM-700	-7	12K+2	
8 RHYTHM GUIT. L	L	M500	-12	—	
9 RHYTHM GUIT. R	R	M500	-12	—	
10 LEAD GUIT.	C	57	-6	3K+2	⏀
11 KEYS L	L	DI	-14	—	
12 KEYS R	R	DI	-14	—	
13 LEAD VOCAL (DON'T USE)	C	414	0	12K+3	⏀
14 LEAD VOCAL 2	C	414	0	12K+3	⏀
15 HARMONY VOC	½ L	414	-6	100-5	⏀
16 HARMONY VOC	½ R	414	-6	100-5	⏀

MIX NOTES

11:30 MUTE VOCALS
11:51 UNMUTE VOCALS 14, 15, 16
12:38 LEAD GUIT -3
12:49 LEAD GUIT -6
14:25 START FADE, OUT AT 14:39

Figure 14.3 A track sheet (multitrack tape log).

5	Drums Overhead R	Shure SM81
6	Hi Toms	Sennheiser MD 421
7	Low Toms	Sennheiser MD 421
8	Electric Lead Guitar	Shure SM57
9	Electric Lead Guitar	Direct
10	Piano L	Crown PZM-30D
11	Piano R	Crown PZM-30D
12	Scratch vocal	Beyer M500

Later you will place this list by the mic snake box and by the mixing console.

Be flexible in your microphone choices—you may need to experiment with various mics during the session to find one giving the best sound with the least console equalization. During lead-guitar overdubs, for example, you can set up a direct box, three close-up microphones, and one distant microphone—then find a combination that sounds best.

Find out what sound the producer wants—a "tight" sound; a "loose, live" sound; an accurate, realistic sound. Ask to hear recordings having the kind of sound the producer desires. Try to figure out what techniques were used to create those sounds, and plan your mic techniques and effects accordingly. Tips on choosing a microphone are given in Chapter 6.

Instrument Layout Chart

Work out an instrument layout chart, indicating where each instrument will be located in the studio, and where baffles and isolation booths will be used (if any). In planning the layout, make sure that all the musicians can see each other and are close enough together to play as an ensemble.

Setting Up the Studio

About an hour before the session starts, clean up the studio to promote a professional atmosphere. Lay down rugs and place AC power boxes according to your layout chart.

Now position the baffles on top of what has gone before. Put out chairs and stools according to the layout. Add music stands and music-stand lights. Run cue cables from each artist's location to the cue panel in the studio.

Place mic stands approximately where they will be used. Wrap one end of a microphone cable around each microphone-stand boom, leaving a few extra coils of cable near the mic-stand base to allow slack for moving. Run the rest of the cable back to the mic input panel or snake box. Plug each cable into the appropriate wall panel or snake box input, according to your mic input list.

Some engineers prefer to run cables in reverse order, connecting to the input panel first and running the cable out to the microphone stand. That procedure leaves less of a confusing tangle at the input panel where connections might be changed.

Now set up the microphones. Check each mic to make sure its switches are in the desired positions. Put the mics in their stand adapters, connect the cables, and balance the weight of the boom against the microphone.

Finally, connect the musicians' headphones for cueing. Set up a spare cue line and microphone for last-minute changes.

Setting Up the Control Room

Having prepared the studio, run through this checklist to make sure the control room is ready for the session:

- Pull all the patch cords from the patch panel.
- If necessary, patch console bus 1 to recorder track 1, bus 2 to track 2, and so on.
- Check out all the equipment to make sure it's working.
- Clean and demagnetize the analog tape machines.
- Thread on or insert some blank tape.
- Label the tape with the artist, date, and reel number.
- Put calibration tones on analog tape.
- Normalize or zero the console by setting all switches and knobs to "off," "zero," or "flat" so as to have no effect.
- Feed a 1 kHz tone to all the console outputs so that the console meters read 0 VU. Adjust the multitrack machine's record levels (if any) so the recorder meters also read 0 VU. This procedure matches the recorder meters to the console meters, so that you have to watch only the console meters while recording. If you're using noise-reduction equipment, refer to Chapter 9 for meter calibration and Dolby-tone recording.
- Switch on phantom powering for condenser microphones.
- Set the console input-selector switches (if any) to "mic" or "line" as needed.

- Attach a **designation strip** of paper leader across the front of the console. Referring to your mic input list, write on the strip the instrument each fader affects (bass, kick, guitar, etc.). Also label the submasters and monitor-mixer knobs according to what is assigned to them.
- Turn up the monitor system. Carefully bring up each fader one at a time and listen to each microphone. You should hear normal studio noise. If you hear any problems such as dead or noisy microphones, hum, bad cables, or faulty power supplies, correct them before the session.
- Verify the mic input list. Have an assistant scratch each mic grille with a fingernail and identify the instrument the microphone is intended to pick up.
- Check all the cue headphones by playing a tone through them and listening while wiggling the cable.

After the musicians arrive, allow them 1/2 hour to 1 hour free setup time for seating, tuning, and mic placement. Show them where to sit, and work out new seating arrangements if necessary to make them more comfortable.

Once the instruments are set up, you may want to listen to their live sound in the studio and do what you can to improve it. A dull-sounding guitar may need new strings; a noisy guitar amp may need new tubes, and so on. Adjust the studio lighting for the desired mood.

Session Overview

This is the typical sequence of events:

1. For efficiency, record the basic rhythm tracks for several songs on the first session.
2. Do the overdubs for all the songs in a dubbing session.
3. Mix all the tunes.
4. Assemble and leader the analog mixdown tape, or edit the DAT mixdown tape.

Recording

Follow the mixer recording procedures described in Chapter 13. Before you start recording, make connections to record a work-print tape of the studio monitor mix, either on two left-over tracks or on a cassette deck. This tape is for the producer to take home to evaluate the performance.

When you're ready to record the tune, briefly play a metronome to the group at the desired tempo, or play a click track (an electronic metronome) through the cue system. Or just let the drummer set the tempo with stick clicks.

Start the tape in record mode. Note the tape-counter time. Hit the slate button and announce the name of the tune and the take number.

Have someone play the keynote of the song (for tuning other instruments later). Then the group leader or the drummer counts off the beat, and the group starts playing.

The producer listens to the musical performance while the engineer watches levels and listens for audio problems. As the song progresses, you may need to make small level adjustments. As stated before, the recording levels are set as high as possible without causing distortion. Balancing the instruments at this time is done with the monitor mixer. The monitor mix affects only what is being heard, not what is going on tape.

The assistant engineer (if any) runs the tape machine and keeps track of the takes on the track sheet, noting the name of the tune, the take number, and whether the take was complete (Figure 14.3). Use a code to indicate whether the take was a false start, nearly completed, a "keeper," and so on.

While the song is in progress, don't use the solo function because the abrupt monitoring change may disturb the producer. The producer should stop the performance if a major fluff (mistake) occurs but should let the minor ones pass.

At the end of the song, the musicians should be silent for several seconds after the last note. Or, if the song ends in a fade-out, the musicians should continue playing for about a minute so there is enough material for a fade-out during mixdown.

After the tune is done, you can either play it back or go on to a second take. Set a rough mix with the aux or monitor-mix knobs. If you connected your multitrack to the insert jacks, use the faders to set a rough mix with EQ and effects. The musicians will catch their fluffed notes during playback; you just listen for audio quality.

Now record other takes or tunes. Pick the best takes. To protect your hearing, try to limit tracking sessions to four hours or less—five hours maximum.

Overdubbing

After recording the basic or rhythm tracks for all the tunes, add overdubs. A musician listens to previously recorded tracks over headphones, and records a new part on an open track. Follow the overdubbing procedures in Chapter 13.

Composite Tracks

If several open tracks are available, you can record a solo overdub in several takes, each on a separate track. This is referred to as recording **composite**

tracks. After recording all the takes, play back the solo (in sync mode) and assign all the overdubbed tracks to a remaining open track set in record mode. You bounce all the solo tracks to a composite track. Match the levels of the different takes. Then switch the overdubbed tracks on and off (using muting), recording just the best parts of each take. Finally, erase the old overdubbed tracks to free them up for other instruments.

Drum Overdubs

Drum overdubs are usually done right after the rhythm session because the microphones are already set up, and the overdubbed sound will match the sound of the original drum track.

Overdubbing in the Control Room

To aid communications among the engineer, producer, and musician, you might have the musician play in the control room while overdubbing. You can patch a synth or electric guitar into the console through a direct box, and feed the direct signal to a guitar amp in the studio via a cue line. Pick up the amp with a microphone, and record and monitor the mic signal.

Breaking Down

When the session is over, tear down the microphones, mic stands, and cables. Put the microphones back in their protective boxes and bags. Wind the mic cables onto a power-cable spool, connecting one cable to the next. Wipe off the cables with a damp rag if necessary. Some engineers hang each cable in big loops on each mic stand. Others wrap the cable "lasso style" with every other loop reversed. You learn this on the job.

Put the labeled tape (stored tail out) in its box. Also in the box, or in a file folder, put the designation strips, track sheet, and take sheet. Label the box and folder. Normally, the studio keeps the multitrack master unless the group wants to buy or rent it.

If you recorded analog, you may want to edit out the out-takes and splice them together on a separate reel. Then put one foot of paper leader between each of the master tape's keeper takes. Write new tape logs indicating the reels' contents.

Log the console settings by writing them in the track sheet or reading them slowly into a portable cassette recorder. At a future session you can play back the tape and reset the console the way it was for the original session. Some consoles can store and recall the control settings.

Mixdown

After all the parts are recorded, you're ready for mixdown. Prepare the console and recorders, record tones, erase noises, and play the multitrack tape through the console while adjusting balances, panning, equalization, reverberation, and effects. Once you've rehearsed the mix to perfection, record it onto a 2-track recorder. Follow the mixdown procedures in Chapter 13.

Record calibration tones on the analog 2-track master tape just before recording the mixes. (See Chapter 13 for more information.)

Assembling Analog Tape Master Reels

Now you're ready to assemble the 2-track open-reel tape into a finished format for cassette or CD duplication. It will contain the songs in the desired order, plus leader tape for separating the songs with silence. It also has calibration tones. The cassette/CD mastering engineer will use those tones to align the playback deck for flat response from your tape.

Sequencing the Songs

Discuss the order of the songs on the recording. For the first song, use a strong, accessible, up-tempo tune. Follow it with something quieter. Alternate keys or tempos from song to song. The last tune should be as good as or better than the first to leave a good final impression.

If you intend cassette release, try to keep the total times for Side A and Side B about equal to conserve cassette tape. Side A should be slightly longer than Side B so that any blank cassette tape is at the end of Side B.

Leader Length

The length of leader between songs depends on how long a pause you want between them. Four seconds is typical. Use longer leader if you want the listener to get out of the mood of the piece he or she just heard before going on to the next. Use shorter leader either to change the mood abruptly or to make similar songs flow together. Short leader also works well after a long fade-out because the fade-out itself acts like a long pause between songs. Each second of leader is 15 inches long (for 15 ips) or 30 inches long (for 30 ips).

What if you want to crossfade between two songs? Either use a digital audio workstation, or follow these steps:

1. Use two 2-track recorders. Put song 1 on one recorder and song 2 on another.

2. Feed both recorders into your mixer, and connect the mixer output to a third 2-track recorder. Match the levels and noise reduction on all the recorders.

3. Play song 1. Near the end of the song before the crossfade, start recording the mix on the third machine.

4. As song 1 fades out, start song 2 and fade it up.

5. After recording this crossfade on the third machine, edit the crossfade into the rest of the program.

Master Reel Assembly

If the recording is too long to fit on a single reel, make a separate reel for Side 1 and Side 2 of the cassette. To assemble the master tapes, wind onto the take-up reel the following material in this order:

1. At least 15 seconds of leader or blank recording tape

2. Tones

3. At least ten seconds of leader

4. First song

5. Leader

6. Second song

7. Leader, etc.

8. Last song on Side 1 of album

9. At least 15 seconds of leader

Then rewind the tape. Play it and time it from the beginning of the first song to the end of the last song (including the leaders between songs). This is called the **running time**. Also note the start and stop time of each selection.

Using a piece of masking tape, fasten the leader tail to the reel and print "TAIL OUT" on the masking tape. Or use blue leader tape for tails out, red for heads out. Type or print a neat label for the tape reel including title, artist, "Side 1," and the running time.

Using another take-up reel, assemble Side 2 of the album (but without tones). Time Side 2 and label the reel.

Be sure to make a safety copy of the master before sending the master tape, in case it is lost or damaged. When copying the master, set the calibration tone from the playback machine to read the same level on the recording machine. There's no need to reset the program levels because you already set them while recording the master tape.

Digitally Editing Your DAT Mixdown Tape

You can edit and re-sequence your DAT mixdown master by using a personal computer, digital editing software, and a sound card. Chapter 10 describes how to edit with your computer. Use the editing software to adjust the level of each song by ear. If the program will be duplicated on cassette, leave a minute of silence between the Side A songs and Side B songs.

You might prefer to send out your mixdown tape for editing/mastering. A mastering engineer can suggest processing for your tape to make it sound more commercial and more uniform from song to song.

Transferring the Edited Program to DAT

If you edited your DAT mixdown tape, now you will copy the edited program from hard disk to a blank DAT. After loading the DAT tape, fast-forward to the end and rewind to the top. This loosens the tape pack, distributes the tape lubricants more evenly, and aligns the tape with the tape guides. You might want to clean the DAT recorder with a cleaning tape. Turn off "automatic start ID."

To prevent DAT drop-outs, record 2 minutes of silence at the head of the DAT tape. When the DAT ABS time reads 2:00 (2 minutes), play the edited program from the hard disk. The editing software tells the start and stop times of each song. Add 2 minutes to these times to get the song start/stop times on the DAT tape.

After the program has copied to DAT, rewind the DAT tape. Refer to your list of DAT start/stop times. Fast-forward to the beginning of each song, and write a manual Start ID there about 1/2 second before each song. When done, have the DAT machine renumber the program numbers.

Make a safety copy by doing another disk-to-tape transfer. Or run two DATs in parallel during the transfer. Hold onto the safety; you'll need it in case the master tape is lost or damaged. You can send a CD-R instead of a DAT.

If you want to make a few cassette copies of your master DAT tape—say, for demo tapes—copy the DAT tape several times with a stereo cassette deck. For larger runs, you'll need an outside duplicating house.

Tape Log

Type or print a **tape log** describing the DAT master tape (Figure 14.4). For analog tape, also include the tape-head format, stereo/mono, tape speed,

```
                        DAT TAPE LOG

       PROJECT TITLE: Out of Your Head
       ARTIST: Hal Klee
       PROJECT NUMBER: 960716

       CLIENT: Hal Klee
               1061 College Ave.
               LaSuer, IN 46596
               Phone 219-555-1049

       RECORDING ENGINEER: Hal Klee
       EDITING ENGINEER: Bruce Bartlett, 56657 CR 40, Fenfen,IN 42596
                219-555-1388 (work)  219-555-9366 (home)

       TAPE CONFIGURATION: TASCAM DA-P20 DAT, 44.1K

       MASTERING INSTRUCTIONS: Just cut flat, no EQ.  Add 6 dB of
       overall compression.  Make chrome cassettes, Dolby B.  Please
       send a proof DAT to Bruce Bartlett.

       There are lots of intentional noises, outtakes, etc.  Distortion
       at 08:32.

       Side A running time: 23:47    Side B running time: 23:08

       ABS times of highest level (0 dBFS):  03:38, 12:43

       Side A
       PGM NO.         TITLE                         START  STOP
          [HEAD OF TAPE]   .......................   00:00
       1  David....................................  02:00  03:43
       2  Burl White In Manhattan..................  03:47  04:46
       3  Wake Up Samantha.........................  04:50  07:23
       4  Wrench and a Bad Man.....................  07:27  09:39
       5  Down Rockaway............................  09:43  12:24
       6  Getting Up...............................  12:28  15:22
       7  Foont...................................   15:26  17:06
       8  Lord of the Bytes.......................,  17:10  18:46
       9  Little Bummer Boy........................  18:50  24:33
       10 Jovian Boo Boos..........................  24:37  25:47

       64 seconds of silence here separates Sides A and B.

       Side B
       PGM NO.         TITLE                         START  STOP
       11 Trumpechello.............................  26:51  29:42
       12 Climbin' Keds............................  29:46  31:12
       13 Rumble Machine...........................  31:16  34:45
       14 Ninety Hats..............................  34:49  37:24
       15 It Pays to Be Silent.....................  37:28  39:50
       16 Ormes....................................  39:54  42:31
       17 Frumplestein.............................  42:35  49:59
```

Figure 14.4 Example of a DAT tape log.

playback equalization (usually NAB or IEC), noise reduction, tail-out designation, the location of the test tones (usually at the beginning of Reel 1), tone frequencies, and levels.

Also include the packaging text such as song lyrics, instrumentation, composers, arrangers, publishers, artwork, and so on. Stay in contact with the duplication house, especially about artwork.

Note that the master tape doesn't leave the studio until all studio time is paid for! When this is done, send the tape or CD-R (PMCD pre-master compact disc) to the duplication house.

It's amazing how the long hours of work with lots of complex equipment have been concentrated into that little tape—but it's been fun. You created a craftsmanlike product you can be proud of. When played, it recreates a musical experience in the ears and mind of the listener—no small achievement.

15

RECORDING
THE SPOKEN WORD

Recording music is fun, but recording speech is the main income producer for many studios. Speech is the major sound element in commercials, promo videos, documentaries, drama, educational programs, and books on tape.

In this chapter you learn ways to record the spoken word effectively. It's not as simple as it seems.

Be Consistent

An important quality of a speech recording is consistency. The tone quality, level, average pitch, and tempo of the voice should not change throughout the recording (except for dramatic effect).

Why is this so important? Often, the voice track is assembled from takes recorded on different days, so the sound of the takes has to match. Otherwise, you hear jarring changes in the voice quality as the recording plays.

Typically, you record a complete script in a single session. Then you send a proof copy of the tape to the script's publisher for error checking. After you get the corrections back, you call the announcer back into your studio to record **inserts**—corrected sentences and paragraphs. You edit the inserts into the original recording. If the sound of the inserts doesn't match the sound of the original, the edited program doesn't sound like a single take. So you need to duplicate the recording setup exactly each time you record the announcer.

Several factors can vary from one session to the next: recording level, mic choice and placement, text position, announcer's position, EQ, noise reduction, and even the voice itself. You need to keep all these factors constant by documenting your setup.

Take notes on the mic used, its switch positions (if any), and its position and distance from the announcer. Also note any EQ or noise reduction used. To reduce the number of variables, you might record without EQ. Settle on a standard setup so you can record a predictable sound.

Microphones

For voice recording, three types of mics are in common use:

- A flat-response cardioid condenser mic.
- A "multiple-D" dynamic mic.
- A top-quality lavalier condenser mic.

A cardioid condenser mic provides a luxurious, big-budget sound—one with full lows and detailed highs. Unfortunately, the mic's bass response varies with the announcer's distance. The closer the announcer is to the mic, the bassier is the recording. Unless the announcer can remain a constant distance away, the voice tone quality will vary.

This close-up bass boost (proximity effect) occurs in **single-D** unidirectional mics. A **multiple-D** mic is designed to compensate for proximity effect: its bass varies only a little with distance. A single-D mic has holes or slots only in the mic capsule; a multiple-D mic (like the EV RE-20) has holes or slots in the handle. As an alternative, you might use an omni mic, which has no proximity effect.

A less popular choice is a lavalier microphone. It's a miniature unit (like TV newscasters use) that clips onto the announcer's tie or shirt. There are excellent models in the $200 price range. Don't skimp on this microphone, or the sound quality will suffer. Even with a good mic, though, the sound is not as natural as with a conventional mic. Lavaliers work well for video shoots where the mic should be inconspicuous.

A lavalier has some advantages in script recording. Since the user wears the lavalier, it remains a constant distance from the mouth, which aids consistency. There are no breath pops unless the user exhales through the nose. If the announcer sits still, cable noise is not a problem. However, if the announcer's head tilts up and down, the voice gets brighter and duller.

Mic Placement

Once you've chosen a mic, place it at various distances from the mouth until you hear the best sound quality. A typical distance is 8 to 12 inches. If the mic is too far, you'll hear too much room reverb. Generally you don't want to hear room sound in a narration recording. If the mic is too close, you'll hear lip and tongue noises, and the voice level will vary when the announcer moves. Find a workable distance somewhere in the middle and stick with it.

Some studios set the miking distance with a spacer or ruler. The announcer can set the spacing with the hand. Ask announcers to spread their fingers, put their thumb on their mouth and their pinky on the mic.

In drama recording, mic distance is less critical. The actors often change their distance to the mic for special effect.

You need to place the mic to avoid breath popping. When a person says words with the letters "p," "t," or "b," a turbulent puff of air shoots out of the mouth. If this air puff hits a mic grille, you hear a thump or pop. Since a pop disturbance leaves the mouth within a narrow angle, you can prevent pops by putting the mic above the mouth at eye height. It also helps to put a foam pop filter on the mic, or use a disk-type pop screen on a gooseneck.

To prevent table thumps, mount the mic on a boom stand and place the stand on the floor. You might want to use a shock mount too. Cover the script stand with felt or foam to absorb noises.

A typical placement for lavalier mics is 8 inches below the chin. Tape the cable to the shirt in order to prevent rubbing noises. Try not to cover the mic with clothing, or the sound will be muffled.

Minimizing Sound Reflections

Angle the script so that the mic does not pick up sound reflections from the script. These reflections can combine at the mic with the direct sound from the announcer. This causes phase interference—a filtered, colored tone quality which changes as the announcer moves.

Figures 15.1 and 15.2 show some right and wrong ways to place scripts. In Figure 15.1A, sound reflects off the script stand into the mic. But in Figure 15.1B, the script is tilted more vertically so reflections bounce away from the mic. In Figure 15.2A, sound reflects off the script into the lavalier mic. But in Figure 15.2B, the script is tilted horizontally to prevent reflections.

If you record more than one announcer at once, seat them at least four feet apart to prevent phase interference between mics.

283

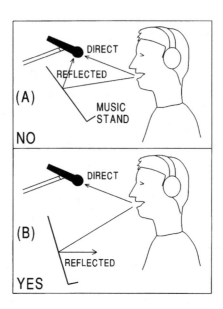

Figure 15.1
Angle the script stand to avoid reflections into the microphone.

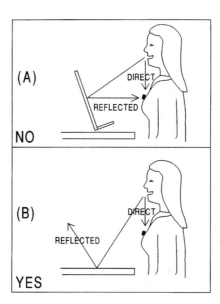

Figure 15.2
Angle the script to avoid reflections into the lavalier microphone.

Reducing Sibilance

Another factor to keep under control is sibilance: the emphasis of "s" and "sh" sounds. These sounds have strong high frequencies around 5 to 10 kHz, which can saturate the tape of duplicated cassettes. If you need to reduce sibilance, use a flat-response mic, mike off-axis to the mouth, or cut EQ a little around 5 to 10 kHz.

A better solution is to use a **de-esser**. This signal processor removes sibilance without affecting tone quality. A de-esser is a compressor with a high-frequency boost before compression and a complementary cut after compression. With a de-esser, only excessive high frequencies (sibilant sounds) are compressed.

Reducing Print-Through

An analog tape of speech might have unwanted echoes or pre-echoes in the pauses between words. These echoes are caused by print-through. In the tape reel, the magnetic signal transfers or "prints" from one layer of tape to the next, causing an echo.

Print-through is a major problem with narration recording because speech has many silent pauses where print-through can be heard. To reduce print-through, use 1 1/2 mil low-print tape, use noise reduction, and store tapes tail out. Better yet, record digitally on DAT or hard disk, which have no print-through.

At the Session

Give the announcer a comfortable, fixed-frame chair with a back to prevent motion. Ask the announcer to move as little as possible, and not to slump over the table.

If the announcer's head moves, the tone quality will change. Why? Low frequencies radiate from the mouth in all directions, but high frequencies radiate mostly straight out. So, if an announcer's head moves while talking, the high frequencies ("s" sounds) will miss the mic sometimes.

Suppose you're ready to record a script. The announcer is seated the proper distance from the mic, and the script is on a music stand. The announcer has folded up the bottom corners of the script pages to form a handle for turning pages without noise. You have an identical script on which to mark **edit points**: spots where the announcer misreads a sentence.

After a level check, start recording. The announcer reads the script, and you follow along in your own script, marking edit points with a pen. Just leave the tape rolling when mistakes occur.

If the announcer misreads a word, or makes a paper noise, have the announcer read the whole sentence over. The announcer should not try to correct the error in mid-sentence, because a mid-sentence edit tends to sound unnatural. If you edit between sentences, the edit will be undetectable.

Figure 15.3 shows a typical script with edit marks. Instead of marking the words that were misread, the producer marked the beginning of the sentences with errors. Those marks correspond to the edit points on tape. Two marks indicate a second retake.

EDITOR'S COPY

SCRIPT FOR "COSMIC ORDER" FILM

Client: Impossible Productions, Inc.

A branch consists of two or more radiations from a common point. Its basic shape is the letter "Y."

Each branch has its own branches. For example, a tree trunk branches into large limbs. Each limb branches into smaller limbs. Each smaller limb branches into sticks. Each stick branches into twigs. Each twig branches into leaf veins. /If a tree were a symphony, you'd hear a basic melody repeated with ever-increasing refinement and complexity.

→ remove cough

Each small branch is a variation on the theme of branching. That is, each part resembles the whole. /For example, the veins in a leaf resemble a tree in shape; capillaries resemble veins and arteries.

→ Noise ?

//Branches are also visible in tree roots, rivers, the circulatory system, the nervous system, the lungs, roads, audio systems, single-point grounding systems, organization of knowledge, distribution of knowledge, and more.

Let's consider that last example.///Knowledge is gathered and distributed like a tree: The research material is the ground the tree is rooted in; the roots represent research; the trunk represents the compilation of research into a new synthesis, and the branches represent the dissemination of the new knowledge to various students via publishing, broadcasting, or education in classrooms.

To digress a bit, consider this: If life seems too complicated, remember it's all happening to you here and now. "Here" is small and "now" is short. You can handle it.

↖ THUMP?

/On the other hand, "here" follows you wherever you go, and "now" is continuous. Never mind!

Figure 15.3 A script with edit marks.

Analog Tape Editing

After the recording is done, you're ready to edit out the mistakes. Follow this procedure:

1. Play the tape and follow along in the marked-up script. When you come to an edit point, stop the tape.
2. Put the recorder in cue or edit mode.
3. Rock the tape back and forth over the playback head to find the exact beginning of the misread sentence. Mark the tape at the playback-head gap with a grease pencil.
4. Using a razor blade and splicing block, cut the tape about 1/2 inch to the right of the mark.
5. Put the feed-reel tape back into the tape-path slot.
6. Pull the tape past the heads. You'll hear the misread sentence. The announcer will stop, then restart the sentence. Mark and cut the tape at the start of the corrected sentence.
7. After removing the misread section of tape, splice the two remaining tape ends together.
8. Play the edited part to check it. You should hear no double breaths between sentences. Edit out paper noises, coughs, and table thumps.

For an alternative method of editing:

1. Before cutting the tape, mark the two edit points: the start of the misread sentence and the start of the good retake.
2. Align both edit marks in your splicing block, cut both at once, and splice the ends together.

Once the tape is edited, add leader tape, label the reel and tape box, and store the tape tail out.

Digital Editing

You might prefer to edit speech with a digital audio workstation (DAW). If you recorded the speech on DAT or analog tape, copy it to your hard disk. (You could record directly to hard disk instead.) When the copy is done, follow this procedure:

1. You'll see the waveform or the soundtrack of the narration on your computer monitor screen. Play the soundtrack and follow along in the marked-up script.

2. When you come to the beginning of a misread sentence, mark it on screen as the beginning of a **region** or sound segment.

3. Play the misread sentence up to the start of the corrected sentence. Mark that as the end point of the region. The region you marked will highlight.

4. Select EDIT and DELETE. The misread sentence will disappear, and the following audio will slide in to fill the gap.

Another method:

1. Mark the start and end points of sentences and paragraphs that you want to keep. Each start and stop point should be at the beginning of a sentence.

2. Define these sound segments as regions or zones.

3. Put them in order in a playlist, and play the playlist.

Check your edits. Are there any double-breaths between sentences? Are any sentences read twice? Do the pauses between sentences sound unnatural? If so, correct these problems by changing the segment start/end points.

Proof Cassettes and Inserts

You might want to make a proof cassette of the edited tape to send to the script publisher for approval. The publisher may notice reading errors that you missed during the recording session. They'll send back a marked-up script showing the errors. The next time the announcer is back in your studio, rerecord the sentences or paragraphs needing corrections.

Be sure to match the recording level, mic position, and so on with those of the original tape. Otherwise the inserts may sound like another person talking. Play some of the original tape aloud so that the announcer can duplicate the pitch and tempo. You may need to equalize the inserts to match the original take.

After editing in the inserts, send another proof tape to the publisher. If it's error-free you can add sound effects or music as needed.

Sound Effects and Music

Many scripts call for sound effects or music to accompany the narration. Start by copying the edited speech to one track of a multitrack recorder (tape or DAW). On other tracks, record sound effects and music where noted in the script. Then mix the tracks.

Libraries of sound effects and mood music are advertised in recording magazines and audio-visual publications. Some publishers require royalty payments for each use of material. To avoid royalties, you can order sound-effects CDs from record stores, or record your own effects and music.

Sound effects can also be prerecorded and sequenced on a computer hard disk or a digital cart machine. You trigger the effects to play while you record the narration.

Many productions have a musical introduction, which you start and fade down just before the announcer starts talking. There's often an outro too. An outro is music that begins near the end of the narration, then fades up and out to conclude the program.

Backtiming a Musical Outro

You might want the outro to end when the narration ends. You'll perform a technique called **backtiming**. Follow these steps:

1. Find the spot in the script where you want the outro to start fading up. Play the speech starting from there, and time it to end of the narration. For this example, it's 20 seconds from outro fade-up to end.

2. In your CD player, put in a CD of outro music and go to the end of the piece you want to use. Press Rewind on the CD player and go back 21 seconds. Now the outro is cued up 21 seconds from the end.

3. At the fade-up point in the script, press Play on the CD player and fade it up. The music should end just after the narration ends. You also can record the outro music on a separate DAW track, and slide the track left or right on-screen until its timing is correct.

This chapter gave some advice about recording the spoken word with a consistent, clean sound, and described how to add sound effects and music. Don't neglect this type of studio work, because it's a potential source of income. And it's another skill to add to your resume.

16

THE MIDI STUDIO: EQUIPMENT AND RECORDING PROCEDURES

The tapeless studio is the computerized world of MIDI equipment—synthesizers, sampling keyboards, drum machines, and sequencers. Because other texts explain this equipment in detail, brief definitions serve the purpose here.

MIDI stands for Musical Instrument Digital Interface. It's a standard connection between electronic musical instruments and computers that allows them to communicate with each other. Some of the things you can do with MIDI are:

- Combine the sounds of two electronic musical instruments by playing them both with one keyboard.
- Create the effect of a band playing. To do this, you record keyboard performances into a computer memory, edit the recording note by note if you wish, and have the recording play through synthesizers and a drum machine in sync.
- Automate a mixdown, or automate effects changes.
- Make a keyboard, electric guitar or breath controller sound like any instrument.
- Automate the playback of sound effects and music for video productions.

The MIDI signal is a stream of digital data—not an audio signal—running at 31,250 bits per second. It sends information about the notes you play on a MIDI controller, such as a keyboard or drum pads. Up to 16 channels of information can be sent on a single MIDI cable.

MIDI is a whole subject in itself. For books on MIDI, see Appendix C.

MIDI Studio Components

The following equipment typically is used in a MIDI studio:

- Synthesizer
- Sampler
- Drum machine
- Sequencer
- Power amplifier and speakers
- Personal computer (optional)
- MIDI computer interface (optional)
- Recorder-mixer (optional)
- Tape synchronizer (optional)
- Mixer (optional)
- 2-track recorder
- Effects
- Audio cables
- MIDI cables
- Power outlet strip
- Equipment stand

You have learned about most of these in previous chapters, but a review might help at this point.

A **synthesizer** is a keyboard musical instrument that creates sounds electronically with oscillators (Figure 16.1). Your studio might have more than one of these.

A **multitimbral synthesizer** can play two or more patches at once. A **patch** is a sound preset (an instrumental timbre), such as a synthesized piano, bass, snare drum, etc. A **polyphonic synthesizer** can play several notes at once (chords) with a single patch.

A **sound generator** or **sound module** is a synthesizer without a keyboard. It is triggered by a sequencer or a controller.

A **sampler** is a device that records sound events, or samples, into computer memory. A **sample** is a digital recording of one note of a real sound

Figure 16.1 A synthesizer.

source: a flute note, a bass pluck, a drum hit, etc. A sample also can be a digital recording of a short segment of another recording. The sampling process is described in Chapter 10.

Often a sampler is built into a sample-playing keyboard, which resembles an electronic piano. It contains samples of several different musical instruments. When you play on the keyboard, the sample notes are heard. The higher the key you press, the higher the pitch of the reproduced sample.

A **drum machine** is device that plays built-in samples of all the sounds of a drum set and percussion (Figure 16.2). It also records and plays back drum patterns that you play or program with built-in keys or drum pads. Some units can sample sounds.

A **sequencer** is a device that records MIDI information about the notes you play into computer memory. Unlike a tape recorder, a sequencer does not record audio. Instead, it records the key number of each note you play, note-on signals, note-off signals, and other parameters such as velocity, pitch-bend, and so on.

Figure 16.2 A drum machine.

Figure 16.3 A sequencer.

The sequencer can be a stand-alone unit (Figure 16.3), a circuit built into a keyboard instrument, or a computer running a sequencer program. Like a multitrack tape recorder, a sequencer can record 8 or more tracks, with each track containing a performance of a different instrument.

Power amplifier and speakers (or **powered speakers**) let you hear what you're performing and recording. Usually these are small monitor speakers set up in a Nearfield arrangement (about 3 feet apart and 3 feet from you).

You might want a computer system and perhaps a printer. The computer is used mainly to run a sequencer program, which replaces the sequencer. Compared to a stand-alone sequencer with an LCD screen, the computer monitor screen displays much more information at a glance, making editing easier and more intuitive (Figure 16.4).

The computer can run other useful programs. A **MIDI/digital audio program** records both MIDI sequences and digital audio tracks on your hard disk, and keeps them in sync. A **librarian program** manipulates patches or samples and stores them on computer disks. A **voice editor program** lets you create your own patches. A **notation program** converts your performance to standard musical notation and prints it. A **film-sound program** lets you enter a list of sound effects or music with the time each occurs, and runs through the list automatically, triggering the effects and music at the right times.

A **MIDI computer interface** plugs into a user port in your computer, and converts MIDI signals into computer signals and vice versa. You need this only if you're using a computer in your system.

A combination **recorder and mixer** records vocals and acoustic instruments. A more elaborate studio might use a separate mixer and multitrack recorder.

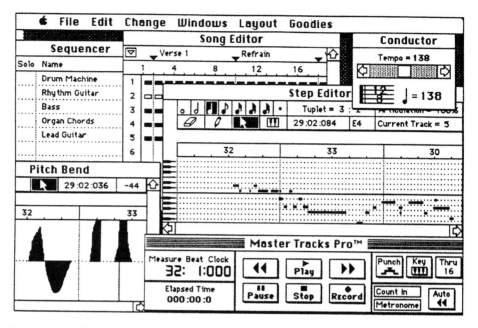

Figure 16.4 A monitor-screen shot of Passport Master Tracks Pro MIDI software (Courtesy Passport Designs, Inc.).

A **tape synchronizer** synchronizes tape tracks with sequencer tracks, and makes the sequencer start at the same place in the song that you start the tape.

If you have two or more synthesizers, or a synth and a drum machine, you need a **mixer** to blend their audio outputs into a single stereo signal.

A **2-track recorder** records the stereo mix of all your sound sources. The tape made on this recorder is the final product. The recorder can be open-reel, cassette, or DAT.

Effects include compression, reverb, gating, and so on.

Audio cables carry audio signals, and typically have a 1/4-inch phone plug on each end. They connect synths, sound modules, and drum machines to your mixer line inputs.

MIDI cables carry MIDI signals, and are used to connect synths, drum machines, and computers together so that they can communicate with each other. MIDI cables have a 5-pin DIN plug on each end.

A **power outlet strip** is a row of electrical outlets to power all your equipment. It's a good idea to have surge protection in the strip.

An **equipment stand** is a system of tubes, rods, and platforms that supports all your equipment in a convenient arrangement. It provides user comfort, shorter cable lengths, and more floor area for other activities.

The Keyboard Workstation

A keyboard workstation includes several MIDI components in one chassis: a keyboard, a sample player, a sequencer, and perhaps a synthesizer and disk drive. That's everything you need to compose, perform, and record instrumental music. Some workstations include drum sounds so that you can get by without a separate drum machine.

If you want to record songs with vocals, you also need a MIDI tape synchronizer plus a multitrack tape recorder or recorder-mixer. An alternative is a computer running multitrack recording software, or MIDI/digital audio software (explained later). If you want a permanent copy of your stereo mixes, you also need a 2-track recorder.

MIDI Recording Procedures

The rest of this chapter describes recording procedures for several different MIDI studio setups, from simple to complex. Read your instruction manuals thoroughly and simplify them into step-by-step procedures for various operations. Note that each piece of MIDI gear has its own idiosyncrasies, and the instructions may have errors or omissions. If you have questions, call the technical service people at the manufacturer of your equipment.

Recording with a Polyphonic Synthesizer

This is the simplest method of recording (Figure 16.5). You plug a MIDI interface into your computer, plug your synth into the interface, and run a sequencer program on the computer. You play chords and melody, record this MIDI data with your sequencer, and play back the sequence through your synthesizer. The basic steps follow:

1. Set your sequencer to record on one track.

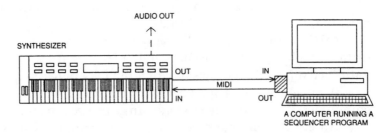

Figure 16.5   A synthesizer connected via a MIDI interface to a computer running a sequencer program.

2. Play a tune on your keyboard.

3. Play back the sequencer recording to hear it. Your performance will be duplicated by the synthesizer.

4. Quantize the track if desired.

5. Punch in/out to correct mistakes.

6. Arrange the song by combining various sequences.

7. Enter any program changes (changes in timbre).

8. Play the composition and set recording levels.

9. Start the sequencer playing, and record the synth output on a 2-track recorder. That recording is the final product. Here are the details for each step.

Set Your Sequencer to Record On One Track

Choose a tempo in your sequencer and choose the track you want to record on. Using your mouse, click on the RECORD key on-screen. You'll hear a metronome ticking at the tempo you set.

Play Music On Your Keyboard

Listen to the sequencer's metronome and play along with its beat. The sequencer keeps track of the measures, beats, and pulses. When the song is done, click on STOP. The sequencer stops recording and should go to the beginning of the sequence (the top of the tune).

Another way to record your performance is in **step-time**, one note at a time. If the part is difficult to play rapidly, you also can set the sequencer tempo very slow, record while playing the synth at that tempo, and then play back the sequence at a faster tempo.

Play Back the Sequence

Click on PLAY. You'll hear the sequence playing through your synthesizer.

Quantize the Track

Quantizing is the process of correcting the timing of each note to the nearest note value (quarter note, eighth note, and so on). If you wish, quantize the performance by the desired amount.

Punch In/Out to Correct Mistakes

To correct mistakes, you can punch into record mode before the mistake, record a new performance, and then punch out of record mode. Here's one way to do it:

1. Go to a point in the song a few bars before the mistake.

2. Just before you get to the mistake, punch into record mode and play a new, correct performance.

3. As soon as you finish the correction, punch out of record mode.

Alternatively, you can use **autopunch**. With this feature, the computer punches in and out automatically at preset measures; all you have to do is play the corrected musical part. Perform an autopunch as follows:

1. Using the computer keyboard or mouse, set the punch-out point (the measure, beat, and pulse where you want to go out of record mode).

2. Set the punch-in point (just before the part you want to correct).

3. Set the cue point (where you want the track to start playing before the punch).

4. Click on PLAY.

5. When the screen indicates punch-in mode, or when the appropriate measure comes up, play the corrected part.

6. The sequencer punches out automatically at the specified point in the song.

These punch-in routines were done in real-time. You can also punch-in/out in step-time:

1. Go to a point in the song just before the mistake.

2. Set the sequencer to step-time mode.

3. Step through the sequence pulse by pulse, and punch into record mode at the proper point.

4. Record the proper note in step-time.

5. Punch-out of record mode.

Arrange the Song by Combining Sequences

Now your sequenced performance is perfect, so you can put together your composition. Many songs have repeated sections: the verse and chorus are each repeated several times. If you wish, you can play the verse and chorus once each and save them as a separate sequence, which you can copy for all the places each occurs in the song.

You can rearrange song sections, and append one section to another, by pressing a few keys on the computer. You can also have any section repeated. In this way, you might build a song by having the computer play sections A, A, B, A, C, A, B, B.

Enter Program Changes

To add variety to the song, you might want to have the synth play different programs (patches) at different parts of the song. For example, play a piano on the first verse, organ on the second, and marimba on the chorus. One way is to press different presets (program numbers) as you record the sequence.

Another way is to record these program changes on another track, which is called the controller track. Be sure to set the controller track to the same channel as the performance track, and turn off any patch on the controller track. Enter the appropriate program numbers at the right time on your synthesizer. Putting the program changes on a separate track makes it easy to edit them. You can punch in new program changes just as you can punch in new performances. When all the program changes are correct, you can bounce them to the performance track if you wish.

Some sequencers do not record the program settings on your synth. They record only program **changes**. Consequently, when you play the sequence into a synth you just turned on, you might hear the wrong sounds. To prevent this problem, insert a few blank measures at the beginning of the tune and record your initial program changes there, according to the sounds you want to hear at the beginning of the tune. Follow this procedure:

1. Insert two or four blank measures at the beginning of your composition.
2. Set each track's patch to the wrong program number. If you want patch #17, for example, set it to #16. This way, you can key in a program change later.
3. Set your sequencer to punch-in mode so that you record only on the blank measures at the beginning of the tune.
4. When the punch-in starts, key in the correct program numbers. You can perform these program changes in several passes, one track at a time.
5. When the sequence plays back, it sets the synth automatically to the correct patches at the beginning of the tune.

An alternative to this procedure is to record a **system exclusive** or **sysex dump**—data about patch settings and so forth—into the sequencer. This works only if your synth and sequencer implement the sysex dump.

Play the Composition and Set Recording Levels

Plug your synthesizer's audio output (mono or stereo) into the line inputs of your 2-track tape recorder. Hit the PLAY key on your computer keyboard, and set the recording level for your recorder according to these guidelines: cassette: +3 dB maximum, open-reel: +3 VU maximum, DAT: 0 dB maximum.

Record the Synth Output

Once your levels are set, put your 2-track recorder in record mode, and start the sequencer. This produces the finished product—a stereo recording of your song.

Recording with a Multitimbral Synthesizer

With this method, you play the parts for several different instruments (patches) on the same keyboard, and record each performance on a separate track of a sequencer. During playback, the sequencer plays the desired patches (instruments) in your synth. It sounds like a band playing. You record the synthesizer's output, and that recording is the final product.

Each track and patch are set to corresponding MIDI channels. For example, suppose both track 1 and the bass patch are set to channel 1. Then track 1's performance in the sequencer plays the bass patch in the synthesizer. Track 2 will play another patch (piano, flute, or whatever).

Some sequencers are designed so that, on power-up, track 1 goes to channel 1, track 2 goes to channel 2, and so on. You select a track to record on and select a patch for that track. The channel assignments are already taken care of.

Refer back to Figure 16.4 to see the connections. Here is an outline of the steps for recording:

1. Start recording on the first sequencer track.
2. Play music on your keyboard.
3. Play back the recording.
4. Punch in/out to correct mistakes.
5. Record overdubs on other tracks.
6. Bounce tracks.
7. Edit the composition.
8. Mix the tracks.
9. Record the mix.

Again, these steps require a closer look. The following sections present details for each one.

Record the First Synthesizer Track

Adjust the metronome tempo on your sequencer as desired. Set the sequencer to record on track 1. If necessary, set sequencer track 1 to MIDI channel 1.

Select the first patch you want to hear on your synthesizer (in this case, bass guitar). You might want to adjust the timbre of the patch with the parameter controls on the synthesizer in order to create unusual sounds. If necessary, set the synthesizer patch to MIDI channel 1.

Click on RECORD on your computer screen.

Play Music on Your Keyboard

Listen to the sequencer's metronome and play along with its beat, or record in step-time. The sequencer keeps track of the measures, beats, and pulses. When you click on STOP, the sequencer stops recording and goes to the beginning of the sequence (the top of the tune).

Play Back the Recording

Click on PLAY on the computer screen. You'll hear the sequence playing through your synthesizer.

Punch In/Out to Correct Mistakes

As described in the previous section, you can correct mistakes by punching into record mode before the mistake, recording a new performance, and then punching out of record mode. You can also quantize the track to make it correct rhythmically.

Record Overdubs on Other Tracks

With your first track recorded and corrected, you're ready to record other tracks. Set the sequencer to record on track 2. If necessary, set sequencer track 2 to MIDI channel 2.

On your synthesizer, select the next patch (instrument timbre) you want to use (in this case, piano). If necessary, set it to MIDI channel 2.

Then click on RECORD on your computer screen. Play the piano on the synthesizer while listening to your prerecorded bass on track 1. This track is played by the synthesizer.

Another example may help. Imagine you just recorded a drum part into the sequencer on track 1. You can go back to the start of the sequence, play the drum part, and add a bass line on track 2 in sync with the drums. Then you can go back to the top and add a piano on track 3.

In short, you record the performance of a different patch on each track in the sequencer, and play back the recording through the multitimbral synthesizer, which plays all the patches simultaneously. Or you can use several synths, one for each part, if necessary. Set each track to a different MIDI channel, and set each instrument or patch to the same channel that its track is set to.

Bounce Tracks

What if your sequencer records eight tracks, but you want to play ten patches at once with several synths? You can make more tracks available by bouncing tracks. If you bounce track 5 into track 4, for example, track 5 can be erased, so that you can record a new instrument on track 5.

To do this, click on BOUNCE on your computer screen. Type the source track and destination track (indicate to which track you want to bounce). In a few seconds, the bounce is accomplished.

You can bounce only one track at a time. You can bounce a track into a prerecorded track, however, without erasing the prerecorded track. The prerecorded track and bounced track then merge. Unlike bouncing with a tape deck, there is no generation loss (no loss of sound quality) when you bounce with a sequencer.

You can record program changes on a separate track set to the same channel as the performance track. Later, bounce the program-change track to the performance track. The performance track and program-change track merge into one. Be sure both tracks are set to the same channel, and that the program-change track has its patch turned off.

Edit the Composition

Now your sequenced performance is satisfactory, so you can put together your composition. As described in the previous section, you can rearrange song sections and append one section to another by making selections on your computer screen. You also can have any section replayed. Key in program changes at the beginning of the song and anywhere else you want the sounds to change.

Mix the Tracks

Now that your song is recorded and arranged, you want to adjust the relative volumes of the tracks to achieve a pleasing balance.

If your multitimbral synthesizer doesn't have separate outputs for each patch, you have to adjust the mix at the sequencer. To do this, adjust the volume (key-velocity scaling) of each track with your computer. This only works if your keyboard is velocity-sensitive.

After you adjust the volume of each track in this way, click PLAY on your computer screen to play the sequence. The desired mix of patches play on your synth. Some synths let you add internal effects to the overall mix.

If your synth has several individual outputs—one for each patch—connect them to a mixer and set up a stereo mix with panning and effects.

Record the Mix

If your synth has a single output (mono or stereo), use your 2-track recorder to record the mix off that output. Plug your synthesizer's audio output into the line inputs of your 2-track recorder. Or, if your synth has several individual outputs connected to a mixer, record off the mixer stereo outputs.

Click PLAY on your computer screen, and set the recording level for your recorder. Then put your 2-track recorder in record mode, and restart the sequencer. This produces the finished product—a stereo recording of your song.

Recording with a Keyboard Workstation

Each workstation operates in a different way, but here is a typical recording procedure:

1. Set up for recording a song.
2. Record the first musical part in real time, or do step recording.
3. Overdub more parts.
4. Punch in.
5. Set effects.
6. Store the song.

The following sections present more detailed instructions for each step.

Set Up for Recording a Song

1. Press SEQ (for SEQUENCER) on the front panel.
2. A sequencer menu appears on the LCD screen. You can move a cursor to select various parameters, and press the up or down buttons to set the value of each parameter.
3. Set the time signature (in the Initialize menu).
4. Select the song number.
5. Set the tempo (in the Sequencer play/Real-time record menu).
6. Select the track number (track 1 to start).
7. Select the program number for the desired sound (for example, a drum set).

Record the First Musical Part

1. Press REC (record) and START. Listen to two measures of metronome clicks and then start playing.
2. When you finish, press STOP.
3. To hear what you just played, press START.
4. If you want to rerecord the part, press REC and START, and play the part again. You also could edit the performance, do punch ins, and so on.

Step Recording

Instead of performing a musical part in real-time, you might prefer to enter the notes one at a time, in step-time. Here's the basic procedure:

1. Select the track number and the measure number where you want to start.
2. Press REC and START.
3. Set the length of the first note (1/32 to 1/1).
4. If necessary, specify triplets, dotted notes, key dynamics, style of playing, and rests.
5. Press the desired note or chords on the keyboard.
6. Release all the keys; the recording proceeds to the next step.
7. After entering all the notes, press STOP.

Overdub More Parts

1. To record the next track, set the track number to the desired track (in this case, track 2).
2. Select the program number for the desired sound (for example, a bass).
3. Press REC and START. As you listen to track 1 playing the drum part, play a bass part on track 2.
4. Continue this procedure (steps 1–3) up to 8 tracks, adding a new instrument each time.

Punch-In

You can correct mistakes easily in each track by punching in, either manually or automatically. The procedure follows:

1. Play the song to find the measures needing correction.
2. Select punch-in mode.

3. Page up one page; set the punch-in measure and the punch-out measure; page down one page.
4. Set the measure number to a point a few bars before the punch-in.
5. Press REC and START. You'll hear the song playing.
6. When the punch-in measure comes up, play the corrected part.
7. Press STOP when done.

Set Effects

You can page up to the effects menus to set overall effects: hall reverb, chorus, flanging, echo, distortion, and so on. Press the correct number on the numeric keypad to get to the effects menus. (Note that these are built-in keyboard effects, not outboard studio effects.)

Save the Song File

Save the completed song in multitrack form to a plug-in RAM card or to an external sequencer and disk drive. To prevent data overload, you might have to do the external sequencer recording one track at a time with other tracks muted. If you're satisfied with the final results, record the stereo output signals of the workstation to a 2-track recorder. In addition to these basic operations, you can:

- Bounce tracks
- Edit each note event
- Create and copy patterns—for a drum or bass part, for example
- Modify track and song parameters
- Insert/delete/erase measures
- Modify sounds and effects (in great detail)
- Change the instrument (patch) that each track plays

Recording with a Drum Machine and a Synthesizer

This system combines a synthesizer with a drum machine. Figure 16.6 shows how to connect the cables. Figure 16.7 shows a preferred setup if your synth has a MIDI THRU port.

Understanding Synchronization

The drum machine has a built-in sequencer that records what you tap on its pads. Suppose you record a drum pattern with its built-in sequencer, and you record a synthesizer melody with an external sequencer. How do you

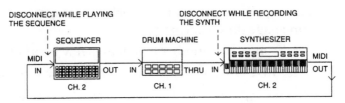

Figure 16.6 Connections to make a sequencer drive a drum machine and a synthesizer.

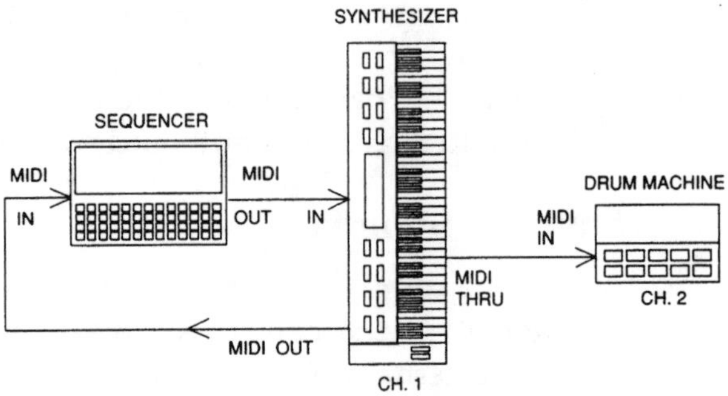

Figure 16.7
Connections to make a sequencer drive a drum machine if your synth has a MIDI THRU port.

synchronize the drum patterns in the drum machine with the synthesizer melody in the sequencer? In other words, how do you get the two devices to play in sync, when both have different patterns recorded in different memories?

To synchronize the machines, use a single **MIDI clock** (timing reference) that sets a common tempo for all the equipment. The MIDI clock is a series of bytes in the MIDI data stream that conveys timing information. The clock is like a conductor's baton movements, keeping all the performers in sync at the same tempo. The clock bytes are added to the MIDI performance information in the MIDI signal. The clock signal is 24, 48, or 96 pulses per quarter note (ppq). That is, for every quarter note of the performance, 24 or more clock pulses (bytes) are sent in the MIDI data stream.

Decide which device you want to be your master timing reference— the sequencer or the drum machine. Set the master device to internal clock and set the slave device to external clock or MIDI clock. Then the slave will follow the tempo of the master.

To make this happen, the master sends clock pulses from its MIDI OUT connector. The slave receives those clock pulses at its MIDI IN connector. The slave also passes clock pulses through its MIDI THRU connector to other slave devices down the chain.

If a slave device lacks a MIDI THRU, enable "Echo MIDI In" in the slave device. Then the incoming pulses are echoed or repeated at the MIDI OUT connector.

306

In the setups shown in Figures 16.6 and 16.7, the sequencer's clock drives both the drum machine and the synthesizer. In other words, the sequencer is the master tempo setter, and the drum machine and synth follow along. The drum machine's internally recorded patterns play in sync with the synthesizer's sequencer-recorded melody.

Basic Recording Procedure

Once you work out the synchronization problem, you are ready to begin recording with this system. There are two basic methods. The first uses the following steps:

1. Record drum patterns into the drum machine.
2. While listening to the drum patterns, record a synth part with your sequencer.
3. Sync the drums and synth by setting MIDI clocks and channels.
4. Press the PLAY key on the sequencer.

Use these steps for the second method:

1. Record drum patterns into the drum machine.
2. Copy these patterns onto one track of your sequencer.
3. While listening to the drum track, record a synth part on another track.
4. Play both sequencer tracks.

Detailed explanations follow for recording drum patterns and synchronizing the drums with the synth.

Recording Drum Patterns

The first step in composing a song is to record a drum pattern. There are many ways to do this; the following is one suggested procedure:

1. On the drum machine, set the tempo, time signature, and pattern length in measures. For this example, the pattern is 2 bars long.
2. Start recording, and play the hi-hat key in time with the metronome beat.
3. At the end of 2 bars, the hi-hat pattern you tapped repeats over and over (loops).
4. While this is happening, you can add a kick drum beat.
5. While the hi-hat and kick drum are looping, add a snare drum back beat, and so on.
6. Mix the recording by adjusting the faders on the drum machine for each instrument.

Next, you repeat the process for a different rhythmic pattern—say, a drum fill—and store this as Pattern 2. Then develop other patterns. Finally, you make a song by repeating patterns and chaining them together as described in the drum machine's instruction manual. A song is a list of patterns in order.

It's a good idea to add a count-off (a few measures of clicks) at the beginning so that later overdubs can start at the correct time. Some musicians like to program a simple repeating drum groove first. While listening to this, they improvise a synth part. After recording the synth part, they redo the drum part in detail, adding hand claps, tom-tom fills, accents, and so on.

Synching Drums and Synth

Now you're ready to add a synth part and synchronize it with the drum track. The following procedure refers to a sequencer; it could also be a computer running a sequencer program.

1. Record a synth part with the sequencer.
2. Set the drum machine to external clock or MIDI clock.
3. Set the sequencer to internal clock.
4. Set MIDI channels: set the drum machine to channel 1; set the sequencer synth track and the synthesizer to channel 2. In this way, the sequencer's recorded performance will play only the synthesizer. The MIDI clock still controls both devices, even though they are set to different channels.
5. Press the PLAY key on the sequencer. As the sequencer plays its recorded synth melody, the sequencer's clock pulses drive the drum machine and synthesizer at the same tempo. The drum machine plays its internally recorded patterns, while the synth plays the sequencer track.

Another way to synchronize a drum machine and a synthesizer is to record the drum patterns on one track of your sequencer. The advantage is that, whenever you rearrange parts of the music in the sequencer, you also rearrange the drum part. So you don't have to change drum patterns each time you repeat or delete a verse or a chorus. Follow this procedure to record the drum patterns into your sequencer:

1. Record a drum pattern with the drum machine's internal sequencer.
2. Enable the drum machine's clock out and MIDI data out.
3. On your sequencer, turn off the MIDI-THRU feature (if it has one).
4. Set the sequencer to external clock or MIDI clock mode, and set an open track in record mode.
5. Hit the PLAY key on the drum machine. The sequencer records the drum pattern on the open track.

To play back the drum patterns you just recorded, follow this procedure:

1. Set the drum machine to external clock mode (slave).
2. Set the sequencer to internal clock (master).
3. Set the drum machine's track and the drum machine to the same MIDI channel.
4. Load an empty pattern into the drum machine so that the machine plays only the sequencer track.
5. Put the sequencer in play mode. The drum machine plays its sequencer track at the sequencer's tempo, and other synths connected to the sequencer play their tracks on their channels.

Recording with a MIDI System Plus Tape Sync

You might have a complete MIDI workstation: a synthesizer, sound module, drum machine, sequencer, and recorder-mixer. Here is a suggested procedure for recording, overdubbing, and mixdown with this system:

1. Connect the system.
2. Record drum patterns; record synth tracks into your sequencer.
3. Record the sync tone.
4. Check the sync-tone playback.
5. Overdub acoustic parts.
6. Mix down all the tracks.

These six steps are the basic outline of the procedure. Detailed explanations follow.

Connect the System

Figure 16.8 shows the hookup. Connect the tape-sync output connector of the sequencer (or MIDI computer interface) to your tape recorder's sync-track input connector. This is an outside track—track 4 of a 4-track recorder, or track 8 of an 8-track recorder. Connect the sync track's output to the tape-sync input of the sequencer or interface. Plug the MIDI equipment audio outputs and the tape-track outputs into a mixer (not shown), and monitor the mixer's output. If your recorder-mixer has enough inputs, you can use it as the mixer.

If your sequencer lacks a tape-sync connector, you need a smart tape-sync box (converter). Connect this system as shown in Figure 16.9.

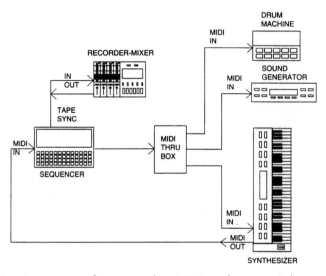

Figure 16.8 Connections for a complete MIDI workstation with tape sync.

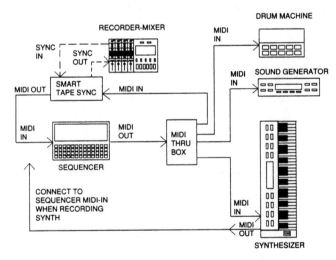

Figure 16.9 Connections for a complete MIDI workstation with smart tape sync.

Compose and Record MIDI Tracks

Now that the system is set up, record drum and synth tracks. Here is a suggested procedure:

1. Set the sequencer, drum machine, and/or converter to internal sync (master) mode.

2. Program the drum tracks into the drum machine. Be sure to set the tempo as desired, because once you record the sync tone, you can't change the tempo. Also consider programming the drum part longer than you need, since the sync tone can be made shorter but not longer.

3. Develop your instrumental arrangement using the synthesizer and its sequencer, set to the same tempo as the drum machine. You can have tempo changes during the song if you enter them before recording the sync track. You might want to record the drum sequence into your sequencer so that changes in the main sequence (e.g., new arrangements, tempo changes) affect the drum track as well.

Record the Sync Tone

Either the drum machine or the sequencer can be used to generate the sync tone. In this example, use the sequencer.

1. Clean and demagnetize the tape heads.

2. Start playing the sequence. A sync tone is generated from the sync output jack.

3. Put the recorder in record-ready mode and set the recording level of the sync tone. Use a level that produces correct synchronization on playback (usually around –4 VU).

4. If you can switch off the noise reduction for the sync track, do so. Dolby is rarely a problem, but dbx can make the sync tone unreadable.

5. Stop the sequencer and reset it to the top of the tune.

6. Tape sync works best if you record the sync track before you record any other tape tracks. Start the tape recorder in record mode. A few seconds later, press play on your sequencer. You'll be recording (**striping**) the FSK or SMPTE tape-sync tone on track 4 of the recorder-mixer. This will be your master clock.

7. Let the sequencer run for the duration of the song. While striping the tone, do not record any musical material because it may not sync with material recorded later.

Check the Sync Tone Playback

Now you play the sync tone you just recorded on tape, and it activates the drum machine and sequencer. First, reset the drum machine and sequencer to the top of the tune, and set them and the converter to external clock mode.

1. Rewind the recorder a little before the beginning of the sync tone and take the sync track out of record-ready mode.

2. Start playing the tape. (You may need to press the PLAY key on the drum machine or sequencer first.) The sequencer should start playing when the tone starts. The tone drives the sequencer, which in turn plays the synth and drives the drum machine—all at the same tempo. They stop when the tone stops.

3. If you have sync problems, try rerecording the tone at a different level. If a tape drop-out causes loss of sync, re-record the tone on a different section of tape or on a new tape or track. You can rerecord the sync track before recording other tracks, but not after—you'll lose sync.

Overdub Acoustic Parts

Now you're ready to overdub vocals and acoustic instruments on tape.

1. Plug a mic into your recorder and assign it to an open track. If possible, don't assign percussive or bass parts to a track adjacent to the sync track because they can interfere with the sync tone.

2. Set the microphone input trim and recording level as described in Chapter 13 on recorder-mixer operation.

3. Rewind the tape to just before the beginning of the sync tone, and hit play. The drum machine and sequencer should start playing. (You may need to press the PLAY key on the drum machine or sequencer first.)

4. While listening to the synth and drum machine playing through headphones, record any vocals, and non-MIDI instruments onto tape. Because you have three tracks available, you could record, for example, lead vocal on track 1, harmony vocal on track 2, and sax on track 3.

Mix Down All the Tracks

After all your tracks are recorded, use the mixer to set up a mix of the tape tracks and MIDI instruments. If your mixer doesn't have enough inputs for all the tape tracks, you can set a stereo mix of the tape tracks with the recorder-mixer, and combine this stereo mix with the audio signals from your MIDI instruments.

1. Adjust levels, panning, and effects. You'll have three tape tracks, probably eight tracks from the synth and sound module, plus stereo drum tracks, with stereo effects—all first generation!

2. Play the song several times to perfect the mix.

3. When you're satisfied with the mix, rewind the tape just before the beginning of the sync tone and hit play. Record the mix onto the 2-track recorder.

Refer to this information for more detail: Chapter 13, the section on mixing procedures; Chapter 14, the section on master-tape assembly; and later in this chapter, the section on automated mixing.

Recording with a MIDI/ Digital Audio Recording System

The tape sync system just described can be cumbersome. Instead, you can record MIDI sequences and digital audio tracks, edit them and mix them, all with your computer. You need MIDI/Digital Audio software, a sound card and a MIDI interface.

This system lets you add several tracks of vocals, sax, guitar, or any audio signal to your MIDI sequences. While playing MIDI tracks, you record the audio digitally to a hard disk. The software synchronizes the sequencer's MIDI data with the audio tracks.

The sound quality is as good as that of a compact disc (16-bit quantization, 44.1 kHz sampling rate). Plus, you can edit the digital audio performances—cut and paste, copy, rearrange, and so on. Because the disk drive is random access, you don't have to wait for shuttling tape.

Be sure your hard drive has enough capacity for digital recording. One track of one minute of CD-quality audio consumes about 5 megabytes. So a 600 megabyte drive can record 30 minutes of 4 audio tracks. Some software removes data created during silent portions of tracks. In this way, one track-minute of audio might use only 1 megabyte instead of 5.

Here's how such a system might work (Figure 16.10):

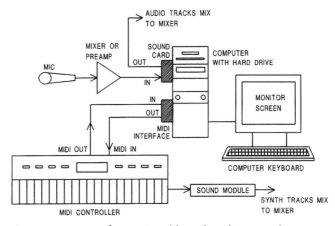

Figure 16.10 Connections for a MIDI/digital audio recording system (mixer not shown).

1. Use the sequencer part of the software to record the MIDI parts, which might play a synth, sound module, sampler, or drum machine.
2. Set up a microphone to record a vocal.
3. While listening to the MIDI tracks, record the vocal onto hard disk as digital audio.
4. You might record a few takes of the vocal part and then cut and paste selected portions to create a perfect take. Record one good chorus and copy it in each chorus section in your song.
5. Add effects to various tracks using DSP effects plug-ins.

   Several things happen when you play back what you recorded:

1. The sequence's MIDI signal comes from the MIDI OUT connector in the MIDI interface.
2. The sequence's MIDI signal plays notes through a synth, sound module, and/or drum machine.
3. The digital audio—a multitrack mix—plays from the sound-card output.
4. A mixer combines the MIDI-equipment audio with the sound-card audio, and you record the mix on DAT (using the software's auto-mated-mixing feature).

Using Effects

No matter how you record with MIDI, effects are an important part of the mix. To keep the sound lively, try to vary the effects throughout the song, or use several types of effects at once.

For example, suppose you have a multitimbral synth, and you want to add a different effect to each patch. Whether or not you can do this depends on your synth. If it has a separate output for each patch, you can use a different effect on each patch. But if your synth has only a single output (mono or stereo) and you run it through an effects device, the same effect is on all the patches.

If your song includes program changes (patch changes), you can have the effects change when the patch changes. Set up a MIDI multi-effects processor so that each synth program change corresponds to the desired effect. When the synth program changes, the effect changes also.

What if you want the effect, but not the synth patch, to change during a mix? Reserve a track and channel just for effects program changes. You don't hear these program changes in your synth, but you do hear the effects change. During a mixdown, it's usually easier to change effects automatically with your sequencer, rather than manually.

314

If your synth is a sampling keyboard, each sample could have reverberation or some other effect already on it; in that case each sample can have a different effect. The effect is not recorded in the sequencer; rather, the effect is part of the sampled sound. Note that the sampled reverberation cuts off every time you play a new note. Although this sounds unnatural, you can use it for special effect.

Because effects are audio signals, tape recorders can record effects but sequencers can't. If an effect is an integral part of the sound of an instrument, it's probably best to record it with the instrument on the multitrack tape. If the effect is overall ambience or reverb (to put the band in a concert hall), however, then it's best to add it to almost everything during mixdown.

XMIDI Stands for Extended MIDI

It is a proposed compatible upgrade to MIDI with enhanced features and capabilities, and more functions:

- 324 addressable devices per cable, equal to 20 MIDI cables
- Better than 12 bit resolution, instead of 7 bit
- Bidirectional transfer
- Speed limited only by the technology
- Automatic setup configuration (channel assignment, patches retrieve, and so on)
- New commands
- Simple cabling of complex setups
- Up to 400 Mbit baud rate

Automated Mixing

Chapter 13 began a discussion of automated mixing, which uses computer memory to store and reset your mix changes. Many types of automation use MIDI devices. Now that you understand these devices, we can explain automation in more detail.

Types of Automation Systems

Four types of automation systems are:

- Automated mixer
- Automated fader/VCA unit

- Automated-mixing software
- Moving fader automation

Each of these is worth a closer look.

Automated Mixer

An automated mixer has a built-in circuit to do automated mixdowns. In the mixer, volume levels and mutes are controlled either by Voltage Controlled Amplifiers (VCAs) or Digitally Controlled Amplifiers (DCAs). Control signals are MIDI, FSK, or SMPTE.

Automated Fader/VCA Unit

You also can automate a standard mixer by adding an external **automated fader/VCA unit**, a box with several faders that control VCA's or DCA's. The VCAs/DCAs plug into your mixer's access jacks or insert jacks (Figure 16.11). These are in series with the signal of each channel. One VCA or DCA is needed for each channel (input, master, or return) you want to automate. Some boxes have one fader or mute button per channel; others have a single fader that you set to the desired channel. MIDI controls this type of system.

Automated-Mixing Software

This software program runs on a personal computer. The monitor screen shows **virtual faders** that you adjust with a mouse. The mix is controlled in

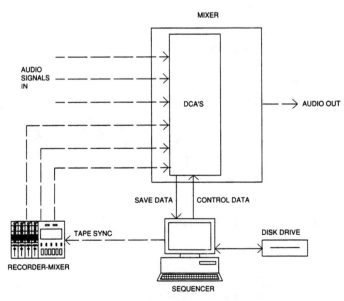

Figure 16.11 Typical connections for MIDI-automated mixing.

two ways: (1) adjust the gain of audio channels, or (2) adjust the MIDI volume or key velocity of sequencer tracks. Some programs allow cut-and-paste edits of mixes. You can append the chorus of mix 1 to the verse of mix 2, for example. Lower-cost programs let you control volume only; higher-cost programs let you control all parameters (e.g., EQ, panning, and effects). Many sequencing programs and digital editing programs have an automated-mixing function.

Moving Fader Automation

With this sophisticated system, SMPTE time code is used to control motorized faders. They move up and down as if controlled by a ghost. There are no VCAs to degrade the signal. In some systems, all control settings can be remembered and recalled.

Storing Mix Moves

A MIDI sequencer can record your mix moves via MIDI program changes. The sequencer is either built into the automated mixer or is external (stand-alone or computer program). When you play the sequencer recording, MIDI program changes control the DCAs to mix the tracks as you did manually. Or the program changes move motorized faders.

Snapshot vs. Continuous Automation

Two types of automation are **snapshot** and **continuous (dynamic)**. With snapshot automation, a computer memory in the mixer takes a "snapshot" or "scene" of the mixer settings and stores them as MIDI program changes. To reset the mixer to any of these stored settings, you punch up the appropriate number (MIDI program change). Alternatively, your sequencer recording can reset the mixer at the correct times as a song plays.

With continuous or dynamic automation, the motion of the mixer controls is recorded. Continuous automation costs more than snapshot and consumes more memory, but permits finer resolution of mix moves. Some digital mixers do both snapshot and continuous automation with their internal memory. Some snapshot units let you program fade times so that you can fade between snapshots to simulate continuous control.

Controls

Some automated-mixing systems have several knobs and faders: one per parameter. Others have just a few "soft" controls: you assign the control one function at a time and adjust its value on a screen.

Simulating Automatic Mixing With MIDI

If you lack an automated-mixer system, you can simulate one by recording MIDI controller data as volume changes on an open sequencer track. This lets you vary the balance among sequencer tracks, but not tape tracks. You might proceed as follows:

1. Choose a continuous controller on your master keyboard, such as a pitch-bend wheel.

2. Set the keyboard to transmit on the MIDI channel of the track you want to automate.

3. In your sound module, assign or map the controller to affect MIDI volume for level changes.

4. Set your sequencer to record on a new track, called the controller track. Set the controller track to the same channel as your master keyboard. Be sure to turn off any patch on the controller track.

5. Start recording. While listening to the music, move the controller to adjust the volume of the track. Later, you can edit the controller track, and perhaps bounce it to the musical track.

6. Repeat this procedure for all the tracks you want to automate. (You can assign the controller to affect filtering or panning as well, but this takes up a lot of sequencer memory.)

MIDI studio equipment, keyboard workstations, and digital audio workstations bring new procedures into the studio with them. Typical recording procedures for MIDI can range from simple to complex. This relatively new technology has changed the way recordings are made.

17

ON-LOCATION RECORDING OF POPULAR MUSIC

Sooner or later you'll want to record a band—maybe your own—playing in a club or concert hall. Many bands want to be recorded in concert because they feel that's when they play best. Your job is to capture that performance on tape and bring it back alive.

There are several ways to record live:

- Record with two mics out front into a 2-track recorder.
- Using the PA mixer, record off a spare MAIN output.
- Record a stereo mic and the PA mixer output into a 4-track recorder-mixer.
- Feed the PA mixer INSERT jacks to a multitrack.
- Feed the PA mixer DIRECT OUTS or INSERT jacks to a recording mixer, and from there to a 2-track or a multitrack recorder.
- Use a mic splitter on stage to feed the PA snake and recording snake. Record to multitrack or 2-track.

We'll start by explaining simple two microphone techniques and work our way up to elaborate multitrack setups.

Two Mics Out Front

Let's start with the simplest, cheapest technique—two mics and a 2-track recorder. The sound will be distant and muddy compared to using a mic on each instrument and vocal. Not exactly CD quality! But you'll hear how your band sounds to an audience.

Recording this way is much simpler, faster, and cheaper than a multi-mic, multitrack recording. Still, if time and budget permit, you'll get better sound with a more-elaborate setup.

Equipment

Here's what you need for 2-mic recording:

- Two mics of the same model number. Your first choice might be cardioid condenser mics. The cardioid pickup pattern cuts down on room reverb and noise. The condenser type generally sounds more natural than the dynamic type. Another option is a pair of boundary mics such as PZMs. Simply put them on the floor or on the ceiling in front of your group.

- A DAT recorder, 2-track open-reel recorder, or cassette deck. DAT has the least distortion and hiss, and lets you record up to two hours non-stop. Either use a unit with mic inputs, or use a separate mic preamp.

- Blank recording tape. Buy the best you can afford. Bring enough tape to cover the duration of the recording.

- Two long mic cables.

- Two mic stands.

- Headphones. You could use speakers in a separate room, but headphones are more portable and sound consistent in any environment. Closed-cup headphones partly block out the live sound of the band so you can better hear what's going on tape. Ideally, you'd set up in a different room than the band is in, so you can clearly hear what you're recording.

Mic Placement

Use a pair of mic stands or hang the mics out of the reach of the audience. Aim the two mics at the group about 12 feet away, and space them about 5 to 15 feet apart (Figure 17.1). Place the mics far apart (that is, close to the PA speakers) to make the vocals louder in the recording. Do the opposite to make them quieter. The stereo imaging will be pretty vague, but at least you can control the balance between instruments and vocals.

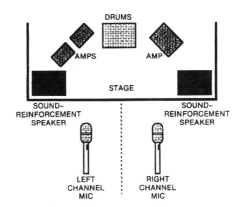

Figure 17.1
Recording a musical group with two spaced microphones.

To record a small folk group or acoustic jazz group, set up two mics of the same model number in a stereo arrangement of your choice. Place the mics about 3 to 10 feet from the group. The balance may not be the best, but the method is simple.

Recording

After setting the recording level, leave it alone as much as possible. If you must change the level, do so slowly and try to follow the dynamics of the music. Switch tapes during pauses or intermissions.

If the playback sounds distorted—even though you did not exceed a normal recording level—the mics probably overloaded the mic preamps in the tape deck. A mic preamp is a circuit in the tape recorder that amplifies a weak mic signal up to a usable level. With loud sound sources such as rock groups, a mic can put out a signal strong enough to cause distortion in the mic preamp.

Some recorders have a pad or input attenuator. It reduces the mic signal level before it reaches the preamp, and so prevents distortion. You can build a pad (Figure 17.2), or buy some plug-in pads from your mic dealer. Some condenser mics have a switchable internal pad that reduces distortion in the mic itself. If you have to set your record-level knobs very low (less than 1/3 up) to get a 0 recording level, that shows you probably need to use a pad.

Recording from the PA Mixer

You can get a fairly good recording by plugging into the main output of the band's PA mixer. (PA stands for public address, but here it means sound

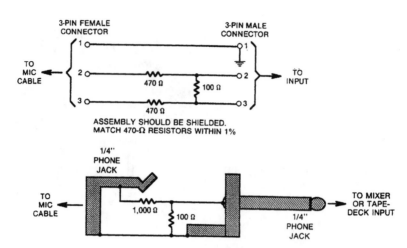

Figure 17.2 Balanced and unbalanced microphone pads.

reinforcement.) Connect the main output(s) of the mixer to the line or aux input(s) of a 2-track recorder. Use the mixer output that is ahead of any graphic equalizer that is used to correct the speakers' frequency response (Figure 17.3).

Mixers with balanced outputs can produce a signal that is too high in level for the recorder's line input, causing distortion. This is probably occurring if your record-level controls have to be set very low. To reduce the output level of the mixer, turn it down so that its signal peaks around –12 VU on the mixer meters, and turn up the PA power amplifier to compensate. That practice, however, degrades the mixer's S/N.

A better solution is to make a 12 dB pad (Figure 17.4.)The output level of a balanced-output mixer is 12 dB higher than the normal input level of a recorder with an unbalanced input.

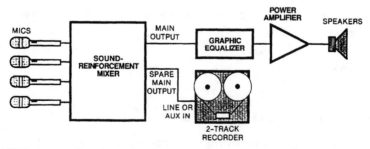

Figure 17.3 Recording from the sound-reinforcement mixer.

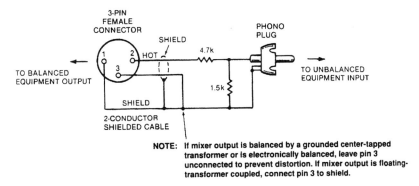

NOTE: If mixer output is balanced by a grounded center-tapped transformer or is electronically balanced, leave pin 3 unconnected to prevent distortion. If mixer output is floating-transformer coupled, connect pin 3 to shield.

Figure 17.4   12 dB pad for matching a balanced output to an unbalanced input.

Drawbacks

The recorded mix off the PA might be poor. The operator of the band's mixer hears a combination of the band's live sound and the reinforced sound through the house system, and tries to get a good mix of both these elements. That means the signal is mixed to augment the live sound—not to sound good by itself. A recording made from the band's mixer is likely to sound too strong in the vocals and too weak in the bass.

Taping with a 4-Tracker

A 4-track portable studio can do a good job of capturing a band's live sound. With this method, you place a stereo mic (or a pair of mics) at the FOH position (Front Of House—the PA mixer location). Record the mic on tracks 1 and 2. Also plug into a spare main output on the PA mixer, and record it on tracks 3 and 4 (Fig. 17.5). After the concert, mix the four tracks together.

Figure 17.5
Recording two mics and a PA mix on a 4-track recorder-mixer.

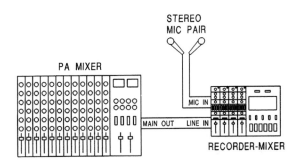

The FOH mics pick up the band as the audience hears it: lots of room acoustics, lots of bass, but rather muddy or distant. The PA mixer output sounds close and clear, but typically is thin in the bass. When you mix the FOH mics with the PA-mixer signal, the combination has both warmth and clarity.

Consider using a stereo mic at the FOH position. Good stereo mics cost over $500, but they provide great stereo in a portable, convenient package. You can also set up two mics of the same model number in a stereo arrangement. For example, angle two cardioid mics 90 degrees apart and space them 1 foot apart horizontally. Or place two omni mics 2 feet apart.

Plug the mics into your 4-track's mic inputs 1 and 2. Adjust the trim to prevent distortion, and set the recording level. Find a spare main output on the PA mixer, and plug it into your 4-track's line inputs 3 and 4. You may need to use the 12 dB pad described earlier. Set recording levels and record the gig.

Back home, mix the four tracks to stereo. Tracks 1 & 2 provide ambience and bass; tracks 3 & 4 provide definition and clarity.

You might hear an echo because the mics pick up the band with a delay (sound takes time to travel to the mics). A typical delay is 20 to 100 msec. To remove the echo, delay the PA mix by the same amount. Patch a delay unit into the access jacks for tracks 3 and 4. As you adjust the delay time up from zero, the echo will get shorter until the signals are aligned in time. You may be surprised at the quality you get with this simple method.

Recording Off the PA Mixer Aux Outputs

On the PA mixer, find an unused aux send output. Plug in a Y-cord: one end fits the PA mixer connector; the other end has two connectors that mate with your 2-track recorder's line inputs (Figure 17.6). If you have two spare aux busses, you could plug a cable into both of them, and set up a stereo mix.

Put on some good closed-cup headphones and plug them into your 2-track recorder to monitor the recording. Adjust the aux-send knob for each instrument and vocal to create a good recording mix. Record the gig.

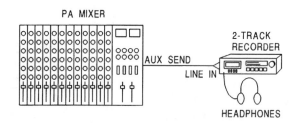

Figure 17.6
Recording off the PA mixer aux outputs.

The advantages of this method are:

- It's simple. All you need is a recorder, a cable, and headphones.
- The recorded sound is close-up and clear.
- If the mix is done well, the sound quality can be very good.

The disadvantages are:

- It's hard to hear what you're mixing. You may need to do several trial recordings. Set up a mix, record, play back, and evaluate. Redo the mix and try another recording
- As you adjust the aux knobs, you might get in the way of the PA operator.
- The recording will be dry (without effects or room ambience). However, you could plug two room mics into the PA mixer and add them to the recording mix. Do not assign these mics to the PA output channels.
- If the aux send is pre-EQ, there will be no EQ on the mics. If the aux send is post-EQ, there will be EQ on the mics, but it may not be appropriate for recording.

Recording an aux mix works best where the setup is permanent and you have time to experiment. Some examples are recording a church service, and recording a regularly scheduled show in a fixed venue.

Feed the PA Mixer Insert Sends to a Modular Digital Multitrack

This is an easy way to record, and it offers excellent sound quality. Plug one or more MDMs into the insert-send jacks on the back of the PA mixer. Set levels with the PA mixer input trims. After the concert, mix the tape tracks back in your studio.

Connections

Suppose you want to record one instrument or vocal on each track. In each PA mixer input channel, locate the INSERT or ACCESS jacks. Connect that jack to a tape track input (Figure 17.7). INSERT jacks are usually pre-fader, pre-EQ. So any fader or EQ changes that the PA operator does will not show up on your tape.

If the PA mixer has separate Send and Receive INSERT jacks, connect the Send to the tape track input, and connect the tape track output to the Receive. Some boards use a single stereo INSERT jack with TRS (tip/ring/sleeve)

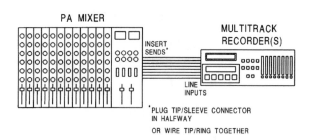

Figure 17.7
Feeding an MDM from the
PA mixer insert jacks.

connections. Usually the tip is send and the ring is receive. In the stereo phone plug you plug into the INSERT jack, wire tip and ring together, and also to the cable hot conductor.

On some PA mixers with a tip/ring/sleeve insert jack, you can use a mono (tip/sleeve) phone plug. Plug it in halfway to the first click so you don't break the signal path—the signal still goes through the PA mixer. If you plug in all the way to the second click, the signal does not go through the PA mixer—just to tape.

What if you want to record several instruments on one track, such as a drum mix? Assign all the drum mics to one or two output busses in the PA mixer. Plug the BUS OUT insert jack to the tape track input. Use two busses for stereo.

Monitor Mix

You may want to set up a monitor mix over headphones so you can hear what you're recording. Here are some ways to do this:

- Connect all the MDM outputs to unused line inputs on your mixer. Use those faders to set up a monitor mix. Assign them to an unused bus, and monitor that bus with headphones.

- Use the aux 1 or aux 2 knobs to set up a monitor mix. Monitor the aux send bus over headphones.

- Monitor the stereo mix and set up a monitor mix with the faders, EQ, etc.

Setting Levels

Set recording levels with the PA mixer's TRIM or INPUT ATTEN knobs. This affects the levels in the PA mix, so be sure to discuss your trim adjustment in advance with the PA mixer operator. If you turn down an input trim, the PA operator must compensate by turning up that channel's fader and monitor send.

Set recording levels before the concert during the sound check (if any!). It's better to set the levels a little too low than too high because during mixdown you can reduce noise but not distortion. A suggested starting level for live recording is –10 dBFS, which allows for surprises. You don't want to exceed 0 dBFS.

Keep a tape log as you record, noting the counter times of tunes, sonic problems, and so on. Refer to this log when you mix.

Mixdown

After the recording is finished, mix down the tracks back in the studio, spending as much time as you need to perfect the mix. You even can overdub parts that were flubbed during the live performance, taking care to match the overdubbed sound to the original recording.

Feed PA Mixer Direct Outs
or Insert Jacks to a Recording Mixer

The previous method has a drawback: You have to adjust the PA mixer's trim controls, and this changes the PA mix slightly. A way around this is to connect the PA mixer direct outs or insert sends to the line inputs of a separate recording mixer. Connect the recording mixer insert sends to the MDM(s) (Figure 17.8). Set MDM recording levels with the recording mixer.

This method has some compromises. You need more cables and another mixer. Also, the signal goes through more electronics, so it is not quite as clean as connecting straight to the MDM.

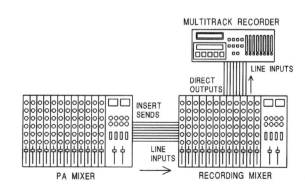

Figure 17.8
Connecting PA mixer to recording mixer to MDM.

Using the recording mixer's faders, you can set up a monitor mix. If you can hear the monitor mix well enough over headphones, you can even omit the multitracks, and attempt a live mix to 2-track.

In a live mix, never turn off a mic completely unless you know for sure that it's not going to be used. Otherwise, you'll invariably miss cues. Turn down unused mics about 10 dB.

Splitting the Microphones

With this method, you plug each mic into a microphone splitter on stage. The mic splitter sends each mic's signal to two paths: the PA mixer snake and the recording mixer snake. Some splitters have a third output to feed a stage monitor mixer. Back at the recording mixer where the snake is plugged in, assign each mic to a different track.

In most mic splitters, the signals are transformer isolated to prevent ground loops and interaction between the mixers. There is a ground-lift switch on each channel (Figure 17.9). Set it to the position where you monitor the least hum. Usually the mic-cable shields are grounded only to the direct-fed console, which supplies phantom power. The cable shields going to the other mixers (getting isolated feeds) are usually floated (disconnected) at the splitter with ground-lift switches.

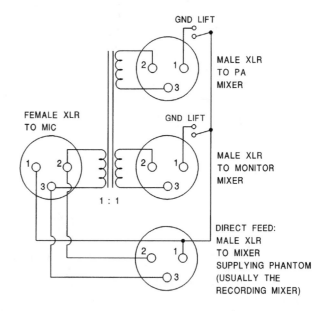

Figure 17.9
Transformer-isolated microphone splitter.

Multitrack Recording in a Van

Here's the ultimate setup. Each mic is split three ways to feed the snake boxes for the recording, reinforcement, and monitor consoles. A long multiconductor snake is run to a recording truck or van parked outside the concert hall or club.

In the van, the snake connects to a mixing console, which is used to submix groups of mics and route their signals to a multitrack tape machine. Sometimes two tape machines are run in parallel to provide a backup in case one fails. Or two analog machines can be synchronized with SMPTE time code to increase the number of tracks available. SMPTE time code is explained in Appendix B.

Mic Placement

For on-location work, you have to place each microphone within a few inches of its source to reject feedback, leakage, and room acoustics. Some suggested mic techniques are covered in Chapter 8.

Audience Microphones

If you have enough mic inputs, you can add one or two audience mics to pick up the room acoustics and audience sounds. This helps the recording to sound "live." Without audience mics, the recording may sound too dry, as if it were done in a studio.

One technique is to tape two boundary mics about 6 feet apart on the front face of the stage. You don't have to hang any microphones, but the mics may pick up conversation in the first few rows of seating.

Another method is to aim two cardioid mics at the audience. Hang them high over the front row of the audience aiming at the back row. You might be able to set up a stereo pair on a mic stand, and tape off the seats around the stand so that it doesn't get bumped.

If the audience mics are far back in the hall—100 feet from the stage, for example—they pick up the sound with a delay. When mixed with the close mics, the audience mics add an echo. Prevent this by placing the audience mics fairly near the stage.

In the PA mixer, leave the audience mics unassigned to prevent feedback. Audience mics can muddy the sound if mixed in too loudly. Keep them down in level, just enough to add some "atmosphere." Bring them up gently to emphasize crowd reactions. You can reduce muddiness by rolling off the lows in the audience mics.

So far this chapter gave an overview of several on-location recording methods. The rest of this chapter explores the details of on-location presession procedures.

Power and Grounding Practice

The following are suggestions for making AC power connections on location. You may want to review Chapter 5 on hum prevention, especially the section on connections to electric guitar amps.

Find a source of power for the remote truck (if any) that can handle the truck's power requirements. Find out whether you'll need a union electrician to make those connections.

Find the circuit breakers for your power source and label them. Ask the custodian not to lock the circuit-breaker box the day of the recording.

Check that your AC power source is not shared with lighting dimmers or heavy machinery; these devices can cause noises or buzzes in the audio. Measure the AC line voltage. Know what your equipment can do under widely varying voltages. Use a line voltage regulator if the AC voltage varies widely. Check AC power on stage with a circuit checker. Are grounded outlets actually grounded? Is there low resistance to ground? Are the outlets correct polarity? There should be a substantial voltage between hot and ground, and no voltage between neutral and ground.

If the AC power is noisy, you might need a power isolation transformer with an electrostatic shield. If possible, get AC power from the same place as the sound-reinforcement company. Run a thick (14 or 16 gauge) extension cord from that point to your recording system. Plug AC outlet strips into the extension cord, then plug all your equipment into the outlet strips.

Interfacing with Telephone Lines

If you're doing a live remote for broadcast, you probably send your signal to the transmitter via rented telephone lines. The Telco (telephone company) noise level of a telephone line is specified in dBrn. A level of 0 dBrn is the "absolutely quiet" reference: 0 dBrn = −90 dBm. Thus, if the noise level is 30 dBrn, the signal-to-noise ratio is 90/30 or 60 dB.

Telco zero level is +8 dBm. You don't have to feed +8 dBm from your console into a phone line; +4 dBm gives 4 dB more headroom. Telco test level is 0 dBm for tones above 400 Hz.

You may want to ask for lossless lines (with unity gain). Otherwise your signal may be down about 20 dB after transmission through the phone lines.

You need a 600-ohm source impedance, achieved by putting a 600-ohm resistor in series with the console output connector (300 ohms per leg of the balanced line). Have a terminated transformer on the sending end. To make a receiving line 600 ohms, put a 600 ohm resistor across pins 2 and 3.

In addition to the program lines, rent a nonequalized private line for communications. Order program lines two or three days in advance. Order a standard nonequalized line for communications about a week in advance. For stereo programs, specify phase-matched lines.

Cables and Connectors

In XLR-type cable connectors, do not connect pin 1 to the shell, or you may get ground loops when the shell contacts a metallic surface. Number the cables near their connectors. You may want to cover these labels with clear heat-shrink tubing. Label both ends of each cable with the cable length. Put a drop of glue on each connector screw to temporarily lock it in place.

To reduce hum pickup and ground-loop problems associated with cable connectors, try to use a single mic cable between each mic and its snake-box connector. Avoid bundling mic cables, line-level cables, and power cables together. If you must cross mic cables and power cables, do so at right angles and space them vertically.

Plug each mic cable into the stage box, then run the cable out to each mic and plug it in. This leaves less of a mess at the stage box. Leave the excess cable at each mic stand so you can move the mics. Don't tape the mic cables down until the musicians are settled. Have an extra microphone and cable offstage ready to use if a mic fails.

Preproduction Meeting

Have a preproduction meeting with the sound-reinforcement company and the production company putting on the event. Find out the date of the event, location, phone numbers of everyone involved, when the job starts, when you can get into the hall, when the second set starts, and other pertinent information. Decide who will provide the split, which system will be plugged in first, second, and so on. Draw block diagrams for the audio system and communications system.

If you're using a mic splitter, the mixer getting the direct side of the split provides phantom power for condenser mics that are not powered on stage. If the house system has been in use for a long time, give them the direct side of the split.

Overloud stage monitors can ruin a recording, so work with the sound-reinforcement people toward a workable compromise. Ask them to start with the monitors quiet because the musicians always want them turned up louder.

Make copies of the meeting notes for all participants. Don't leave things unresolved. Know who is responsible for supplying what equipment.

Figure 17.10 shows a typical equipment layout worked out at a preproduction meeting. There are three systems in use: sound-reinforcement, recording, and monitor mixing. The mic signals are split three ways to feed these systems.

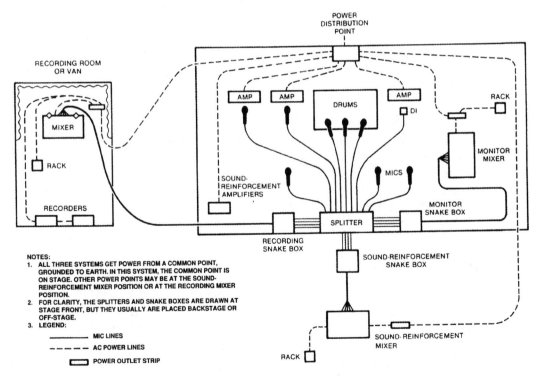

Figure 17.10 Typical layout for an on-location recording of a live concert.

Site Survey

Visit the recording site in advance and go through the following checklist:

- Check the AC power as described earlier.
- Listen for ambient noises—ice machines, coolers, 400-Hz generators, heating pipes, air conditioning, nearby discos, etc. Try to have these noise sources under control by the day of the concert.
- Sketch dimensions of all rooms related to the job. Estimate distances for cable runs.
- Turn on the sound-reinforcement system to see if it functions okay by itself (no hum, and so on). Turn the lighting on at various levels with the sound system on. Listen for buzzes. Try to correct any problem so that you don't document bad PA sound on your tape.
- Determine locations for any audience/ambience mics. Keep them away from air-conditioning ducts and noisy machinery.
- Plan your cable runs from stage to recording mixer.
- If you plan to hang mic cables, feel the supports for vibration. You may need microphone shock mounts. If there's a breeze in the room, plan on taking windscreens.
- Make a file on each recording venue including the dimensions and the location of the circuit breakers.
- Determine where the control room will be. Find out what surrounds it—any noisy machinery?
- Visit the site when a crowd is there to see where there may be traffic problems.

After doing the site survey, draw a complete system block diagram including all cables and connectors. Use this to generate an equipment list. Keep a file of system block diagrams for various recording venues.

Recording Console Setup

Here's a suggested procedure for setting up the recording system efficiently:

1. Turn up the monitor system and verify that it is clean.
2. Plug in one mic at a time and monitor it to check for hums and buzzes. Troubleshooting is easier if you listen to each mic as you connect it, rather than plugging them all in, and trying to find a hum or buzz.

3. Check and clean up one system at a time: first the sound-reinforcement system, then the stage-monitor system, then the recording system. Again, this makes troubleshooting easier because you have only one system to troubleshoot.

4. Use as many designation strips as you need for complex consoles. Label the input faders bottom and top. Also label the monitor-mix knobs and the meters.

5. Monitor the reverb returns (if any) and check for a clean signal.

6. Make a short test recording and listen to the playback.

7. Verify that left and right channels are correct, and that the pan-pot action is not reversed audibly.

8. Do a preliminary pan-pot set-up. Panning similar instruments to different locations helps you identify them.

Miscellaneous Tips

Here are some helpful hints for successful on-location recordings:

- Hook up and use unfamiliar equipment before going on the road. Don't experiment on the job!

- Arrive several hours ahead of time for setup. Expect failures—there's always something going wrong, something unexpected. Allow 50 percent more time for troubleshooting than you think you'll need. Have backup plans if equipment fails. Leave as little to chance as possible. Consider recording with redundant (double) systems so you have a backup if one fails.

- Learn the names of your crew members, and be friendly. These people can be your assets or your enemies. Think before you comment to them!

- Don't be caught without the little things, like spare tape reels, spare tape, spare cables, hub adapters, pencil and paper, flashlight, and electrical 3-to-2 adapters.

- Bring a tool kit with screwdrivers, pliers, soldering iron, connectors, adapters, cables, 9-volt batteries, guitar cords, guitar strings, AC-outlet checkers, fuses, a pocket radio to listen for interference, ferrite beads of various sizes for RFI suppression, canned air to shoot out dirt, cotton swabs and pipe cleaners, and De-Oxit from Caig Labs to remove oxide from connectors.

- If a concert will be longer than the running time of a reel of tape, switch reels at intermissions. Another method is to feed two identical tape machines the same signal in parallel. Record on one machine. As the reel of tape nears the end, start recording on the second machine so that none of the performance is lost. Edit the two tapes together back in the studio.

- Walkie-talkies are okay for preshow use, but don't use them during the performance because they cause RF interference. Use hard-wired communications headsets. Assistants can relay messages to and from the stage crew while you're mixing.

- During short set changes, use a closed-circuit TV system and light table (or fax machines) to show what set changes and mic-layout changes are coming up next; transmit this information to the monitor mixer and sound-reinforcement mixer.

- Don't unplug mics plugged into phantom power because this will make a popping noise in the sound-reinforcement system.

- After the gig, note equipment failures and fix broken equipment as soon as possible.

- Don't put tapes through airport X-ray machines because the transformer in these machines is not always well shielded. Have the tapes inspected by hand.

- Hand-carry your mics on airplanes. Arrange to load and unload your own freight containers, rather than trusting them to airline freight loaders. Expect delays here and at security checkpoints.

- Get a public-liability insurance policy to protect yourself against lawsuits.

- In general, plan everything in advance so you can relax at the gig and have fun!

By following these suggestions, you should improve your efficiency—and your recordings—at on-location sessions.

Most of the information in the second half of this chapter (starting with Power and Grounding Practice) was derived from two workshops presented at the 79th convention of the Audio Engineering Society in October, 1985. These workshops were titled "On the Repeal of Murphy's Law—Interfacing Problem Solving, Planning, and General Efficiency On-Location," given by Paul Blakemore, Neil Muncy, and Skip Pizzi; and "Popular Music Recording Techniques," given by Paul Blakemore, Dave Moulton, Neil Muncy, Skip Pizzi, and Curt Wittig.

18

ON-LOCATION
RECORDING
OF CLASSICAL MUSIC

Perhaps your civic orchestra or high school band is giving a concert, and you would like to make a professional recording. Or maybe there's an organist or string quartet playing at the local college, and they want you to record them.

This chapter explains how to make professional-quality recordings of these ensembles. It describes the necessary equipment, microphone techniques, and session procedures.

Incidentally, recording classical music ensembles is a great way for the beginning recording engineer to gain experience. With just two microphones and a 2-track recorder, much can be learned about acoustics, microphone placement, level setting, and editing—all essential skills in the studio.

Equipment

You need the following equipment to record classical music on-location:

- 2-track recorder
- Microphones
- Cables

- Mic stands and stand adapters
- Headphones (or powered speakers)
- Mic preamps (optional)
- Mixer (optional)

If you plan to record overseas, you may need a power converter that converts 50 Hz, 220V AC power to 60 Hz, 110V. Or power your equipment from batteries, and recharge DAT batteries overnight with a 220V/110V converter. You also need some AC power outlet adapters.

The 2-Track Recorder

A good cassette, MiniDisc or open-reel recorder can be used for live recording, but for highest quality a DAT recorder is preferred. Chapter 10 covers DAT machines in detail. DAT tapes can record up to 2 hours nonstop, making them ideal for recording live concerts. DAT tape cannot be edited unless you copy from one DAT recorder to another, or edit the tape with a digital audio workstation.

Microphones

Next on your list of equipment are some quality microphones. You need two or three of the same model number, or a stereo mic. Good mics are essential, for the microphones—and their placement—determine the sound of your recording. You should spend at least $250 per microphone, or rent some good ones, for professional-quality sound.

For classical music recording, the preferred microphones are condenser types with a wide, flat frequency response, and very low self-noise (less than 21 dB equivalent SPL, A-weighted). (Self-noise is explained in Chapter 6.)

These mics are available with an omnidirectional or unidirectional pickup pattern. An omnidirectional mic is equally sensitive to sounds arriving from any direction, so it helps to add liveness (reverberation) to a recording made in an acoustically dead hall. Omni condenser mics have excellent low-frequency response, so they are a good choice for recording pipe organ or bass drum.

A unidirectional microphone (such as a cardioid) is most sensitive to sounds approaching the front of the microphone, and partly rejects sounds approaching the sides and rear. It helps reduce excessive reverberation in the recording. You need a pair of unidirectional mics if you want to do coincident or near-coincident stereo miking (see Chapter 7).

Stands vs. Hanging

You can mount the microphones on stands or hang them from the ceiling with nylon fishing line. Stands are much easier to set up, but are more visually distracting at live concerts. Stands are more suitable for recording rehearsals or sessions with no audience present.

The mic stands should have a tripod folding base, and should extend at least 14 feet high. You can purchase "baby booms" to extend the height of regular mic stands. Many camera stores have telescoping photographic stands that are lightweight and compact.

A useful accessory is a **stereo bar** or **stereo microphone adapter**. This device mounts two microphones on a single stand for stereo recording.

Hiding Microphones

In some live concerts—especially those that are videotaped—the microphones must not be seen. You might be able to hang some miniature condenser mics, or place boundary mics on the stage floor. If the musical ensemble is large (e.g., an orchestra), and you lay the mics on the stage floor, this placement usually overemphasizes the front row of the ensemble and results in a muffled sound. But if the ensemble is small (a string quartet or small choir, for example), floor placement can work very well. You can also mount boundary mics on the ceiling or on the front edge of a balcony. These placements tend to sound too distant, but they may be your only option.

Monitors

For monitoring, you can use either high-quality loudspeakers or headphones. The headphones should be closed-cup, circumaural (around the ear) types to block out the sound of the musicians. You want to hear only what's being recorded. Of course, the headphones should have a wide-range, smooth response for accurate monitoring.

Loudspeakers give more accurate stereo imaging than headphones. So you might want to set up monitor speakers in a control room separate from the concert hall. Place a pair of Nearfield monitors about 3 feet apart and 3 feet from you, on stands behind the mixer. An alternative is to use high-end consumer or professional loudspeakers placed several feet from the walls to weaken early reflections. You could add absorptive material such as acoustic foam to the walls behind and to the side of the speakers. For the best stereo imaging, sit exactly between the speakers, and as far from them as they are spaced apart.

Mic Cables

You have to sit far from the musicians to clearly monitor what you're recording. To do that, you need a pair of 50-foot mic cables. Longer extensions are needed if the mics are hung from the ceiling, or if you want to monitor in a separate room.

Mic Preamp or Mixer

You need a mixer when you want to record more than one source—an orchestra and a choir, for instance, or a band and a soloist. You might put a pair of microphones on the orchestra and another pair on the choir. The mixer blends the signals of all four mics into a composite stereo signal. It also lets you control the balance (relative loudness) among microphones. You also need a mixer if you want to use **spot microphones** (**accent microphones**) placed close to each orchestra section or soloist.

For the cleanest sound, consider using some high quality stand-alone mic preamps instead of the preamps built into recorders and mixers. Place each preamp on stage, then run its line-level output signal back to your recording gear. Other miscellaneous equipment you may need includes a power extension cord, an outlet strip, DAT dry cleaning tape, spare mic cables, pen and notebook, and duct tape or vinyl mats to keep cables in place.

Stereo Microphone Techniques

As a starting point, you place two or three mics several feet in front of the group, raised up high (Figure 18.1). The mic placement controls the acoustic perspective or sense of distance to the ensemble, the balance among instruments, and the stereo imaging.

Recall from Chapter 7 that there are four mic techniques commonly used for stereo recording: coincident pair, near-coincident pair, spaced-pair, and baffled omni techniques. Below is a review:

- Coincident-pair: Two directional mics angled apart with their grilles nearly touching and their diaphragms aligned vertically.

- Near-coincident pair: Two directional mics angled apart and spaced a few inches apart horizontally.

- Spaced pair: Two or three matched microphones of any pattern aiming straight ahead toward the ensemble and spaced several feet apart horizontally.

- Baffled omni: Two omnidirectional mics that are ear spaced and separated by a padded baffle or a hard-surface sphere.

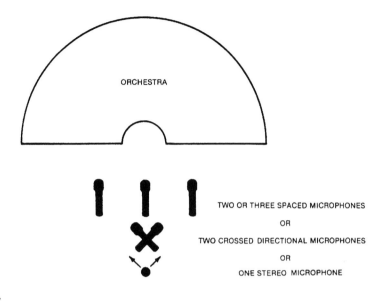

(A) Top view.

ORCHESTRA

TWO OR THREE SPACED MICROPHONES

OR

TWO CROSSED DIRECTIONAL MICROPHONES

OR

ONE STEREO MICROPHONE

(B) Side view.

Figure 18.1 Typical microphone placement for on-location recording of a classical music ensemble.

Recall from Chapter 6 that boundary microphones can be mounted on clear plastic panels about 2 feet square. You can space these panels apart for spaced-pair stereo, or place them with one edge touching to form a "V." Aim the point of the V at the ensemble. This near-coincident arrangement provides excellent stereo imaging. Also available is a stereo boundary microphone that is smaller than the two panels, and provides excellent stereo imaging, extended low-frequency response, and mono-compatibility.

Preparing for the Session

Once you have the equipment, you are ready to go on-location. First, ask the musical director what groups and soloists will be playing, where they will be located, and how long the program is.

If possible, plan to record in a venue with good acoustics. It should have low background noise and adequate reverberation time for the music being performed. This is very important, because it can make the difference between an amateur-sounding recording and a commercial-sounding one. Try to record in an auditorium or spacious church rather than in a band room or gymnasium. If you're forced to record in a hall that is relatively dead, you might want to add artificial reverberation.

Next, get all your equipment ready. Fast-forward and rewind the blank DAT tape and clean the heads with a dry cleaning tape. Label the tape with the name of the artist, location, and date of the session. Check all cables and equipment for proper operation.

Keep your equipment inside your home or studio until you're ready to leave. A DAT recorder left outside in a cold car may become sluggish if the lubricant stiffens, and batteries may lose some voltage.

Session Setup

Allow an extra hour or so for setup and for fixing broken cables, etc. There's always something unexpected in any new recording situation.

When you first arrive at the recording venue, locate some AC power outlets where you want to set up. Check that these outlets are "live." If not, ask the custodian to turn on the appropriate circuit breaker. Always check in with union technicians if the venue is a union one.

Find a table or folding chairs on which to set your equipment. Plug into the AC outlets and let your equipment warm up. Leave a few turns of AC cord near the outlet, and tape down the cord so that it isn't pulled out accidentally.

Then take out your microphones and place them in the desired stereo miking arrangement. As an example, suppose you are recording an orchestra rehearsal with two crossed cardioids on a stereo bar (the near-coincident method). Screw the stereo bar onto a mic stand, and mount two cardioid microphones on the stereo bar. For starters, angle them 110 degrees apart and space them 7 inches apart horizontally. Aim them down so that they point at the orchestra when raised. You may want to mount the microphones in shock mounts or put the stands on sponges to isolate the mics from floor vibration.

As a starting position, place the mic stand behind the conductor's podium, about 12 feet in front of the front-row musicians. Connect mic cables. Raise the microphones about 14 feet off the floor. This prevents overly loud pickup of the front row relative to the back row of the orchestra.

Mic techniques for piano recitals and other solo instruments are covered in Chapter 8.

Leave some extra turns of mic cable at the base of each stand so you can reposition the stands. This slack also allows for people pulling on the cables accidentally. Try to route the mic cables where they won't be stepped on, or cover them with mats.

Make connections in one of the following ways:

- If you are using 2 mics, and your DAT recorder has high-quality mic preamps, plug the mics directly into the recorder mic inputs.
- If you prefer to use an outboard mic preamp, plug the mics into the preamp, and plug the preamp output into the recorder line inputs.
- If you're using multiple mics and a mixer, plug the mics into the mixer mic inputs, and plug the mixer stereo outputs into the recorder line inputs.

Now put on your headphones, turn up the recording-level controls, and monitor the signal. When the orchestra starts to play, set the recording levels to peak around –12 dBFS.

Microphone Placement

Nothing has more effect on the production style of a classical music recording than microphone placement. Miking distance, stereo positioning, and spot miking all influence the recorded sound character.

Distance

The microphones must be placed closer to the musicians than a good live listening position would be. If you place the mics out in the audience where the live sound is good, the recording probably will sound muddy and distant when played over speakers. That's because the recorded reverberation is condensed into the space between the playback speakers, along with the direct sound of the orchestra. Close miking (5 to 20 feet from the front row) compensates for this effect by increasing the ratio of direct sound to reverberant sound.

The closer the mics are to the orchestra, the closer it sounds in the recording. If the instruments sound too close, too edgy, too detailed—if the recording lacks hall ambience—the mics are too close to the ensemble. Move the mic stand 1or 2 feet farther from the orchestra and listen again.

If the orchestra sounds too distant, muddy, or reverberant, the mics are too far from the ensemble. Move the mic stand a little closer to the musicians and listen again.

Eventually you'll find a "sweet spot" where the direct sound of the orchestra is in a pleasing balance with the ambience of the concert hall. Then the reproduced orchestra will sound neither too close nor too far.

Stereo-Spread Control

Now concentrate on the stereo spread. If the spread heard over headphones is too narrow, that means the mics are angled or spaced too close together. Increase the angle or spacing between mics until localization is accurate. Angling the mics farther apart makes the instruments sound farther away; spacing the mics farther apart does not, but may make the images less focused.

If the instruments that are slightly off-center are heard far-left or far-right in your headphones, your mics are angled or spaced too far apart. Move them closer together until localization is accurate.

You localize sounds differently with headphones than with speakers. For this reason, coincident-pair recordings have less stereo spread over headphones than over loudspeakers. Take this into account when monitoring.

Surround miking was covered at the end of Chapter 7.

You can test the stereo localization accuracy of your chosen stereo miking method. If you have time, record yourself speaking from various positions on stage while announcing your position: far-left, half-left, center, half-right, and far-right. Listen to the monitor system to check whether the image of your voice is reproduced in corresponding positions. Generally, the far-left and far-right positions should be reproduced at the left and right loudspeakers, respectively.

Soloist Pickup and Spot Microphones

Sometimes a soloist plays in front of the orchestra. By raising or lowering the stereo mic pair, you can control the balance between soloist and ensemble. If the soloist is too loud relative to the orchestra (as monitored), raise the mics. If the soloist is too quiet, lower the mics. You may want to add a spot mic about 3 feet from the soloist and mix it with the other microphones.

Many recording companies prefer to use several spot mics and a multitrack recorder when taping classical music. Such a method gives more control of balance and definition, and is necessary in many situations. If you use spot or accent mics on various instruments or instrumental sections, mix them at a low level relative to the main pair—just loud enough to add definition, but not loud enough to destroy depth. Operate the spot-mic faders subtly or leave them untouched. Otherwise the close-miked instruments may "jump forward" when you bring up the fader, and then "fall back in" when you bring down the fader.

Spot mics sound more natural if you delay their signals to match the acoustic delay of the main pair. Typical delays are 15 to 25 milliseconds.

Solo the main stereo pair, and note the image locations of instruments that you are spot-miking. Pan the spot mics so their image locations coincide with those of the main mic pair.

Recording

Now that the mics are positioned properly, you're ready to record. At a live concert, you might want to set your recording levels to read about –15 dBFS with the opening applause. This procedure should result in approximately correct recording levels when the musicians start playing. Or set the record-level controls where they were at previous sessions.

Start recording a few seconds before the music starts. Once the recording is in progress, let the recording-level meters peak at –3 dB maximum for a DAT recorder. This allows a little headroom for surprises. Leave the recording level alone as much as possible. If you must adjust the level, do so slowly and try to follow the dynamics of the music.

If there is applause at the end of a musical piece, you can fade it out over 3 to 5 seconds by slowly turning down the recording-level controls or the mixer master volume control. Or leave it alone for a later fade-out during editing.

Most classical recording sessions are done in several takes. Keep track of the DAT ABS times for these takes, and slate each one. Mark the keeper takes for later editing.

After the concert, pack the mics away first. Otherwise, they may be stolen or damaged.

Editing

Once you have your tapes home, you may want to edit them to make a tight presentation. Currently most editing is done with a digital audio workstation, described in Chapter 10.

Put about 4 seconds of silence between each selection. Or you may want to insert an interval of recorded "room sound" or "room tone," especially between movements of a symphony.

Some on-location recordings have a fair amount of background noise. If you insert silence between pieces, there will be an abrupt cutoff of this noise at the end of each piece. To prevent that, either fade the noise down to silence, or replace the silence with room tone.

If you recorded a live concert with applause, fade down the applause over several seconds. At the point where the applause fades to silence, edit that to a point a few seconds before the start of the next piece. After copying the edited program to another blank DAT, add Start IDs as described in Chapter 14.

Congratulations! You now have your finished product—a realistic, professional recording of a classical music ensemble.

19

JUDGING SOUND QUALITY

Seat an engineer behind a mixing console and ask him or her to do a mix. It sounds great. Then seat another engineer behind the same console and again ask for a mix. It sounds terrible. What happened?

The difference lies mainly in their ears—their critical listening ability. Some engineers have a clear idea of what they want to hear and how to get it. Others haven't acquired the essential ability to recognize good sound. By knowing what to listen for, you can improve your artistic judgments during recording and mixdown. You are able to hear errors in microphone placement, equalization, and so on, and correct them.

To train your hearing, try to analyze recorded sound into its components—such as frequency response, noise, reverberation—and concentrate on each one in turn. It's easier to hear sonic flaws if you focus on a single aspect of sound reproduction at a time. This chapter is a guide to help you do this.

Classical vs. Popular Recording

Classical and popular music have different standards of "good sound." One goal in recording classical music (and often folk music or jazz) is to accurately reproduce the live performance. This is a worthy aim because the sound of an orchestra in a good hall can be quite beautiful. The music was composed and the instruments were designed to sound best when heard live in the concert hall. The recording engineer, out of respect for the music, should always try to translate that sound to tape with as little technical intrusion as possible.

By contrast, the accurate translation of sound to tape is not always the goal in recording popular music. Although the aim may be to reproduce the original sound, the producer or engineer may also want to play with that sound to create a new sonic experience, or to do some of both.

In fact, the artistic manipulation of sounds through studio techniques has become an end in itself. Creating an interesting new sound is as valid a goal as re-creating the original sound. There are two games to play, each with its own measures of success.

If the aim of a recording is realism or accurate reproduction, the recording is successful when it matches the live performance heard in the best seat in the concert hall. The sound of musical instruments is the standard by which such recordings are judged.

When the goal is to enhance the sound or produce special effects (as in most pop-music recordings), the desired sonic effect is less defined. The live sound of a pop group could be a reference, but pop music recordings generally sound better than live performances. Recorded vocals are clearer and less harsh, the bass is cleaner and tighter, and so on. The sound of pop music reproduced over speakers has developed its own standards of quality apart from accurate reproduction.

Good Sound in a Pop Music Recording

Currently, a good-sounding pop recording might be described as follows (there are always exceptions):

- Well-mixed
- Wide-range
- Tonally balanced
- Clean
- Clear
- Smooth
- Spacious

  It also has:

- Presence
- Sharp transients
- Tight bass and drum

- Wide and detailed stereo imaging
- Wide but controlled dynamic range
- Interesting sounds
- Suitable production

The next sections explore each one of these qualities in detail so that you know what to listen for. Assume that the monitor system is accurate, so that any colorations heard are in the recording and not in the monitors.

A Good Mix

In a good mix, the loudness of instruments and vocals are in a pleasing balance with each other. Everything can be clearly heard, yet nothing is obtrusive. The most important instruments or voices are loudest; less-important parts are in the background.

A successful mix goes unnoticed. When all the tracks are balanced correctly, nothing sticks out and nothing is hidden. Of course, there's a wide latitude for musical interpretation and personal taste in making a mix. Dance mixes, for example, can be very severe sonically.

Sometimes you don't want everything to be clearly heard. In rare occasions you may want to mix in certain tracks very subtly for a subconscious effect.

The mix must be appropriate for the style of music. For example, a mix that's right for loud rock music usually won't work for a pop ballad. A rock mix typically has the drums way up front and the vocals only slightly louder than the accompaniment. In contrast, a pop ballad has the vocals loudest, with the drums used just as "seasoning" in the background.

Level changes during the mix should be subtle, or should make sense. Otherwise, instruments jump out for a solo and fall back in afterwards. Move faders slowly, or set them to preset positions during pauses in the music. Nothing sounds more amateurish than a solo that starts too quietly then comes up as it plays. You can hear the engineer working the fader.

Wide Range

Wide range means extended low-frequency and high-frequency response. Cymbals should sound crisp and distinct, but not sizzly or harsh; kick drum and bass should sound deep, but not overwhelming or muddy. Wide-range sound results from using high-quality microphones and adequate EQ. Analog tape recorders need proper alignment, good tape, high tape speed, and clean tape heads.

Tonal Balance

The overall tonal balance of a recording should be neither bassy nor trebley. That is, the perceived spectrum should not emphasize low frequencies or high frequencies. Low bass, midbass, midrange, upper midrange, and highs should be heard in equal proportions (Figure 19.1). Emphasis of any one frequency band over the other eventually causes listening fatigue. Dance club mixes, however, are heavy on the bass end to get the crowd moving.

Recorded tonal balance is inversely related to the frequency response of the studio's monitor system. If the monitors have a high-frequency roll-off, the engineer will compensate by boosting highs in the recording to make the monitors sound right. The result is a bright recording.

Before doing a mix, it helps to play over the monitors some commercial recordings whose sound you admire. This helps you become accustomed to a commercial spectral balance. After your mix is recorded, play it back and alternately switch between your mix and a commercial recording. This comparison indicates how well you matched a commercial spectral balance. Of course, you may not care to duplicate what others are doing.

In pop music recordings, the tonal balance or timbre of individual instruments does not necessarily have to be natural. Still, many listeners want to hear a realistic timbre from acoustic instruments, such as the guitar, flute, sax, or piano. The reproduced timbre depends on microphone frequency response, microphone placement, the musical instruments themselves, and equalization.

Clean Sound

Clean means free of noise and distortion. Tape hiss, hum, and distortion are inaudible in a good recording. Distortion in this case means distortion added by the recording process, not distortion already present in the sound

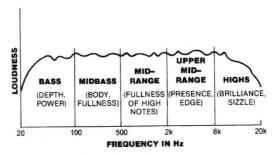

Figure 19.1
Loudness vs. frequency of a good-sounding pop recording.

of electric-guitar amps or Leslie speakers. There are exceptions to this guideline; some popular recordings have noise or distortion added intentionally.

Clean also means "not muddy"—free of low-frequency overhang and leakage. A clean mix is one that is uncluttered, free of excess instrumentation. This is achieved by arranging the music so that similar parts don't overlap, and not too many instruments play at once in the same frequency range. Usually, the fewer the instruments, the cleaner the sound. Too many overdubs can muddy the mix.

Clarity

In a clear-sounding recording, instruments do not crowd or mask each other's sound. They are separate and distinct. As with a clean sound, clarity arises when instrumentation is sparse, or when instruments occupy different areas of the frequency spectrum. For example, the bass provides low frequencies; keyboards might emphasize midbass, lead guitar provides upper midrange, and cymbals fill in the highs.

In addition, a clear recording has adequate reproduction of each instrument's harmonics. That is, the high-frequency response is not rolled off.

Smoothness

Smooth means easy on the ears, not harsh, uncolored. Sibilants sounds are clear but not piercing. A smooth, effortless sound allows relaxation; a strained or irritating sound causes muscle tension in the ears or body. Smoothness is a lack of sharp peaks or dips in the frequency response, as well as a lack of excessive boost in the midrange or upper midrange.

Presence

Presence is the apparent sense of closeness of the instruments—a feeling that they are present in the listening room. Synonyms are clarity, detail, and punch.

Presence is achieved by close miking, overdubbing, and using microphones with a presence peak or emphasis around 5 kHz. Using less reverb and effects can help. Upper-midrange boost helps too. Most instruments have a frequency range which, if boosted, makes the instrument stand out more clearly or become better defined. Presence sometimes conflicts with smoothness because presence often involves an upper-midrange boost, while a smooth sound is free of such emphasis. You have to find a tasteful compromise between the two.

Spaciousness

When the sound is spacious or airy, there is a sense of air around the instruments. Without air or ambience, instruments sound as if they are isolated in stuffed closets. (Sometimes, though, this is the desired effect.) You achieve spaciousness by adding reverb, recording instruments in stereo, using room mics, or miking farther away.

Sharp Transients

The attack of cymbals and drums generally should be sharp and clear. A bass guitar and piano may or may not require sharp attacks, depending on the song.

Tight Bass and Drums

The kick drum and bass guitar should "lock" together so that they sound like a single instrument—a bass with a percussive attack. The drummer and bassist should work out their parts together so as to hit accents simultaneously, if this is desired.

To further tighten the sound, damp the kick drum and record the bass direct. Equalize them for presence and clarity. Rap music, however, has its own sound—the kick drum usually is undamped and boomy, sometimes with short reverb added.

Wide and Detailed Stereo Imaging

Stereo means more than just left and right. Usually, tracks should be panned to many points across the stereo stage between the monitor speakers. Some instruments should be hard-left or hard-right; some should be in the center; others should be half-left or half-right. Try to achieve a stereo stage that is well balanced between left and right (Figure 19.2). Instruments occupying the same frequency range might be panned to opposite sides of center.

You may want some tracks to be unlocalized. Backup choruses and strings should be spread out rather than appearing as point sources. Stereo keyboard sounds can wander between speakers. A lead-guitar solo can have a fat, spacious sound.

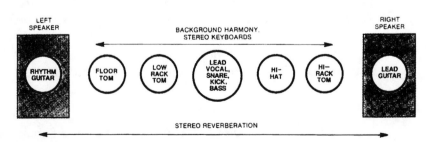

Figure 19.2 An example of image placement between speakers.

There should also be some front-to-back depth. Some instruments should sound close or up front; others should sound farther away.

If you want the stereo imaging to be realistic (for a jazz combo, for example), the reproduced ensemble should simulate the spatial layout of the live ensemble. If you're sitting in an audience listening to a jazz quartet, you might hear drums on the left, piano on the right, bass in the middle, and sax slightly right. The drums and piano are not point sources, but are somewhat spread out. If spatial realism is the goal, you should hear the same ensemble layout between your speakers. Often the piano and drums are spread all the way between speakers—an interesting effect, but unrealistic.

Pan-potted mono tracks often sound artificial in that each instrument sounds isolated in its own little space. It helps to add some stereo reverberation around the instruments to "glue" them together.

Often, TV mixes are heard in mono. Hard-panned signals sound weak in mono relative to center-panned signals. So pan sound sources to 3 and 9 o'clock, not hard-right and hard-left.

Wide but Controlled Dynamic Range

Dynamic range is the range of volume levels from softest to loudest. A recording with a wide dynamic range becomes noticeably louder and softer, adding excitement to the music. To achieve this, don't add too much compression (automatic volume control). An overly compressed recording sounds squashed—crescendos and quiet interludes lose their impact.

Vocals often need some compression or gain-riding because they have more dynamic range than the instrumental backup. A vocalist may sing too loudly and blast the listener, or sing too softly and become buried in the mix. A compressor can even out these extreme level variations, keeping the vocals at a constant loudness. Bass guitar also can benefit from compression.

Interesting Sounds

The recorded sound may be too flat or neutral—lacking character or color. In contrast, a recording with creative production has unique musical instrument sounds, and typically uses special effects. Some of these are equalization, echo, reverberation, doubling, chorus, flanging, compression, and stereo effects. Making sounds interesting or colorful can conflict with accuracy or fidelity, so effects and equalization should be used with discretion.

Suitable Production

The way a recording sounds should imply the same message as the musical style or lyrics. In other words, the sound should be appropriate for the particular tune being recorded.

For example, some rock music is rough and raw. The sound should be, too. A clean, polished production doesn't always work for high-energy rock and roll. There might even be a lot of leakage or ambience to suggest a garage studio or nightclub environment. The role of the drums is important, so they should be loud in the mix. The toms should ring, if that is desired.

New Age, disco, rhythm and blues, contemporary Christian, or pop music is slickly produced. The sound is usually tight, smooth, and spacious. In old-time folk music, instruments are miked at a respectful distance, giving an airy, natural effect.

Actually, each style of music is not locked into a particular style of production. You tailor the sound to complement the music of each individual tune. Doing this may break some of the guidelines of good sound, but that's usually okay as long as the song is enhanced by its sonic presentation.

Good Sound in a Classical Music Recording

As with pop music, classical music should sound clean, wide-range, and tonally balanced. But since classical recordings are meant to sound realistic—like a live performance—they also require good acoustics, a natural balance, tonal accuracy, suitable perspective, and accurate stereo imaging (see Chapter 18).

Good Acoustics

The acoustics of the concert hall or recital hall should be appropriate for the style of music to be performed. Specifically, the reverberation time should be neither too short (dry) nor too long (cavernous). Too short a reverberation time results in a recording without spaciousness or grandeur. Too long a reverberation time blurs notes together, giving a muddy, washed-out effect. Ideal reverberation times are around 1.2 seconds for chamber music or soloists, 1.5 seconds for symphonic works, and 2 seconds for organ recitals. To get a rough idea of the reverb time of a room, clap your hands once, loudly, and count the seconds it takes for the reverb to fade to silence.

A Natural Balance

When a recording is well balanced, the relative loudness of instruments is similar to that heard in an ideal seat in the audience area. For example, violins are not too loud or soft compared to the rest of the orchestra; harmonizing or contrapuntal melody lines are in proportion.

Generally, the conductor, composer, and musicians balance the music acoustically, and you capture that balance with your stereo mic pair. But sometimes you need to mike certain instruments or sections to enhance definition or balance. Then you mix all the mics. In either case, consult the conductor for proper balances.

Tonal Accuracy

The reproduced timbre or tone quality should match that of live instruments. Fundamentals and harmonics should be reproduced in their original proportion.

Suitable Perspective

Perspective is the sense of distance of the performers from the listener—how far away the stage sounds. Do the performers sound like they're eight rows in front of you, in your lap, or in another room?

The style of music suggests a suitable perspective. Incisive, rhythmically motivated works (such as Stravinsky's "Rite of Spring") sound best with closer miking; lush romantic pieces (a Bruckner symphony) are best served by more distant miking. The chosen perspective depends on the taste of the producer.

Closely related to perspective is the amount of recorded ambience or reverberation. A good miking distance yields a pleasing balance of direct sound from the orchestra and ambience from the concert hall.

Accurate Imaging

Reproduced instruments should appear in the same relative locations as they were in the live performance. Instruments in the center of the ensemble should be heard in the center between the speakers; instruments at the left or right side of the ensemble should be heard from the left or right speaker. Instruments halfway to one side should be heard halfway off center, and so on. A large ensemble should spread from speaker to speaker, while a quartet or soloist can have a narrower spread.

It's important to sit equidistant from the speakers when judging stereo imaging, otherwise the images shift toward the side on which you're sitting. Sit as far from the speakers as they are spaced apart. Then the speakers appear to be 60 degrees apart, which is about the same angle an orchestra fills when viewed from the typical ideal seat in the audience (tenth row center, for example).

The reproduced size of an instrument or instrumental section should match its size in real life. A guitar should be a point source; a piano or string section should have some stereo spread. Each instrument's location should be as clearly defined as it was heard from the ideal seat in the concert hall.

Reproduced reverberation (concert hall ambience) should either surround the listener, or at least it should spread evenly between the speakers. Surround-sound technology is needed to make the recorded ambience surround the listener, although spaced-microphone recordings have some of this effect.

There should be a sense of stage depth, with front-row instruments sounding closer than back-row instruments. Accurate imaging is illustrated in Figure 19.3.

Training Your Hearing

The critical process is easier if you focus on one aspect of sound reproduction at a time. You might concentrate first on the tonal balance—try to pinpoint what frequency ranges are being emphasized or slighted. Next listen to the mix, the clarity, and so on. Soon you have a lengthy description of the sound quality of your recording.

Developing an analytical ear is a continuing learning process. Train your hearing by listening carefully to recordings—both good and bad.

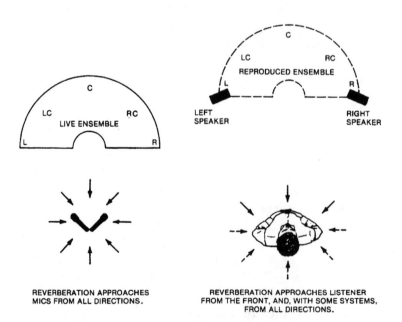

Figure 19.3 With accurate imaging, the sound-source location and size are reproduced during playback, as well as the reverberant field.

Make a checklist of all the qualities mentioned in this chapter. Compare your own recordings to live instruments, and to commercial recordings.

A pop music record that excels in all the attributes of good sound is "The Sheffield Track Record" (Sheffield Labs, Lab 20), engineered and produced by Bill Schnee. In effect, it's a course in state-of-the-art sound—required listening for any recording engineer or producer.

Another record with brilliant production is "The Nightfly" by Donald Fagen (Warner Brothers 23696-2); engineered by Roger Nichols, Daniel Lazerus, and Elliot Scheiner, produced by Gary Katz, and mastered by Bob Ludwig. The sound is razor sharp, elegant, and tasteful; and the music just pops out of the speakers.

The following listings are more examples of outstanding rock production, and set high standards:

"I Need Somebody" by Bryan Adams. Producer, Bob Clearmountain

"The Power of Love" by Huey Lewis & The News. Producer, Huey Lewis & The News

"Synchronicity" by The Police. Producer, Hugh Padgham and The Police

"90125" by Yes. Producer, Trevor Horn

"Dark Side of the Moon" by Pink Floyd. Producer, Alan Parsons

"Thriller" by Michael Jackson. Engineer, Bruce Swedien; Producer, Quincy Jones

"Avalon" by Roxy Music. Engineer, Bob Clearmountain; Producer, Roxy Music

Then there are the incredibly clean recordings of Tom Jung (with DMP records) and George Massenberg. You can learn a lot by emulating these superb recordings, and many others.

Once you're making recordings that are competent technically—clean, natural, and well mixed—the next stage is to produce imaginative sounds. You're in command; you can tailor the mix to sound any way that pleases you or the band you're recording. The supreme achievement is to produce a recording that is a sonic knockout—beautiful or thrilling.

Troubleshooting Bad Sound

Now you know how to recognize good sound, but can you recognize bad sound? Suppose you're monitoring a recording in progress, or listening to a recording you've already made. Something doesn't sound right. How can you pinpoint what's wrong, and how can you fix it?

The rest of this chapter includes step-by-step procedures to solve audio-related problems. Read down the list of "bad sound" descriptions until you find one matching what you hear. Then try the solutions until your problem disappears. Only the most common symptoms and cures are mentioned; console maintenance is not covered.

This troubleshooting guide is divided into four main sections:

- Bad sound on all recordings (including those from other studios).
- Bad sound on playback only (mixer output sounds all right).
- Bad sound in a pop music recording.
- Bad sound in a classical-music recording.

Before you start, check for faulty cables and connectors. Also check all control positions; rotate knobs and flip switches to clean the contacts, clean connectors with De-Oxit from Caig Labs.

Bad Sound on All Recordings

If you have bad sound on all your recordings, including those from other studios, follow this checklist to find the problem:

- Upgrade your monitor system.
- Adjust tweeter and woofer controls on speakers.
- Adjust the relative gains of tweeter and woofer amplifiers in a bi-amped system.
- Relocate speakers.
- Improve room acoustics.
- Equalize the monitor system
- Try different speakers
- Upgrade the power amp and speaker cables.

Bad Sound on Playback Only

You might have bad sound on your playback only, but your mixer output sounds okay. If cassette or open-reel playback has a dull sound or drop-outs, follow these steps:

- Check that the oxide side of the tape is against the heads.
- Clean and demagnetize the tape path.
- Try another brand of tape.
- Align tape heads; calibrate the electronics.

- Do maintenance on the tape transport.
- Check and replace tape heads if necessary.

  If your tape playback is plagued with distortion, try the following:

- Reduce the recording level.
- Increase the bias level.
- Check for bad cables.

  If your tape playback has tape hiss, follow these steps:

- Increase the recording level.
- Use Dolby or dbx noise reduction.
- Use better tape.
- Align the tape recorder.

  If DAT tape playback has glitches or drop-outs, try these steps:

- Clean the recorder with a dry cleaning tape.
- Before recording, fast-forward the tape to the end and rewind it to the top.
- Use better tape.
- Format videocassettes nonstop from start to finish.

Bad Sound in a Pop Music Recording Session

Sometimes you have bad sound in a pop music recording session.

Muddiness (Leakage)

If the sound is muddy from excessive leakage, try the following:

- Place the microphones closer to their sound sources.
- Spread the instruments farther apart to reduce the level of the leakage.
- Place the instruments closer together to reduce the delay of the leakage.
- Use directional microphones (such as cardioids).
- Overdub the instruments.
- Record the electric instruments direct.
- Use baffles (goboes) between instruments.
- Deaden the room acoustics (add absorptive material, flexible panels, or slot absorbers).

- Filter out frequencies above and below the spectral range of each instrument. Be careful or you'll change the sound of the instrument.
- Turn down the bass amp in the studio, or monitor the bass with headphones instead.

Muddiness (Excessive Reverberation)

If the sound is muddy due to excessive reverberation, try these steps:

- Reduce the effects-send levels or effects-return levels. Or don't use effects until you figure out what the real problem is.
- Place the microphones closer to their sound sources.
- Use directional microphones (such as cardioids).
- Deaden the room acoustics.
- Filter out frequencies below the fundamental frequency of each instrument.

Muddiness (Lacks Highs)

If your sound is muddy and lacks highs, or has a dull or muffled sound, try the following:

- Use microphones with better high-frequency response, or use condenser mics instead of dynamics.
- Change the mic placement. Put the mic in a spot where there are sufficient high frequencies. Keep the high-frequency sources (such as cymbals) on-axis to the microphones.
- Use small-diameter microphones, which generally have a flatter response off-axis.
- Boost the high-frequency equalization.
- Change musical instruments; replace guitar strings; replace drum heads. (Ask the musicians first!)
- When bouncing tracks, record bright-sounding instruments last to reduce generation loss.
- Avoid excessive recording levels with bright-sounding instruments. The high-frequency response of cassettes and open-reel gradually rolls off as the recording level is increased.
- Use an enhancer signal processor, but watch out for noise.
- Use a direct box on the electric bass. Have the bassist play percussively or use a pick if the music requires it. When compressing the bass, use a long attack time to allow the note's attack to come through.

(Some songs don't require sharp bass attacks—do whatever's right for the song.)

- Damp the kick drum with a pillow or blanket, and mike it next to the center of the head near the beater. Use a wooden beater if the song and the drummer allow it.

Muddiness (Lacks Clarity)

If your sound is muddy because it lacks clarity, try these steps:

- Consider using fewer instruments in the musical arrangement.
- Equalize instruments differently so that their spectra don't overlap.
- Try less reverberation.
- Using equalizers, boost the presence range of instruments that lack clarity.
- In a reverb unit, add about 30 to 100 msec of predelay.

Distortion

If you have distortion in a pop music recording, try the following:

- Increase input attenuation (reduce input gain), or plug in a pad between the microphone and mic input.
- Readjust gain-staging: set faders and pots to their design centers (shaded areas).
- If you still hear distortion, switch in the pad built into the microphone (if any).
- Check connectors for stray wires and bad solder joints.
- Unplug and plug-in connectors.

Tonal Imbalance

If you have bad tonal balance—the sound is boomy, dull, or shrill, for example—follow these steps:

- Change musical instruments; change guitar strings; change reeds, etc.
- Change mic placement. If the sound is too bassy with a directional microphone, you may be getting proximity effect. Mike farther away or roll off the excess bass.

- Use the 3:1 rule of mic placement to avoid phase cancellations. When you mix two or more mics to the same channel, the distance between mics should be at least three times the mic-to-source distance.
- Try another microphone. If a cardioid mic's proximity effect is causing a bass boost, try an omnidirectional mic instead.
- If you must place a microphone near a hard reflective surface, try a boundary microphone on the surface to prevent phase cancellations or pad the surface.
- Change the equalization. Avoid excessive boost.
- Use equalizers with a broad bandwidth, rather than a narrow, peaked response.

Lifelessness

If your pop music recording has a lifeless sound and is unexciting, these steps might help you solve it:

- Work on the live sound of the instruments in the studio to come up with unique effects.
- Add special effects—reverberation, echo, exciter, doubling, equalization, etc.
- Use and combine recording equipment in unusual ways.
- Try overdubbing little vocal licks or synthesized sound effects.

If your sound seems lifeless due to dry or dead acoustics, try these:

- If leakage is not a problem, put microphones far enough from instruments to pick up wall reflections. If you don't like the sound this produces, try the next suggestion.
- Add reverb or echo to dry tracks. (Not all tracks require reverberation. Also, some songs may need very little reverberation so that they sound intimate.)
- Use omnidirectional microphones.
- Add hard reflective surfaces in the studio, or record in a hard-walled room.
- Allow a little leakage between microphones. Put mics far enough from instruments to pick up off-mic sounds from other instruments. Don't overdo it, though, or the sound becomes muddy and track separation becomes poor.

Noise (Hiss)

Sometimes your pop music recording has extra noise on it. If your sound has hiss, try these:

- Check for noisy guitar amps or keyboards.
- Switch out the pad built into the microphone (if any).
- Reduce mixer input attenuation (increase input gain).
- Use a more sensitive microphone.
- Use a quieter microphone (one with low self-noise).
- Increase the sound pressure level at the microphone by miking closer. If you're using PZMs, mount them on a large surface or in a corner.
- Apply any high-frequency boost during recording, rather than during mixdown.
- If possible, feed tape tracks from mixer direct outs or insert sends instead of group or bus outputs.
- Use a lowpass filter (high-cut filter).
- As a last resort, use a noise gate.

Noise (Rumble)

If the noise is a low-frequency rumble, follow these steps:

- Reduce air-conditioning noise or temporarily shut off the air conditioning.
- Use a highpass filter (low-cut filter) that is set around 40 to 80 Hz.
- Use microphones with limited low-frequency response.
- See Noise (Thumps) below.

Noise (Thumps)

- Change the microphone position.
- Change the musical instrument.
- Use a highpass filter set around 40 to 80 Hz.
- If the cause is mechanical vibration traveling up the mic stand, put the mic in a shock-mount stand adapter. Or use a microphone that is less susceptible to mechanical vibration, such as an omnidirectional mic, or a unidirectional mic with a good internal shock mount.
- Use a microphone with a limited low-frequency response.
- If the cause is piano pedal thumps, also try working on the pedal mechanism.

Hum

Hum is a subject in itself. See Chapter 5 for causes and cures of hum.

Pop

Pops are explosive breath sounds in a vocalist's microphone. If your pop-music recording has pops, try these solutions:

- Place the microphone above or to the side of the mouth.
- Place a foam windscreen (pop filter) on the microphone.
- Stretch a silk stocking over a darning hoop, and mount it on a mic stand a few inches from the microphone (or use an equivalent commercial product).
- Place the microphone farther from the vocalist.
- Use a microphone with a built-in pop filter (ball grille).
- Use an omnidirectional microphone, because it is likely to pop less than a directional (cardioid) microphone.
- Switch in a highpass filter (low-cut filter) set around 80 Hz.

Sibilance

Sibilance is an overemphasis of "s" and "sh" sounds. If you are getting sibilance on your pop-music recording, try these steps:

- Use a de-esser.
- Place the microphone farther from the vocalist.
- Place the microphone toward one side of the vocalist, rather than directly in front.
- Cut equalization in the range from 5 to 10 kHz.
- Change to a duller-sounding microphone.

Bad Mix

If you get a bad mix on your recording, some instruments or voices are too loud or too quiet. To improve a bad mix, try the following:

- Change the mix. (Maybe change the mix engineer!)
- Compress vocals or instruments that occasionally get buried.
- Change the equalization on certain instruments to help them stand out.
- During mixdown, continuously change the mix to highlight certain instruments according to the demands of the music.

Unnatural Dynamics

When your pop-music recording has unnatural dynamics, loud sound don't get loud enough or soft sounds disappear. If this happens, try these steps:

- Check the tracking of noise-reduction units. For example, a 10-dB level increase at the input of the encode unit should appear as a 10-dB level increase at the output of the decode unit.
- Use the same type of noise reduction that was used during recording.
- Use less compression or limiting.
- Avoid overall compression.

Isolated Sound

If some of the instruments on your recording sound too isolated, as if they are not in the same room as the others, follow these steps:

- In general, allow a little crosstalk between the left and right channels. If tracks are totally isolated, it's hard to achieve the illusion that all the instruments are playing in the same room at the same time. You need some crosstalk or correlation between channels. Some right channel information should leak into the left channel, and vice versa.
- Place microphones farther from their sound sources to increase leakage.
- Use omnidirectional mics to increase leakage.
- Use stereo reverberation or echo.
- Pan effects returns to the channel opposite the channel of the dry sound source.
- Pan extreme left-and-right tracks slightly toward center.
- Make the effects-send levels more similar for various tracks.
- To give a lead-guitar solo a fat, spacious sound, use a stereo chorus. Or send its signal through a delay unit, pan the direct sound hard left and pan the delayed sound hard right.

Lack of Depth

If the mix lacks depth, try these steps:

- Achieve depth by miking instruments at different distances.
- Use varied amounts of reverberation on each instrument. The higher the ratio of reverberant sound to direct sound, the more distant the track sounds.

Bad Sound in a Classical Music Recording

Check the following procedures if you have problems recording classical music:

Too Dead

If the sound in your classical recording is too dead—there is not enough ambience or reverberation—try these measures to solve the problem:

- Place the microphones farther from the performers.
- Use omnidirectional microphones.
- Record in a concert hall with better acoustics (longer reverberation time).
- Add artificial reverberation.

Too Close

If the sound is too detailed, too close, or too edgy, follow these steps:

- Place the microphones farther from the performers.
- Place the microphones lower or on the floor (as with a boundary microphone).
- Roll off the high frequencies.
- Use mellow-sounding microphones (many ribbon mics have this quality).

Too Distant

If the sound is distant and there is too much reverberation, these steps might help:

- Place the microphones closer to the performers.
- Use directional microphones (such as cardioids).
- Record in a concert hall that is less live (reverberant).

Stereo Spread Imbalance

If your classical-music recording has a narrow stereo spread, try these steps:

- Angle or space the main microphone pair farther apart.
- If you're doing midside stereo recording, turn up the side output of the stereo microphone.
- Place the main microphone pair closer to the ensemble.

If the sound has excessive stereo spread (or "hole-in-the-middle"), try the following:

- Angle or space the main microphone pair closer together.
- If you're doing midside stereo recording, turn down the side output of the stereo microphone.
- In spaced-pair recording, add a microphone midway between the outer pair, and pan its signal to the center.
- Place the microphones farther from the performers.

Lack of Depth

Try the following to bring more depth into your classical-music recording:

- Use only a single pair of microphones out front. Avoid multimiking.
- If you must use spot mics, keep their level low in the mix.
- Add more artificial reverberation to the distant instruments than to the close instruments.

Bad Balance

If your classical-music recording has bad balance, try the following:

- Place the microphones higher or farther from the performers.
- Ask the conductor or performers to change the instruments' written dynamics. Be tactful!
- Add spot microphones close to instruments or sections needing reinforcement. Mix them in subtly with the main microphones' signals.

Muddy Bass

If your recording has a muddy bass sound, follow these steps:

- Aim the bass-drum head at the microphones.
- Put the microphone stands and bass-drum stand on resilient isolation mounts, or place the microphones in shock-mount stand adapters.
- Roll off the low frequencies or use a highpass filter set around 40 to 80 Hz.
- Record in a concert hall with less low-frequency reverberation.

Rumble

Sometimes your classical-music recording picks up rumble from air conditioning, trucks, and other sources. Try the following to clear this up:

- Check the hall for background rumble problems.
- Temporarily shut off the air conditioning.

- Record in a quieter location.
- Use a highpass filter set around 40 to 80 Hz.
- Use microphones with limited low-frequency response.

Distortion

If your classical-music recording has distortion, try the following:

- Switch in the pads built into the microphones (if any).
- Increase the mixer input attenuation (turn down the input trim).
- Check connectors for stray wires or bad solder joints.

Bad Tonal Balance

Bad tonal balance expresses itself in a sound that is too dull, too bright, or colored. If your recording has this problem, follow these steps:

- Change the microphones. Generally, use flat-response microphones with minimal off-axis coloration.
- Follow the 3:1 rule mentioned in Chapter 7.
- If a microphone must be placed near a hard reflective surface, use a boundary microphone to prevent phase cancellations between direct and reflected sounds.
- Adjust equalization.
- Place the mics at a reasonable distance from the ensemble (too-close miking sounds shrill).
- Avoid mic positions that pick up standing waves or room modes.
- Experiment with small changes in mic position.

This chapter described a set of standards for good sound quality in both popular-music and classical-music recordings. These standards are somewhat arbitrary, but the engineer and producer need guidelines to judge the effectiveness of the recording. The next time you hear something you don't like in a recording, the lists in this chapter will help you define the problem and find a solution.

20

MUSIC: WHY
YOU RECORD

This book focuses on techniques for recording music, without paying much attention to the music itself. Occasionally it's wise to remember that music is the main reason for recording!

Music can be exalting, exciting, soothing, sensuous, and fulfilling. It's wonderful that recordings can preserve it. As a recording engineer, it's to your advantage to better understand what music is all about.

Music starts as musical ideas or feelings in the mind and heart of its composer. Musical instruments are used to translate these ideas and feelings into sound waves. Somehow, the emotion contained in the music—the message—is coded in the vibrations of air molecules. Those sounds are converted to electricity, and stored magnetically. The composer's message manages to survive the trip through the mixing console and tape machines; the signal is transferred to disc or tape. Finally, the original sound waves are reproduced in the listening room, and miraculously the original emotion is reproduced in the listener as well.

Of course, not everyone reacts to a piece of music the same way, so the listener may not perceive the composer's intent. Still, it's amazing that anything as intangible as a thought or feeling can be conveyed by tiny magnetic patterns on a cassette tape, or by pits on a compact disc.

The point of music lies in what it's doing now, in the present. In other words, the meaning of "Doo wop she bop" is "Doo wop she bop." The meaning of an Am7 chord followed by a Fmaj7 chord is the **experience** of Am7 followed by Fmaj7.

Increasing Your Involvement in Music

Sometimes, to get involved in music, you must relax enough to lie back and listen. You have to feel unhurried, to be content to sit between your stereo speakers or wear headphones, and listen with undivided attention. Actively analyze or feel what the musicians are playing.

Music affects people much more when they are already feeling the emotion expressed in the song. For example, hearing a fast Irish reel when you're in a party mood, or hearing a piece by Debussy when you're feeling sensuous, is more moving because your feelings resonate with those in the music. When you're falling in love, any music that is meaningful to you is enhanced.

If you identify strongly with a particular song, that tells you something about yourself and your current mood. And the songs that other people identify with tell you something about them. You can understand individuals better by listening to their favorite music.

Different Ways of Listening

There are so many levels on which to listen to music—so many ways to focus attention. Try this. Play one of your favorite records several times while listening for these different aspects:

- Overall mood and rhythm
- Lyrics
- Vocal technique
- Bass line
- Drum fills
- Sound quality
- Technical proficiency of musicians
- Musical arrangement or structure
- Reaction of one musician to another's playing
- Surprises vs. predictable patterns

By listening to a piece of music from several perspectives, you'll get much more out of it than if you just hear it as background. There's a lot going on in any song that usually goes unnoticed. Sometimes you play an old familiar record and listen to the lyrics for the first time. The whole meaning of the song changes for you.

Most people react to music on the basic level of mood and rhythmic motivation. But as a recording enthusiast, you hear much more detail because your focus demands sustained critical listening. The same is true of trained musicians focusing on the musical aspects of a performance.

It's all there for anyone to hear, but you must train yourself to hear selectively, to focus attention on a particular level of the multidimensional musical event. For example, instead of just feeling excited while listening to an impressive lead-guitar solo, listen to what the guitarist is actually playing. You may hear some amazing things.

Here's one secret of really involving yourself in recorded music: Imagine yourself playing it! For example, if you're a bass player, listen to the bass line in a particular record, and imagine that you're playing the bass line. You'll hear the part as never before. Or respond to the music visually; see it as you do in the movie "Fantasia."

Follow the melody line and see its shape. Hear where it reaches up, strains, relaxes. Hear how one note leads into the next. How does the musical expression change from moment to moment?

There are times you can almost touch music. Some music has a prickly texture (many transients, emphasized high frequencies); some music is soft and sinuous (sine-wave synthesizer notes, soaring vocal harmonies); some music is airy and spacious (much reverberation).

Two Different Production Styles

Suppose you are in the control room working on a pop music mix, and you're aiming for a realistic, natural sound. Listen to the reproduced instruments and try to make them sound as if they're really playing in front of you. That is, instead of trying to make a pleasant mix or a sonically interesting recording, try to control the sound you hear to simulate real instruments—to make them believable.

It's like an artist trying to draw a still-life as realistically as possible. The artist compares the drawing to the real object, notes the difference, and then modifies the drawing to reduce the difference.

When you're striving for a natural sound, compare the recorded instrument with your memory of the real thing. How does it sound different? Turn the appropriate knob on the console that reduces the difference.

Alternatively, when you're mixing, imagine you're creating a sonic experience between the monitor speakers, rather than just reproducing instruments. Sometimes you don't want a recording to sound too realistic. If a recording is very accurate, it sounds like musical instruments, rather than just music itself.

This approach contradicts the basic edict of high fidelity—to reproduce the original performance as it sounded in the original environment.

Some songs seem to require unreal sounds. That way, you don't connect the sounds we hear with physical instruments, but with the music behind the instruments—the composer's dream or vision.

Here's one way to reproduce pure music rather than reproducing instruments playing in a room: Mike closely or record direct to avoid picking up studio ambience and then add reverb. Also add EQ, double-tracking, sampling, and special effects to make the instrument or voice slightly unreal. The idea is to make a production, rather than a documentation—a record, rather than a recording.

Try to convey the musician's intentions through the recorded sound quality. If the musician has a loving, soft message, translate that into a warm, smooth tone quality. Add a little mid-bass or slightly reduce the highs. If the musical composition suggests grandeur or space, add reverberation with a long decay time. Ask the musicians what they are trying to express through the music, and try to express that through the sound production as well.

Why Record?

Recording is a real service. Without it, people would be exposed to much less music, limited to the occasional live concert or to their own live music, played once, and forever gone.

With recordings, you can preserve a performance for thousands of listeners. You can hear an enormous variety of musical expressions whenever you want. Unlike a live concert, a record can be played over and over for analysis. Tapes or discs are also a way to achieve a sort of immortality. The Beatles may be gone, but their music lives on.

Records can even reveal your evolving consciousness as you grow and change. A tape or disc stays the same physically, but you hear it differently over the years as your perception changes. Recordings are a constant against which you measure change in yourself. Be proud that you are contributing to the recording art—it is done in the service of music.

A

DB OR NOT DB

In the studio, you must know how to set and measure signal levels, and match equipment levels. You also need to evaluate microphones by their sensitivity specs. To learn these skills, you must understand the decibel, the unit of measurement of audio level.

Definitions

In a recording studio, **level** originally meant power, and amplitude referred to voltage. Nowadays, many audio people also define "level" in terms of voltage or sound pressure, even though this terminology is not strictly correct. You should know both definitions in order to communicate.

Audio level is measured in decibels (dB). One dB is the smallest change in level that most people can hear—the just-noticeable difference. Actually, the just-noticeable difference varies from 0.1 dB to about 5 dB, depending on bandwidth, frequency, program material, and the individual. But 1 dB is generally accepted as the smallest change in level that most people can detect. A 6 to 10 dB increase in level is considered by most listeners to be "twice as loud." Sound pressure level, signal level, and change in signal level all are measured in dB.

Sound Pressure Level

Sound pressure level (SPL) is the pressure of sound vibration measured at a point. It's usually measured with a sound level meter in dB SPL (decibels of sound pressure level).

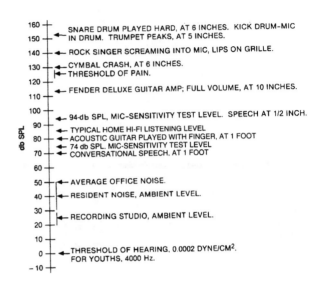

Figure A.1
A chart of sound pressure
levels.

The higher the sound pressure level, the louder the sound (Figure A.1). The quietest sound you can hear, the threshold of hearing, is 0 dB SPL. Average conversation at one foot is 70 dB SPL. Average home-stereo listening level is around 85 dB SPL. The threshold of pain—so loud that the ears hurt and can be damaged—is 125 to 130 dB SPL.

Sound pressure level in decibels is 20 times the logarithm of the ratio of two sound pressures:

$$\text{dB SPL} = 20 \log\frac{P}{P_{ref}}$$

where P is the measured sound pressure in dynes/cm², and P_{ref} is a reference sound pressure: 0.0002 dyne/cm² (the threshold of hearing).

Signal Level

Signal level also is measured in dB. The level in decibels is 10 times the logarithm of the ratio of two power levels:

$$\text{dB} = 10 \log\frac{P}{P_{ref}}$$

where P is the measured power in watts, and P_{ref} is a reference power in watts.

Recently it's become common to use the decibel to refer to voltage ratios as well:

$$dB = 20 \log\frac{V}{V_{ref}}$$

where V is the measured voltage, and V_{ref} is a reference voltage.

This expression is mathematically equivalent to the previous one, because power equals the square of the voltage divided by the circuit resistance:

$$dB = 10 \log\frac{P_1}{P_2}$$

$$= 10 \log\left(\frac{V_1^2/R}{V_2^2/R}\right)$$

$$= 10 \log\left(\frac{V_1^2}{V_2^2}\right)$$

$$= 20 \log\frac{V_1}{V_2}$$

The resistance R (or impedance) in this equation is assumed to be the same for both measurements, and thus divides out.

Signal level in decibels can be expressed in various ways, using various units of measurement:

- dBm: decibels referenced to 1 milliwatt
- dBu or dBv: decibels referenced to 0.775 volt (dBu is preferred)
- dBV: decibels referenced to 1 volt

dBm

If you're measuring **signal power,** the decibel unit to use is dBm, expressed in the equation:

$$dBm = 10 \log\frac{P}{P_{ref}}$$

where P is the measured power, and P_{ref} is the reference power (1 milliwatt).

For an example of signal power, use this equation to convert 0.01 watt to dBm:

$$= 10 \log \frac{P}{P_{ref}}$$

$$= 10 \log \frac{0.01}{0.001}$$

$$= 10$$

So, 0.01 watt is 10 dBm (10 decibels above 1 milliwatt).
Now convert 0.001 watt (1 milliwatt) into dBm:

$$dBm = 10 \log \frac{P}{P_{ref}}$$

$$= 10 \log \frac{0.001}{0.001}$$

$$= 0$$

So, 0 dBm = 1 milliwatt. This has a bearing on voltage measurement as well.
Any voltage across any resistance that results in 1 milliwatt is 0 dBm. This relationship can be expressed in the equation:

$$0 \text{ dBm} = \frac{V^2}{R} = 1 \text{ milliwatt}$$

where V = the voltage in volts, and R is the circuit resistance in ohms.
For example, 0.775 volt across 600 ohms is 0 dBm. One volt across 1000 ohms is 0 dBm. Each results in 1 milliwatt.
Some voltmeters are calibrated in dBm. The meter reading in dBm is accurate only when you're measuring across 600 ohms. For an accurate dBm measurement, measure the voltage and circuit resistance, then calculate:

$$dBm = 10 \log \frac{(V^2/R)}{0.001}$$

dBv or dBu

Another unit of measurement expressing the relationship of decibels to voltage is dBv or dBu. This means decibels referenced to 0.775 volt. This figure comes from 0 dBm, which equals 0.775 volt across 600 ohms (because 600 ohms used to be a standard impedance for audio connections):

$$dBu = 20 \log \frac{V}{V_{ref}}$$

where V_{ref} is 0.775 volt

dBV

Signal level also is measured in dBV (with a capital V), or decibels referenced to 1 volt:

$$dBV = 20 \log \frac{V}{V_{ref}}$$

where V_{ref} is 1 volt.
For example, use this equation to convert 1 millivolt (0.001 volt) to dBV:

$$dBV = 20 \log \frac{V}{V_{ref}}$$
$$= 20 \log \frac{0.001}{1}$$
$$= -60$$

So, 1 millivolt = –60 dBV (60 decibels below 1 volt).
Now convert 1 volt to dBV:

$$dBV = 20 \log \frac{1}{1}$$
$$= 0$$

So, 1 volt = 0 dBV.
To convert dBV to voltage, use the formula

$$Volts = 10^{(dBV/20)}$$

Change in Signal Level

Decibels also are used to measure the change in power or voltage across a fixed resistance. The formula is:

$$dB = 10 \log \frac{P_1}{P_2}$$

or

$$dB = 20 \log \frac{V_1}{V_2}$$

where P_1 is the new power level, P_2 is the old power level, V_1 is the new voltage level, and V_2 is the old voltage level.

For example, if the voltage across a resistor is 0.01 volt, and it changes to 1 volt, the change in dB is

$$dB = 20 \log \frac{V_1}{V_2}$$

$$= 20 \log \frac{1}{0.01}$$

$$= 40 \, dB$$

Doubling the **power** results in an increase of 3 dB; doubling the **voltage** results in an increase of 6 dB.

The VU Meter, Zero VU, and Peak Indicators

A VU meter is a voltmeter of specified transient response, calibrated in volume units or VU. It shows approximately the relative volume or loudness of the measured audio signal.

The VU-meter scale is divided into volume units, which are not necessarily the same as dB. The volume unit corresponds to the decibel only when measuring a steady sine wave tone. In other words, a change of 1 VU is the same as a change of 1 dB **only** when a steady tone is applied.

Most recording engineers use 0 VU to define a convenient "zero reference level" on the VU meter. When the meter on your mixer or recorder reads "0" on a steady tone, your equipment is producing a certain level at its output. Different types of equipment produce different levels when the meter reads 0 (Figure A.2). Zero VU corresponds to:

- +8 dBm in older broadcast and telephone equipment
- +4 dBm in balanced recording equipment
- –10 dBV in unbalanced recording equipment

378

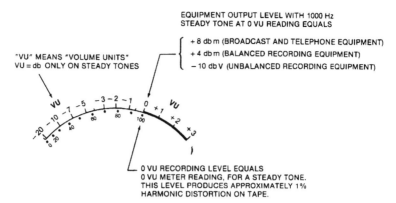

EQUIPMENT OUTPUT LEVEL WITH 1000 Hz
STEADY TONE AT 0 VU READING EQUALS

+ 8 db m (BROADCAST AND TELEPHONE EQUIPMENT)
+ 4 db m (BALANCED RECORDING EQUIPMENT)
− 10 db V (UNBALANCED RECORDING EQUIPMENT)

"VU" MEANS "VOLUME UNITS"
VU = db ONLY ON STEADY TONES

0 VU RECORDING LEVEL EQUALS
0 VU METER READING, FOR A STEADY TONE.
THIS LEVEL PRODUCES APPROXIMATELY 1%
HARMONIC DISTORTION ON TAPE.

Figure A.2 VU-meter scale.

When a tape operator says to a mixing engineer, "Send me a 0 VU tone," it means, "Send me a tone that reads 0 on your VU meter." The signal level itself isn't too important because the tape operator receiving the tone just wants to match the tape-deck meters to those on the console.

A **0 VU recording level** (0 on the record level meter) is the normal operating level of an analog tape recorder; it produces the desired recorded flux on tape. A "0 VU recording level" does not mean a "0 VU signal level."

With a VU meter, 0 VU corresponds to a recording level 8 dB below the level that produces 3 percent third-harmonic distortion on tape at 400 Hz. Distortion at 0 VU typically is below 1 percent.

The response of a VU meter is not fast enough to track rapid transients accurately. In addition, when a complex waveform is applied to a VU meter, the meter reads less than the peak voltage of the waveform. (This means you must allow for undisplayed peaks above 0 VU that use up headroom.)

In contrast, a peak indicator responds quickly to peak program levels, making it a more accurate indicator of recording levels. One type of peak indicator is an LED that flashes on peak overloads. Another is the LED bar-graph meter commonly seen on cassette decks. Yet another is the PPM (peak program meter). It is calibrated in dB, rather than VU. Unlike the VU meter reading, the PPM reading does not correlate with perceived volume.

In a digital recorder, the meter is an LED or LCD bargraph meter that reads up to 0 dBFS (FS means Full Scale). In theory, 0 dBFS means all 16 bits are ON. The OVER indication means that the input level exceeded the voltage needed to produce 0 dBFS, and there is some short-duration clipping of the output analog waveform. Some manufacturers calibrate their meters so that 0 dBFS is less than 16 bits ON; this allows a little headroom.

Balanced vs. Unbalanced Equipment Levels

Generally, audio equipment with balanced (3-pin) connectors works at a higher nominal line level than equipment with unbalanced (phono) connectors. There's nothing inherent in balanced or unbalanced connections that makes them operate at different levels; they're just standardized at different levels.

These are the nominal (normal) input and output levels for the two types of equipment:

- Balanced: +4 dBm (1.23 volts)
- Unbalanced: −10 dBV (0.316 volt)

In other words, when a balanced-output recorder reads 0 VU on its meter with a steady tone, it is producing 1.23 volts at its output connector. This voltage is called +4 dBm when referenced to 1 milliwatt. When an unbalanced-output recorder reads 0 on its meter with a steady tone, it is usually producing 0.316 volt at its output connector. This voltage is called −10 dBV when referenced to 1 volt.

Interfacing Balanced and Unbalanced Equipment

There's a difference of 11.8 dB between +4 dBm and −10 dBV. To find this, convert both levels to voltages:

$$dB = 20 \log\left(\frac{1.23}{0.316}\right)$$

$$= 11.8.$$

So, +4 dBm is 11.8 dB higher in voltage than −10 dBV (assuming the resistances are the same).

A cable carrying a nominal +4 dBm signal has a signal-to-noise ratio (S/N) 11.8 dB better than the same cable carrying a −10 dBV signal. This is an advantage in environments with strong radio frequency or hum fields. But in most studios with short cables, the difference is negligible.

Connecting a +4 dBm output to a −10 dBV input might cause distortion if the signal peaks of the +4 equipment exceed the headroom of the −10 equipment. If this happens, use a pad to attenuate the level 12 dB (Figure A.3). The

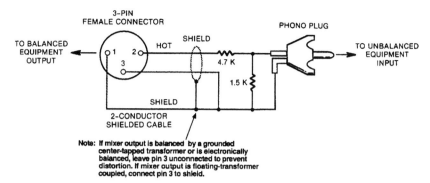

Figure A.3. Use a pad of 12 dB to match a balanced +4dBm output to an unbalanced −10dBV input.

pad converts from balanced to unbalanced as well as reducing the level 12 dB. You may have to substitute a stereo phone plug for the 3-pin connector.

You don't always need that pad. Many pieces of equipment have a **+4/–10 level switch**. Set the switch to the nominal level of the connected equipment.

Microphone Sensitivity

Decibels are an important concern in another area: microphone sensitivity. A high-sensitivity mic puts out a stronger signal (higher voltage) than a low-sensitivity mic when both are exposed to the same sound pressure level.

A **microphone-sensitivity specification** tells how much output (in volts) a microphone produces for a certain input (in SPL). The standard is millivolts per pascal, where one pascal (Pa) is 94 dB SPL.

A typical "open-circuit sensitivity" spec is 5.5 mV/Pa for a condenser mic and 1.8 mV/Pa for a dynamic mic. "Open-circuit" means that the mic is **unloaded** (not connected to a load, or connected to a mic preamp with a very high input impedance). If the spec is 5.5 mV/Pa, that means the mic produces 5.5 mV when the SPL at the mic is 94 dB SPL.

If you put a microphone in a 20 dB louder sound field, it produces 20 dB more signal voltage. For example, if 74 dB SPL in gives 0.18 mV out (−75 dBV), then 94 dB SPL in gives 1.8 mV out (−55 dBV). 150 dB SPL in gives 1.1 volt out (+1 dBV), which is approximately line level! That's why you need so much input padding when you record a kick drum or other loud source.

B

INTRODUCTION
TO SMPTE TIME CODE

Have you ever wished you had more tracks? Suppose you've filled up all the tracks of a 16-track recorder, the band you're recording wants to over-dub several more instruments, and you don't have a 24-track machine.

There is a solution: Synchronize the 16-track machine with an 8-track machine by using SMPTE time code. This is a special signal recorded on tape that can sync together two multitrack recorders so that they operate as one. Time code also can synchronize an audio recorder with a video recorder, or even an automated mixing system to a multitrack recorder.

SMPTE stands for the Society of Motion Picture and Television Engineers. The SMPTE standardized the time code signal for use in video production, and you can use it in audio recording as well. SMPTE time code is something like a digital tape counter, where the counter time is recorded as a signal on tape.

How the Time Code Works

A **time-code generator** creates the time code signal (a 1200 Hz modulated square wave). You record—or **stripe**—this signal onto one track of both recorders. A **time-code reader** reads the code off the two tapes. Then a **time-code synchronizer** compares the codes from the two transports and locks them together in time by varying the motor speed of one of the transports.

The counter time is recorded as a signal on tape. Pictures on a video screen are updated approximately 30 frames per second, where a frame is a still picture made of 525 lines on the screen. SMPTE time code assigns a unique number (address) to each video frame—8 digits that specify HOURS:MINUTES:SECONDS:FRAMES.

Each video frame is identified with its own time code address; for example, **01:26:13:07** means "1 hour, 26 minutes, 13 seconds, and 7 frames." These addresses are recorded sequentially: for each successive video frame, the time code number increases by one frame count. There are approximately 30 frames per second in the American TV system, so the time code counts frames from 0 to 29 each second.

Time Code Signal Details

The SMPTE time code is a data stream that is divided into **code words**. Each code word includes 80 binary digits (or **bits**) that identify each video frame (Figure B.1).

The 80-bit time code word is synchronized to the start of each video frame. The code uses binary 1s and 0s. During each half-cycle of the square wave, the voltage may be constant (signifying a 0) or changing (signifying a 1). That is, a voltage transition in the middle of a half-cycle of the square wave equals a 1. No transition signifies a 0. This is called **biphase modulation** (Figure B.2). It can be read forward or reverse, at almost any tape speed. A time code reader detects the binary 1's and 0's, and converts them to decimal numbers to form the time code addresses.

SMPTE words also can include user information. There are 32 multi-purpose bits (8 digits or 4 characters) reserved for the user's data, for example, the take number.

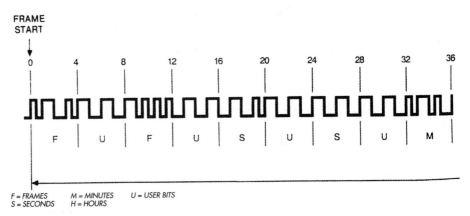

FRAME
START

0   4   8   12   16   20   24   28   32   36

F | U | F | U | S | U | S | U | M

F = FRAMES   M = MINUTES   U = USER BITS
S = SECONDS   H = HOURS

Figure B.1 An 80-bit time code word.

The last 16 bits in the word are a fixed number of 1s and 0s called **sync bits**. These bits indicate the end of the time code word, so that the time code reader can tell whether the code is being read forward or in reverse.

Drop-Frame Mode

SMPTE code can run in various modes depending on the application. One of these, Drop-Frame mode, is needed for specific reasons.

Black-and-white video runs at 30 frames/sec. A time code signal also running at 30 frames/sec will agree with the clock on the wall. Color video, on the other hand, runs at 29.97 frames/sec. If a color program is clocked at 30 frames/sec for one hour, the actual show length will run 3.6 seconds (108 frames) longer than an hour.

The **Drop-Frame mode** causes the time code to count at a rate to match the clock on the wall. Each minute, frame numbers 00 and 01 are dropped, except every 10th minute. (Instead of seeing frames ...27, 28, 29, 00 on the counter; you see frames ...27, 28, 29, 02.) This speeds up the time code counter to match the rate of the video frames. The video frames still progress at 29.97 frames/sec, and the time code progresses at 30 frames/sec, but it drops every few frames—so the effective time code frame rate is 29.97 frames/sec.

You program the time-code generator to operate in **Drop** (Drop-Frame) or **Non-Drop** (Non-Drop-Frame) mode. Non-Drop can be used for audio-only synchronizing, but Drop mode should be used if the audio will be synched to a video tape later on.

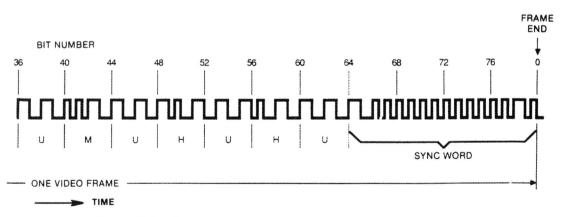

Figure B.1 Continued.

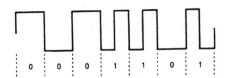

Figure B.2
Biphase modulation used in
SMPTE time code.

Setting Up a Time Code System

To use the SMPTE time code, you need a time-code generator, reader, and synchronizer. These may be all-in-one or separate units. Figure B.3 shows a typical system hookup, in which the generator, reader, and synchronizer are combined in one unit.

Set the generator to Time-of-Day code, or any other convenient starting time. If you are synching to video, feed the generator a sync signal from the video source being recorded. This will lock the generator together in time with the video source. For audio-only applications, use the internal crystal sync.

Select Drop-Frame or Non-Drop-Frame mode, and stay with it for the entire production. Use Drop-Frame mode if you anticipate synching audio to video in the future.

Next, set the frame rate: 29.97, 30, 24, or 25 frames/second, using the following guidelines:

MASTER

SMPTE
IN

SMPTE
OUT

SLAVE

SMPTE
IN

SMPTE
OUT

CONTROL
INTERFACE
CABLE

SMPTE
IN

SMPTE
TIME CODE
OUT

11:14:25:09

SMPTE
IN

SYNCHRONIZER

Figure B.3
Typical hookup for synchronizing two tape transports
with SMPTE time code.

- Color video productions require 29.97 frames/sec.
- Black-and-white video or audio-only productions use 30 frames/sec.
- Film usually runs at 24 frames/sec.
- European TV—using EBU (European Broadcast Union) time code—requires 25 frames/sec.

The time code signal appears at the generator output, which is a standard 3-pin audio connector. Signal level is +4 dBm. The signal is fed through a standard 2-conductor shielded audio cable. To avoid crosstalk of time code into audio channels, separate the time code cables from audio cables. Patch the time code signal into an outside track of the recorders you want to lock together. Then, patch the outputs of those time code tracks to the inputs of the time code reader.

The reader decodes the information recorded on tape and, in some models, displays the time code data in HOURS:MINUTES:SECONDS:FRAMES format. Some readers have an **error bypass** feature which corrects for missing data.

The time-code synchronizer matches bits between two time code signals to synchronize them. The synchronizer compares tape direction, address, and phase to synchronize two SMPTE tracks via servo control of the transport motors. The two tape machines to be synched are called "master" and "slave." The synchronizer controls the slave by making its tape position and speed follow that of the master.

Connect the shielded multipin interface cable between the synchronizer and slave machine to control the slave's tape transport and motors. This interface cable has channels for controlling the capstan motor, tape direction, shuttle modes, and tachometer (more on the tach later).

Because the time code signal becomes very high in frequency when the tape is shuttled rapidly, special playback amplifier cards with extended high-frequency response may be needed to reproduce the SMPTE signal accurately. These cards are available from the recorder manufacturer.

Unfortunately, when the tape is in shuttle mode (fast forward or rewind), the tape usually is lifted from the heads—losing the SMPTE signal. In this case, the recorders are synchronized using tach pulses from the recorders as a replacement for the SMPTE time code. Some synchronizers are fed tach pulses from the slave only.

If **chase mode** is available, the slave follows the shuttle motions of the master. If the master is put in fast-forward, the slave goes into fast-forward, and so on. Without chase mode, the synchronizer notes the address of the master tape when it is stopped and cues the slave to match that location. Chase mode is useful for repetitive overdubs.

How to Use the SMPTE Time Code

Suppose you want to synchronize two multitrack recorders. Follow this procedure:

1. Clean the heads and the tape path.
2. Record the SMPTE time code on an outside track of both recorders at –5 to –10 VU, leaving the adjacent track blank, if possible, to avoid time code crosstalk. Don't put high-transient sounds (such as drums) on that adjacent track; they can cause sync problems in analog tape machines.
3. Start recording, or striping, the code about 20 seconds before the music starts, and continue non-stop with no breaks in the signal. Stripe the two tapes simultaneously. If that is not possible, you need a **time-code editor** to correct or insert an offset.
4. During playback, manually cue the slave to approximately the same point as the master tape, using time code address information as a reference.
5. Engage the synchronizer in Lock and Chase mode, and enable it.
6. Put both recorders in Play mode.
7. Adjust the slave's tape speed to gradually reduce the error between transports to less than one time code frame.

With some synchronizers, this operation is automatic. You set the slave tape to approximately the same point as the master tape. Then put the master in play. When the synchronizer detects master time code, it sets the slave machine in play mode and, in a few seconds, adjusts the slave's speed to synchronize the two recorders. This condition is called **locked up**.

When you record on two synchronized transports, try not to split stereo pairs between two tapes. The slight time differences between machines can degrade stereo imaging. Keep all stereo pairs on the same tape, copying them if necessary onto the other tape.

Restriping Defective Code

You may encounter degraded or erased sections on a time code track. This lost code must be replaced with good code in proper sequence. If you need to rerecord (restripe) a defective SMPTE track, use the **Jam Sync** mode on the time code generator. This feature produces new code which matches the original addresses and frame count.

For example, suppose the slave tape needs to be restriped. Follow this procedure:

1. Patch the slave's time code track into the generator set to Jam Sync mode.
2. Patch the generator output into the time code track input on the slave machine (or into another track).
3. Play the tape. The time code reader built into the generator detects a section of good code and initializes the generator with that information.
4. Start recording the new, regenerated code over the bad data (or on a new track).

Jam Sync also should be used when you copy a tape containing time code. With Jam Sync in operation, the code is regenerated to create a clean copy. This procedure is preferable to copying the time code track directly because each generation can distort the code signal.

Audio-for-Video SMPTE Applications

With the advent of music videos and other audio/video combinations, there's a widespread need to sync audio to video. Studios doing soundtrack work for film or video can use SMPTE time code to synchronize sound and picture for overdubbing narration, dialog, lip-sync, music, environmental sounds, or sound effects.

Synchronizing to Video

Running audio and videotapes in synchronization for TV audio editing is a typical postproduction method. You can edit the audio and video portions of a program independently even though they are locked together in time.

When you sync audio and video, select **Longitudinal Time Code** (LTC) or **Vertical Interval Time Code** (VITC). Longitudinal code records along the length of an audio track on the video tape. Vertical Interval code is combined with the video signal and is placed in the vertical blanking interval—the black bar seen over the TV picture when it is rolling vertically. VITC frees up an audio track for other purposes.

If you record the time code signal on an audio or cue track of the videotape, do not use automatic level control because it may distort the SMPTE waveform. Instead, adjust the time code signal level manually.

When you play an audio tape synched to video, the time code track on the audio tape will be delayed with respect to the video's code due to the spacing between the record and playback heads in the audio recorder. This delay (about 5 frames) can be corrected by the **offset** function in the synchronizer.

Some time code systems include a **character inserter** which displays the address on the video monitor. If desired, these addresses can be recorded with (**burned into**) the picture, a feature called **window dub**.

The Audio-Tape Synchronization Procedure

At a typical on-location video shoot, the video from the camera(s) is recorded on a videocassette recorder, while the audio from the microphones is recorded on a separate high-quality tape recorder (such as a Nagra) or a portable DAT with SMPTE time code capability. Both video and audio tapes are prestriped with SMPTE time code so that they can be synchronized later in postproduction.

Back at the studio, you connect the audio and video decks for SMPTE sync as described earlier. When you play the videotape, the SMPTE time code locks the picture and sound together. You can equalize the audio tape or change levels, and then lay it back (copy it) to the videocassette.

If you sync video to a multitrack tape recorder, you can run the video over and over as you refine the mix. Update your mix moves with an automated mixer or automated mixing program. Finally, when the mix is satisfactory, record the mixer output signal onto the video tape.

This procedure eliminates the dubbing step when transferring the audio soundtrack to videotape. That is, you can mix the multitrack tape master directly to the videotape (keeping sync), rather than mixing down to 2-track and dubbing that to videotape.

The Film/Video Soundtrack Program

SMPTE can be used with a film/video soundtrack program you run on a computer. This software package automates the playback of music and sound-effects cues for motion-picture and video postproductions. Using SMPTE time code, you can synchronize audio events, such as sound effects, to film or videotape. You create a **cue list** (also known as an **Edit Decision List** or **EDL**) of audio events, each with its own time code address. These events are played by MIDI instruments, or are played from recordings on a computer hard disk.

Some programs let you record the video and audio onto hard disk. The video shows up in a window on your computer screen. While watching the video, you note the SMPTE times where you want sound effects or music to occur. In this way you create the sound track and sync it to video, all in your computer. Then dump the edited soundtrack to a prestriped videocassette.

Soundtrack Program Features

The following are some of the many tasks that are offered in a soundtrack program.

- Spot and layback sound effects (turn them on at the proper times and record them onto the videotape).
- Do an automated mix.
- Backtime events.
- Repeat events.
- Name events.
- Print cue lists, libraries, and recording logs.
- Map keyboards (show each effect's location on a piano-style keyboard).
- Cut/copy/paste, insert, and delete events.
- Enter cue locations by tapping on the space bar as the program progresses.
- Display sound-effect cues graphically along with the music.
- Convert sequencer files to a cue sheet (hit list).
- Indicate both SMPTE time code and bar/beat for each event.
- Expand or compress the duration of a soundtrack.
- Enter subtle tempo variations, or introduce time offsets, to make cues fall exactly on the beat.
- Write standard MIDI files with meter, beat, tempo and event data for use in a sequencer.
- Lock to MIDI Time Code and SMPTE.

You also can perform various other tasks found in sequencer programs, such as quantization, tap-in tempo, and so on.

Digital Audio Editing

It's becoming common to record the elements of a soundtrack—dialog, ambience, music, and sound effects—into a digital audio workstation, then edit the soundtrack with that. Either SMPTE or MIDI time code keeps the soundtrack and video in sync. With MIDI time code, you can cue MIDI devices to play music and effects at various SMPTE times relative to the video program. This process can go through various steps.

1. First, you're handed a **work tape**, which is a videotape of the program you're working on. It has SMPTE time code already striped on tape, and the SMPTE time code appears in a window on-screen called a window dub. As stated before, some programs let you copy the video onto your hard disk.

2. Watch the picture. Using a MIDI keyboard workstation, compose musical parts related to the video scenes and their SMPTE start/stop times. These times indicate how long the music needs to be for each musical segment. This music can be recorded with a MIDI sequencer or multitrack recorder synched to the video recorder.

3. Dialog and wild sound are often recorded on a portable 2-track recorder with a center track for time code. (Wild sound is ambient noise recorded on the set while shooting, not synched with the video.) Sync the dialog tape to the videotape.

4. Missing or poorly recoded dialog is replaced during a process called **looping** or **ADR** (**Automatic Dialog Replacement**). Have actors watch the work tape and lip-sync their lines onto a synched multi-track audio recorder.

5. Record sound effects so that they line up with corresponding events in the video. Audition several sound effects from CD libraries, pick the ones you like, and sample them into the digital audio workstation.

6. Using slow motion or freeze frame, go through the video and note the SMPTE times where each sound effect should occur.

7. Using a sequencer program, enter the cue point (SMPTE time) for each sound effect. This creates the EDL.

8. Play the video. The sequencer runs down the cue list. Each cue point (SMPTE time code address) triggers the sampler to play a corresponding sound effect or music at the correct time.

9. If you wish, you redo the SMPTE time cues or change the effects independently.

Other Time Code Applications

SMPTE time code allows video editing under computer control. In editing a video program, you copy program segments from two or more video-tapes onto a third recorder. On a computer you specify the **edit points** (time code addresses) where you want to switch from one video source to another. You can rehearse edits as often as required.

Time code is used also as an index for locating cue points on tape. During a mixdown, you can use these cue points to indicate where to make changes in the mix.

Time code also can be used as a reference for console automation and MIDI instruments. With this latter application, MIDI synthesizers can be cued to any point within a sequence, rather than having to start at the beginning. By using SMPTE time code to lock together multiple audio or video transports, you can greatly expand your operating flexibility.

C

FURTHER EDUCATION

The following books, magazines, and literature are recommended to anyone desiring further education in recording technology.

Books

Music Books Plus

A catalog that offers many audio-related books. Their Web site is http://www.musicbooksplus.com. Address: 2315 Whirlpool St. #132, Niagara Falls, NY 14305, phone 800-265-8481.

Mix Bookshelf

A major audio books catalog. It's at 100 Newfield Ave., Edison, NJ 08837-3817. Phone 800-233-9604, international phone 908-417-9575. Some topics covered are the music business, recording, live sound, audio, electronic music, and so on. Their Web site is http://www.mixbookshelf.com.

For a detailed explanation of stereo theory and stereo mic techniques, see **Stereo Microphone Techniques** by Bruce Bartlett, published by Focal Press.

Recording Magazines

Electronic Musician
6400 Hollis St.
Emeryville, CA 94608
510-653-3307
(Home and project studio recording.)

EQ
Miller Freeman PSN, Inc.
460 Park Ave. South, 9th Floor
New York, NY 10016-7315
212-378-0400
http://www.eqmag.com
(Home, project, and pro recording.)
(Also offers a recording and sound buyer's guide.)
Gig
P.O. Box 0532
Baldwin, NY 11510
212-378-0449
(Musician's magazine with some recording and live sound columns.)
Mix
6400 Hollis St. #12
Emeryville, CA 94608
415-653-3307
(Pro recording.)
(Mix also offers a master directory of the pro audio industry.)
Recording
3849 Ventura Canyon
Sherman Oaks, CA 91423
619-738-5571
(Home and project studio recording.)

In the back of those magazines are ads related to recording products and services.

Pro Audio Magazines

Journal of the Audio Engineering Society (JAES)
60 E. 42nd St., Room 2520
New York, NY 10165-2520
(Pro audio engineering. Theoretical and academic.)
Live Sound
112 Market St.
Sun Prairie, WI 53590-2986
608-837-2200
Pro Audio Review
IMAS Publishing
P.O. Box 1214
Falls Church, VA 22041
703-998-7600
(Reviews of pro audio equipment.)

Radio World Newspaper
IMAS Publishing
P.O. Box 1214
Falls Church, VA 22041
703-998-7600
(Engineering and management tips for radio broadcasters.)

Religion
Box 35
20 Wellington St.
E. Aurora, Ontario, Canada L4G 3KA
416-841-5200
(Audio for churches.)

Consumer Audio Magazines

Audio
1633 Broadway
New York, NY 10019
212-767-6332
(Consumer audio.)

Stereo Review
1633 Broadway
New York, NY 10019
(Consumer audio.)

Guides, Brochures, and Other Literature

Careers in Audio Engineering and the **Journal of the Audio Engineering Society**, from the Audio Engineering Society, 60 E. 42nd St., New York, NY 10165.

Microphone application guides are available from AKG Acoustics Inc., 77 Selleck St., Stamford, CT 06902; Audio-Technica U.S. Inc., 1221 Commerce Drive, Stow, Ohio 44224; Countryman Associates Inc., 417 Stanford Ave., Redwood City, CA 94063; Crown International, 1718 W. Mishawaka Rd., Elkhart, IN 46517; Sennheiser Electronic Corp., 6 Vista Drive, P.O. Box 987, Old Lyme, CT 06371; and Shure Brothers Inc., 222 Hartrey Ave, Evanston, IL 60202.

You can find a lot of valuable information in user manuals and free sales literature provided by manufacturers of recording equipment. Ask your equipment dealer for manufacturer's phone numbers.

The International MIDI Association has MIDI technical information for sale, such as the MIDI 1.0 specification and a 50-page detailed explanation of MIDI. International MIDI Association, 5316 W. 57th St., Los Angeles, CA 90056. Phone 213-649-MIDI. Their Web site is http://www.ima.org.

Guides to Recording Schools

Each July issue of **Mix** magazine contains a comprehensive directory of recording schools, seminars, and programs. Universities and colleges in most major cities have recording-engineering courses. Investigate them thoroughly, however, before making a decision.

A complete reference guide to audio education is the book **New Ears: A Guide to Education in Audio and the Recording Sciences, 2nd Ed.**, edited by Mark Drews, and published by New Ear Productions of Syracuse, New York (1992). It's all here: course descriptions of all the known recording schools; directories of industry-related magazines, journals, and textbooks; professional audio, music, and broadcasting associations; audio research facilities, non-profit studios, and other helpful resources. The book is available from New Ear Productions, 1033 Euclid Ave., Syracuse, NY 13210.

The Internet

A good place to ask questions, besides magazines, is on the Internet. Check out the newsgroup rec.audio.pro. You can ask questions and get answers from pro engineers. You may get conflicting answers, because often there are many ways to do the same thing. Also, some who reply are more expert than others. But you'll often find stimulating debates.

Do an Internet Web search with the keywords audio, recording, DAT, digital audio, and so on. Also search for the name and model number of equipment you have, or are interested in. You'll discover hundreds of audio-related Web sites and links.

Recording Equipment Catalogs

Here are but a few catalogs from which to order recording gear:

American Musical Supply
600 Industrial Ave.
Paramus, NJ 07652
800-458-4076
http://www.americanmusical.com

Bananas At Large
1504 Fourth St.
San Rafael, CA 94901-2713
415-457-7600
http://www.bananas.com

Manny's Mailbox Music
156 W. 48th St.
New York, NY 10036
800-448-8478

Music Emporium
12401 Twinbrook Parkway
Rockville, MD 20852
800-648-8460

Musician's Friend
P.O. Box 4520
Medford, OR 97501
800-776-5173
http://www.musiciansfriend.com

Sweetwater Sound
5335 Bass Rd.
Ft. Wayne, IN 46808
219-432-8176
http://www.sweetwater.com

The Woodwind and the Brasswind
19880 State Line Road
South Bend, IN 46637
800-348-5003

GLOSSARY

A WEIGHTING—*See* Weighted.

A-B—A listening comparison between two audio programs, or between two components playing the same program, performed by switching immediately from one to the other. The levels of the two signals are matched. *See also* Spaced-Pair.

ACCENT MICROPHONE—*See* Spot Microphone.

ACCESS JACKS—Two jacks in a console input module or output module that allow access to points in the signal path, usually for connecting a compressor. Plugging into the access jacks breaks the signal flow and allows you to insert a signal processor in series with the signal.

ACTIVE COMBINING NETWORK—A combining network with gain. *See* Combining Network.

AES—Audio Engineering Society.

AES/EBU—Also called IEC 988 Type 1, an interface format for digital signals, using a balanced 110 ohm mic cable terminated with XLR-type connectors. *See also* S/PDIF.

ALIGNMENT—The adjustment of tape-head azimuth and of tape-recorder circuitry to achieve optimum performance from the particular type of tape being used.

ALIGNMENT TAPE—A prerecorded tape with calibrated tones for alignment of a tape recorder.

AMBIENCE—Room acoustics, early reflections, and reverberation. Also, the audible sense of a room or environment surrounding a recorded instrument.

AMBIENCE MICROPHONE—A microphone placed relatively far from its sound source to pick up ambience.

AMPLITUDE, PEAK—On a graph of a sound wave, the sound pressure of the waveform peak. On a graph of an electrical signal, the voltage of the waveform peak. The amplitude of a sound wave or signal as measured on a meter is 0.707 times the peak amplitude.

ANALOG-TO-DIGITAL (A/D) CONVERTER—A circuit that converts an analog audio signal into a stream of digital data (bit stream).

ASSIGN—To route or send an audio signal to one or more selected channels.

ATRAC—Adaptive Transform Acoustic Coding. A data compression scheme that reduces by 5:1 the storage needed for digital audio. ATRAC is a perceptual coding method, which omits data deemed inaudible due to masking.

ATTACK—The beginning of a note. The first portion of a note's envelope in which a note rises from silence to its maximum volume.

ATTACK TIME—In a compressor, the time it takes for gain reduction to occur in response to a musical attack.

ATTENUATE—To reduce the level of a signal.

ATTENUATOR—In a mixer (or mixing console) input module, an adjustable resistive network that reduces the microphone signal level to prevent overloading of the input transformer and mic preamplifier.

AUTOLOCATE—A recorder function which makes the tape or disk go to a program address (counter time) at the press of a button.

AUTOMATED MIXING—A system of mixing in which a computer remembers and updates console settings. With this system, a mix can be performed and refined in several stages and played back at a later date exactly as set up previously.

AUXILIARY BUS (AUX-BUS)—*See* Effects Bus.

AUXILIARY SEND (AUX-SEND)—*See* Effects Send.

A/V DRIVE—A hard disk drive meant for audio/video use. It postpones thermal recalibration until the disk is inactive, preventing data errors.

AZIMUTH—In a tape recorder, the angular relationship between the head gap and the tape path.

AZIMUTH ALIGNMENT—The mechanical adjustment of the record or playback head to bring it into proper alignment (90 degrees) with the tape path.

BACK-TIMING—A technique of cueing up the musical background or a sound effect to a narration track so that the music or effect ends simultaneously with the narration.

BAFFLED-OMNI—A stereo miking arrangement that uses two ear-spaced omnidirectional microphones separated by a hard or padded baffle.

BALANCE—The relative volume levels of various tracks or instruments.

BALANCED AC POWER—AC power from a center-tapped power transformer. Instead of one 120V line and one 0V line, it has two 60V lines. They are in phase with each other, and sum to 120V. But they are connected to the center-tap ground out of phase (one is +60V; the other is −60V). Any hum and noise on the grounding system cancels out.

BALANCED LINE—A cable with two conductors surrounded by a shield, in which each conductor is at equal impedance to ground. With respect to ground, the conductors are at equal potential but opposite polarity; the signal flows through both conductors.

BANDPASS FILTER—In a crossover, a filter that passes a band or range of frequencies but sharply attenuates or rejects frequencies outside the band.

BASIC TRACKS—Recorded tracks of rhythm instruments (bass, guitar, drums, and sometimes keyboard).

BASS TRAP—An assembly that absorbs low-frequency sound waves in the studio.

BIAMPLIFICATION (BIAMPING)—Driving a woofer and tweeter with separate power amplifiers. An active crossover is connected ahead of these power amplifiers.

BIAS—In tape-recorder electronics, an ultrasonic signal that drives the erase head. This signal is also mixed with the audio signal applied to the record head to reduce distortion.

BIDIRECTIONAL MICROPHONE—A microphone that is most sensitive to sounds arriving from two directions—in front of and behind the microphone. It rejects sounds approaching either side of the microphone. Sometimes called a cosine or figure-eight microphone because of the shape of its polar pattern.

BINAURAL RECORDING—A 2-channel recording made with an omnidirectional microphone mounted near each ear of a human or a dummy head, for playback over headphones. The object is to duplicate the acoustic signal appearing at each ear.

BLUMLEIN ARRAY—A stereo microphone technique in which two coincident bidirectional microphones are angled 90 degrees apart (45 degrees to the left and right of center).

BOARD—*See* Mixing Console.

BOUNCING TRACKS—A process in which two or more tracks are mixed, and the mixed tracks are recorded on an unused track or tracks. Then the original tracks can be erased, which frees them up for recording more instruments.

BOUNDARY MICROPHONE—A microphone designed to be used on a boundary (a hard reflective surface). The microphone capsule is mounted very close to the boundary so that direct and reflected sounds arrive at the microphone diaphragm in phase (or nearly so) for all frequencies in the audible band.

BREATHING—The unwanted audible rise and fall of background noise that may occur with a compressor. Also called pumping.

BULK TAPE ERASER—A large electromagnet used to erase a whole reel of recording tape at once.

BUS—A common connection of many different signals. An output of a mixer or submixer. A channel that feeds a tape track, signal processor, or power amplifier.

BUS IN—An input to a program bus, usually used for effects returns.

BUS MASTER—In the output section of a mixing console, a potentiometer (fader or volume control) that controls the output level of a bus.

BUS OUT—The output connector of a bus.

BUS TRIM—A control in the output section of a mixing console that provides variable gain control of a bus, used in addition to the bus master for fine adjustment.

BUZZ—An unwanted edgy tone that sometimes accompanies audio, containing high harmonics of 60 Hz.

CALIBRATION—*See* Alignment.

CAPACITOR—An electronic component that stores an electric charge. It is formed of two conductive plates separated by an insulator called a dielectric. A capacitor passes AC but blocks DC.

CAPACITOR MICROPHONE—*See* Condenser Microphone.

CAPSTAN—In a tape-recorder transport, a rotating post that contacts the tape (along with the pinch roller) and pulls the tape past the heads at a constant speed during recording and playback.

CARDIOID MICROPHONE—A unidirectional microphone with side attenuation of 6 dB and maximum rejection of sound at the rear of the microphone (180 degrees off-axis). A microphone with a heart-shaped directional pattern.

CD-R—CD-Recordable, a recordable compact disc that cannot be rewritten. Once recorded, it cannot be erased and reused.

CD-RW—CD-Rewritable, a recordable compact disc that can be rewritten. Once recorded it can be erased and reused.

CHANNEL—A single path of an audio signal. Usually, each channel contains a different signal.

CHANNEL ASSIGN—*See* Assign.

CHORUS—1. A special effect in which a signal is delayed by 15 to 35 milliseconds, the delayed signal is combined with the original signal, and the delay is varied randomly or periodically. This creates a wavy, shimmering effect. 2. The main portion of a song that is repeated several times throughout the song with the same lyrics.

CLEAN—Free of noise, distortion, overhang, leakage. Not muddy.

CLEAR—Easy to hear, easy to differentiate. Reproduced with sufficient high frequencies.

COINCIDENT-PAIR—A stereo microphone, or two separate microphones, placed so that the microphone diaphragms occupy approximately the same point in space. They are angled apart and mounted one directly above the other.

COMB-FILTER EFFECT—The frequency response caused by combining a sound with its delayed replica. The frequency response has a series of peaks and dips caused by phase interference. The peaks and dips resemble the teeth of a comb.

COMBINING AMPLIFIER—An amplifier at which the outputs of two or more signal paths are mixed together to feed a single track of a tape recorder.

COMBINING NETWORK—A resistive network at which the outputs of two or more signal paths are mixed together to feed a single track of a tape recorder.

COMPING—Recording composite tracks.

COMPLEX WAVE—A wave with more than one frequency component.

COMPOSITE TRACKS—The process of recording several performances of a musical part on different tracks, so that the best segments of each performance can be played in sequence during mixdown.

405

COMPRESSION—1. The portion of a sound wave in which molecules are pushed together, forming a region with higher-than-normal atmospheric pressure. 2. In signal processing, the reduction in dynamic range or gain caused by a compressor. *See also* Data Compression.

COMPRESSION RATIO (SLOPE)—In a compressor, the ratio of the change in input level (in dB) to the change in output level (in dB). For example, a 2:1 ratio means that for every 2 dB change in input level, the output level changes 1 dB.

COMPRESSOR—A signal processor that reduces dynamic range or gain by means of automatic volume control. An amplifier whose gain decreases as the input signal level increases above a preset point.

CONDENSER MICROPHONE—A microphone that works on the principle of variable capacitance to generate an electrical signal. The microphone diaphragm and an adjacent metallic disk (called a backplate) are charged to form two plates of a capacitor. Incoming sound waves vibrate the diaphragm, varying its spacing to the backplate, which varies the capacitance, which in turn varies the voltage between the diaphragm and backplate.

CONNECTOR—A device that makes electrical contact between a signal-carrying cable and an electronic device, or between two cables. A device used to connect or hold together a cable and an electronic component so that a signal can flow from one to the other.

CONSOLE—*See* Mixing Console.

CONTACT PICKUP—A transducer that contacts a musical instrument and converts its mechanical vibrations into a corresponding electrical signal.

CONTROL ROOM—The room in which the engineer controls and monitors the recording. It houses most of the recording hardware.

CROSSOVER—An electronic network that divides an incoming signal into two or more frequency bands.

CROSSOVER, ACTIVE (ELECTRONIC CROSSOVER)—A crossover with amplifying components, used ahead of the power amplifiers in a biamped or triamped speaker system.

CROSSOVER FREQUENCY—The single frequency at which both filters of a crossover network are down 3 dB.

CROSSOVER, PASSIVE—A crossover with passive (nonamplifying) components, used after the power amplifier.

CROSSTALK—The unwanted transfer of a signal from one channel to another. Crosstalk often occurs between adjacent tracks within a record or playback head in a tape recorder, or between input modules in a console.

CUE, CUE SEND—In a mixing-console input module, a control that adjusts the level of the signal feeding the cue mixer that feeds a signal to headphones in the studio.

CUE LIST—*See* Edit Decision List.

CUE MIXER—A submixer in a mixing console that takes signals from cue sends as inputs and mixes them into a composite signal that drives headphones in the studio.

CUE SHEET—Used during mixdown, a chronological list of mixing-console control adjustments required at various points in the recorded song. These points may be indicated by tape-counter or ABS-time readings.

CUE SYSTEM—A monitor system that allows musicians to hear themselves and previously recorded tracks through headphones.

DAMPING FACTOR—The ability of a power amplifier to control or damp loudspeaker vibrations. The lower the amplifier's output impedance, the higher the damping factor.

DAT (R-DAT)—A digital audio tape recorder that uses a rotating head to record digital audio on tape.

DATA COMPRESSION—A scheme for reducing the amount of data storage on a medium. *See* ATRAC.

DAW—Abbreviation for digital audio workstation.

dB—Abbreviation for decibel. *See also* Decibel.

DEAD—Having very little or no reverberation.

DECAY—The portion of the envelope of a note in which the envelope goes from maximum to some midrange level. Also, the decline in level of reverberation over time.

DECAY TIME—*See* Reverberation Time.

DECIBEL (dB)—The unit of measurement of audio level. Ten times the logarithm of the ratio of two power levels. Twenty times the logarithm of the ratio of two voltages. dBV is decibels relative to 1 volt. dBu is decibels relative to 0.775 volt. dBm is decibels relative to 1 milliwatt. dBA is decibels, A weighted (*See* Weighted).

DECODED TAPE—A tape that is expanded after being compressed by a noise-reduction system. Such a tape has normal dynamic range.

DE-ESSER—A signal processor that removes excessive sibilance ("s" and "sh" sounds) by compressing high frequencies around 5 to 10 kHz.

DELAY—The time interval between a signal and its repetition. A digital delay or a delay line is a signal processor that delays a signal for a short time.

DEMAGNETIZER (DEGAUSSER)—An electromagnet with a probe tip that is touched to elements of the tape path (such as tape heads and tape guides) to remove residual magnetism.

DEPTH—The audible sense of nearness and farness of various instruments. Instruments recorded with a high ratio of direct-to-reverberant sound are perceived as being close; instruments recorded with a low ratio of direct-to-reverberant sound are perceived as being distant.

DESIGN CENTER—The portion of fader travel (usually shaded), about 10 to 15 dB from the top, in which console gain is distributed for optimum headroom and signal-to-noise ratio. During normal operation, each fader in use should be placed at or near design center.

DESIGNATION STRIP—A strip of paper taped near console faders to designate the instrument that each fader controls.

DESK—The British term for mixing console.

DESTRUCTIVE EDITING—In a digital audio workstation, editing that rewrites the data on disk. A destructive edit cannot be undone.

DI—Short for direct injection, recording with a direct box.

DIFFUSION—An even distribution of sound in a room.

DIGITAL AUDIO—An encoding of an analog audio signal in the form of binary digits (ones and zeros).

DIGITAL AUDIO WORKSTATION (DAW)—A computer, sound card, and editing software that allows you to record, edit, and mix audio programs entirely in digital form. Stand-alone DAWs include real mixer controls; computer DAWS have virtual controls on-screen.

DIGITAL RECORDING—A recording system in which the audio signal is stored in the form of binary digits.

DIGITAL-TO-ANALOG CONVERTER—A circuit that converts a digital audio signal into an analog audio signal.

DIM—To reduce the monitor volume temporarily by a preset amount so that you can carry on a conversation.

DIRECT BOX—A device used for connecting an amplified instrument directly to a mixer mic input. The direct box converts a high-impedance unbalanced audio signal into a low-impedance balanced audio signal.

DIRECT INJECTION (DI)—Recording with a direct box.

DIRECT OUTPUT, DIRECT OUT—An output connector following a mic preamplifier, fader, and equalizer, used to feed the signal of one instrument to one track of a tape recorder.

DIRECT SOUND—Sound traveling directly from the sound source to the microphone (or to the listener) without reflections.

DIRECTIONAL MICROPHONE—A microphone that has different sensitivity in different directions. A unidirectional or bidirectional microphone.

DISTORTION—An unwanted change in the audio waveform, causing a raspy or gritty sound quality. The appearance of frequencies in a device's output signal that were not in the input signal. Distortion is caused by recording at too high a level, improper mixer settings, components failing, or vacuum tubes distorting. (Distortion can be desirable—for an electric guitar, for example.)

DOLBY TONE—A reference tone recorded at the beginning of a Dolby-encoded tape for alignment purposes.

DOUBLING—A special effect in which a signal is combined with its 15 to 35 millisecond delayed replica. This process mimics the sound of two identical voices or instruments playing in unison. In another type of doubling, two identical performances are recorded and played back to thicken the sound.

DROP-FRAME—For color video production, a mode of SMPTE time code that causes the time code to match the clock on the wall. Once every minute, frame numbers 00 and 01 are dropped, except every 10th minute.

DROP-OUT—During playback of a tape recording, a momentary loss of signal caused by separation of the tape from the playback head by dust, tape-oxide irregularity, etc.

DRUM MACHINE—A device that plays samples of real drums, and includes a sequencer to record rhythm patterns.

DRY—Having no echo or reverberation. Referring to a close-sounding signal that is not yet processed by a reverberation or delay device.

DSP—Abbreviation for Digital Signal Processing, modifying a signal in digital form.

DVD—Digital Versatile Disc. A storage medium the size of a compact disc which holds much more data. The DVD stores video, audio, or computer data.

DYNAMIC MICROPHONE—A microphone that generates electricity when sound waves cause a conductor to vibrate in a stationary magnetic field. The two types of dynamic microphone are moving coil and ribbon. A moving-coil microphone is usually called a dynamic microphone.

DYNAMIC RANGE—The range of volume levels in a program from softest to loudest.

EARTH GROUND—A connection to moist dirt (the ground we walk on). This connection is usually done via a long copper rod or an all-metal cold-water pipe.

ECHO—A delayed repetition of a signal or sound. A sound delayed 50 milliseconds or more, combined with the original sound.

ECHO CHAMBER—A hard-surfaced room containing a widely separated loudspeaker and microphone, once used for creating reverberation.

EDIT DECISION LIST (EDL)—A list of program events in order, plus their starting and ending times.

EDITING—The cutting and rejoining of magnetic tape to delete unwanted material, to insert leader tape, or to rearrange recorded material into the desired sequence. Also, the same actions performed with a digital audio workstation, hard-disk recorder, or MiniDisc recorder-mixer—without cutting any tape.

EDITING BLOCK—A metal block that holds magnetic tape during the editing/splicing procedure.

EFFECTS—Interesting sound phenomena created by signal processors, such as reverberation, echo, flanging, doubling, compression, or chorus.

EFFECTS BUS—The bus that feeds effects devices (signal processors).

EFFECTS LOOP—A set of connectors in a mixer for connecting an external effects unit, such as a reverb or delay device. The effects loop includes a send section and a receive section. *See* Effects Send, Effects Return.

EFFECTS MIXER—A submixer in a mixing console that combines signals from effects sends, and then feeds the mixed signal to the input of a special-effects device, such as a reverberation unit.

EFFECTS RETURN (EFFECTS RECEIVE)—In the output section of a mixing console, a control that adjusts the amount of signal received from an effects unit. Also, the connectors in a mixer to which you connect the effects-unit output signal. They might be labeled "bus in" instead. The effects-return signal is mixed with the program bus signal.

EFFECTS SEND—In an input module of a mixing console, a control that adjusts the amount of signal sent to a special-effects device, such as a reverberation or delay unit. Also, the connector in a mixer which you connect to the input of an effects unit. The effects-send control normally adjusts the amount of reverberation or echo heard on each instrument.

EFFICIENCY—In a loudspeaker, the ratio of acoustic power output to electrical power input.

EIA—Electrical Industries Association.

EIA RATING—A microphone-sensitivity specification that states the microphone output level in dBm into a matched load for a given Sound Pressure Level (SPL). SPL + dB (EIA rating) = dBm output into a matched load.

ELECTRET-CONDENSER MICROPHONE—A condenser microphone in which the electrostatic field of the capacitor is generated by an electret—a material that permanently stores an electrostatic charge.

ELECTROSTATIC FIELD—The force field between two conductors charged with static electricity.

ELECTROSTATIC INTERFERENCE—The unwanted presence of an electrostatic hum field in signal conductors.

ENCODED TAPE—A tape containing a signal compressed by a noise-reduction unit.

END-ADDRESSED—Referring to a microphone whose main axis of pickup is perpendicular to the front of the microphone. You aim the front of the mic at the sound source. *See* Side-Addressed.

ENVELOPE—The rise and fall in volume of one note. The envelope connects successive peaks of the waves comprising a note. Each harmonic in the note might have a different envelope.

EQUALIZATION (EQ)—The adjustment of frequency response to alter the tonal balance or to attenuate unwanted frequencies.

EQUALIZER—A circuit (usually in each input module of a mixing console, or in a separate unit) that alters the frequency spectrum of a signal passed through it.

ERASE—To remove an audio signal from magnetic tape by applying an ultrasonic varying magnetic field so as to randomize the magnetization of the magnetic particles on the tape.

ERASE HEAD—A head in a tape recorder that erases the signal on tape.

EXPANDER—1. A signal processor that increases the dynamic range of a signal passed through it. 2. An amplifier whose gain decreases as its input level decreases. When used as a noise gate, an expander reduces the gain of low-level signals to reduce noise between notes.

FADE-OUT—To gradually reduce the volume of the last several seconds of a recorded song, from full level down to silence, by slowly pulling down the master fader.

FADER—A linear or sliding potentiometer (volume control), used to adjust signal level.

FEED—1. To send an audio signal to some device or system. 2. An output signal sent to some device or system.

FEEDBACK—1. The return of some portion of an output signal to the system's input. 2. The squealing sound you hear when a PA system microphone picks up its own amplified signal through a loudspeaker.

FEED REEL—The left-side reel on a tape recorder that unwinds during recording or playback.

FILTER—1. A circuit that sharply attenuates frequencies above or below a certain frequency. Used to reduce noise and leakage above or below the frequency range of an instrument or voice. 2. A MIDI Filter removes selected note parameters.

FLANGING—A special effect in which a signal is combined with its delayed replica, and the delay is varied between 0 and 20 milliseconds. A hollow, swishing, ethereal effect like a variable-length pipe, or like a jet plane passing overhead. A variable comb filter produces the flanging effect.

FLETCHER MUNSON EFFECT—Named after the two people who discovered it, the psychoacoustical phenomenon in which the subjective frequency response of the ear changes with program level. Due to this effect, a program played at a lower volume than the original level subjectively loses bass and treble.

FLOAT—To disconnect from ground.

FLUTTER—A rapid periodic variation in tape speed.

FLUTTER ECHOES—A rapid series of echoes that occurs between two parallel walls.

FLUX—Magnetic lines of force.

FLUXIVITY—The measure of the flux density of a magnetic recording tape, per unit of track width.

FLY-IN (LAY-IN)—To copy part of a recorded track onto another recorder, then re-record that copy back onto the original multitrack tape in a different part of the song, in sync with other recorded tracks. For example, copy the vocal track from the first chorus of the song onto an external recorder or sampler. Rerecord (fly-in) that copy onto the multitrack tape at the second chorus. Then the first and second choruses have identical vocal performances.

FOLDBACK (FB)—*See* Cue System.

FREQUENCY—The number of cycles per second of a sound wave or an audio signal, measured in hertz (Hz). A low frequency (for example, 100 Hz) has a low pitch; a high frequency (for example, 10,000 Hz) has a high pitch.

FREQUENCY RESPONSE—1. The range of frequencies that an audio device will reproduce at an equal level (within a tolerance, such as +/–3 dB). 2. The range of frequencies that a device (mic, human ear, etc.) can detect.

FULL TRACK—A single tape track recorded across the full width of a tape.

FUNDAMENTAL—The lowest frequency in a complex wave.

GAIN—Amplification. The ratio, expressed in decibels, between the output voltage and the input voltage, or between the output power and the input power.

GAP—In a tape-recorder head, the thin break in the electromagnet that contacts the tape.

GATE—1. To turn off a signal when its amplitude falls below a pre-set value. 2. The signal-processing device used for this purpose. *See also* Noise Gate.

GATED REVERB—Reverberation with the reverberant "tail" cut off before it fades out.

GENERATION—A copy of a tape or a bounce of a track. A copy of the original master recording is a first generation tape. A copy made from the first generation tape is a second generation, and so on.

GENERATION LOSS—The degradation of signal quality (the increase in noise and distortion) that occurs with each successive generation of a tape recording.

GOBO—A moveable partition used to prevent the sound of an instrument from reaching another instrument's microphone. Short for go-between.

GRAPHIC EQUALIZER—An equalizer with a horizontal row of faders; the fader-knob positions indicate graphically the frequency response of the equalizer. Usually used to equalize monitor speakers for the room they are in. Sometimes used for complex EQ of a track.

GROUND—The zero-signal reference point for a system of audio components.

GROUND BUS—A common connection to which equipment is grounded, usually a heavy copper plate.

GROUND LOOP—1. A loop or circuit formed of ground leads. 2. The loop formed when unbalanced components are connected together via two ground paths—the connecting-cable shield and the power ground. Ground loops cause hum and should be avoided.

GROUNDING—Connecting pieces of electronic equipment to ground. Proper grounding ensures that there is no voltage difference between equipment chassis. An electrostatic shield needs to be grounded to be effective.

GROUP—*See* Submix.

GUARD BAND—The spacing between tracks on a multitrack tape or tape head, used to prevent crosstalk.

HALF-TRACK—A tape track recorded across approximately half the width of a tape. A half-track recorder usually records two such tracks simultaneously in the same direction to make a stereo recording.

HARD DISK—A random-access storage medium for computer data. A hard disk drive contains a stack of magnetically coated hard disks that are read by, and written to by, an electromagnetic head.

HARD DISK RECORDER—A device dedicated to recording digital audio on a hard disk drive. A hard disk recorder-mixer includes a built-in mixer.

HARMONIC—An overtone whose frequency is a whole-number multiple of the fundamental frequency.

HARMONIZER—A signal processor that provides a wide variety of pitch-shifting and delay effects.

HEAD—An electromagnet in a tape recorder that either erases the audio signal on tape, records a signal on tape, or plays back a signal that is already on tape.

HEAD GAP—*See* Gap.

HEADPHONES—A head-worn transducer that covers the ears and converts electrical audio signals into sound waves.

HEADROOM—The safety margin, measured in decibels, between the signal level and the maximum undistorted signal level. In a tape recorder, the dB difference between standard operating level (corresponding to a 0 VU reading) and the level causing 3 percent total harmonic distortion. High-frequency headroom increases with analog tape speed.

HERTZ (Hz)—Cycles per second, the unit of measurement of frequency.

HIGHPASS FILTER—A filter that passes frequencies above a certain frequency and attenuates frequencies below that same frequency. A low-cut filter.

HISS—A noise signal containing all frequencies, but with greater energy at higher octaves. Hiss sounds like wind blowing through trees. It is usually caused by random signals generated by microphones, electronics, and magnetic tape.

HOT—1. A high recording level causing slight distortion, maybe used for special effect. 2. A condition in which a chassis or circuit has a potentially dangerous voltage on it. 3. Referring to the conductor in a microphone cable which has a positive voltage on it at the instant that sound pressure moves the diaphragm inward.

HUM—An unwanted low-pitched tone (60 Hz and its harmonics) heard in the monitors. The sound of interference generated in audio circuits and cables by AC power wiring. Hum pickup is caused by such things as faulty grounding, poor shielding, and ground loops.

HYPERCARDIOID MICROPHONE—A directional microphone with a polar pattern that has 12 dB attenuation at the sides, 6 dB attenuation at the rear, and two nulls of maximum rejection at 110 degrees off axis.

IMAGE—An illusory sound source located somewhere around the listener. An image is generated by two or more loudspeakers. In a typical stereo system, images are located between the two stereo speakers.

IMPEDANCE—The opposition of a circuit to the flow of alternating current. Impedance is the complex sum of resistance and reactance. Abbreviated as Z.

INPUT—The connection going into an audio device. In a mixer or mixing console, a connector for a microphone, line-level device, or other signal source.

INPUT ATTENUATOR—*See* Attenuator.

INPUT MODULE—In a mixing console, the set of controls affecting a single input signal. An input module usually includes an attenuator (trim), fader, equalizer, effects send, cue send, and channel-assign controls.

INPUT SECTION—The row of input modules in a mixing console.

INPUT/OUTPUT (I/O) CONSOLE (IN-LINE CONSOLE)—A mixing console arranged so that input and output sections are aligned vertically. Each module (other than the monitor section) contains one input channel and one output channel.

INSERT JACKS—*See* Access Jacks.

JACK—A female or receptacle-type connector for audio signals into which a plug is inserted.

KEYBOARD WORKSTATION—Several MIDI components in one chassis—a keyboard, a sample player, a sequencer, and perhaps a synthesizer and disk drive.

KILO—A prefix meaning one thousand. Abbreviated k.

LAY-IN—*See* Fly-In.

LEADER TAPE—Plastic or paper tape without an oxide coating, used for a spacer between takes (for silence between songs).

LEADERING—The process of splicing leader tape between program selections.

LEAKAGE—The overlap of an instrument's sound into another instrument's microphone. Also called bleed or spill.

LED INDICATOR—A recording-level indicator using one or more Light Emitting Diodes.

LEDE—Abbreviation for Live-End/Dead-End, a type of control room acoustic treatment in which the front half of the control room prevents early reflections to the mixing position, while the back half of the control room reflects diffused sound to the mixing position.

LEVEL—The degree of intensity of an audio signal—the voltage, power, or sound pressure level. The original definition of level is the power in watts.

LEVEL SETTING—In a recording system, the process of adjusting the input-signal level to obtain maximum level on the recording media without distortion. A VU meter or other indicator shows recording level.

LIMITER—A signal processor whose output is constant above a preset input level. A compressor with a compression ratio of 10:1 or greater, with the threshold set just below the point of distortion of the following device. Used to prevent distortion of attack transients or peaks.

LINE LEVEL—In balanced professional recording equipment, a signal whose level is approximately 1.23 volts (+4 dBm). In unbalanced equipment (most home hi-fi or semipro recording equipment), a signal whose level is approximately 0.316 volt (–10 dBV).

LIVE—1. Having audible reverberation. 2. Occurring in real-time, in person.

LIVE RECORDING—A recording made at a concert. Also, a recording made of a musical ensemble playing all at once, rather than overdubbing.

LOCALIZATION—The ability of the human hearing system to tell the direction of a real or illusory sound source.

LOOP—In a sampling program, to play the sustain portion of a sound's envelope repeatedly. In a sequencer or DAW, to play a few measures of drum patterns or music repeatedly.

LOUDSPEAKER—A transducer that converts electrical energy (the signal) into acoustical energy (sound waves).

LOWPASS FILTER—A filter that passes frequencies below a certain frequency and attenuates frequencies above that same frequency. A high-cut filter.

M—Abbreviation for mega, or one million (as in megabytes).

MAGNETIC RECORDING TAPE—A recording medium made of magnetic particles (usually ferric oxide) suspended in a binder and coated on long strip of thin plastic (usually Mylar).

MASK—To hide or cover up one sound with another sound. To make a sound inaudible by playing another sound along with it.

MASTER FADER—A volume control that affects the level of all program buses simultaneously. It is the last stage of gain adjustment before the 2-track recorder.

MASTER TAPE—A completed tape used to generate tape copies or compact discs.

MD—Abbreviation for MiniDisc.

MDM—Abbreviation for Modular Digital Multitrack.

MEMORY—A group of integrated circuit chips used to store digital data temporarily or permanently (such as an audio signal in digital format).

MEMORY REWIND—A tape-recorder function that rewinds the tape to a preset tape-counter position.

METER—A device that indicates voltage, resistance, current, or signal level.

MIC—An abbreviation for microphone.

MIC LEVEL—The level or voltage of a signal produced by a microphone, typically 2 millivolts.

MIC PREAMP—*See* Preamplifier.

MICROPHONE—A transducer or device that converts an acoustical signal (sound) into a corresponding electrical signal.

MICROPHONE TECHNIQUES—The selection and placement of microphones to pick up sound sources.

MIDI—Abbreviation for Musical Instrument Digital Interface, a specification for a connection between synthesizers, drum machines, and computers that allows them to communicate with and/or control each other.

MIDI CHANNEL—A route for transmitting and receiving MIDI signals. Each channel controls a separate MIDI musical instrument or synth patch. Up to 16 channels can be sent on a single MIDI cable.

MIDI CONTROLLER—A musical performance device (keyboard, drum pads, breath controller, etc.) that outputs a MIDI signal designating note numbers, note on, note off, and so on.

MIDI IN—A connector in a MIDI device that receives MIDI messages.

MIDI INTERFACE—A circuit that plugs into a computer, and converts MIDI data into computer data for storage in memory or on hard disk. The interface also converts computer data into MIDI data.

MIDI OUT—A connector in a MIDI device that transmits MIDI messages.

MIDI THRU—A connector in a MIDI device that duplicates the MIDI information at the MIDI-In connector. Used to connect another MIDI device in the series.

MID-SIDE—A coincident-pair stereo microphone technique using a forward-facing unidirectional, omnidirectional, or bidirectional mic and a side-facing bidirectional mic. The microphone signals are summed and differenced to produce right- and left-channel signals.

MIKE—To pick up with a microphone.

MILLI—A prefix meaning one thousandth, abbreviated m.

MINIDISC (MD)—A rewritable, magneto-optical storage medium that is read by a laser. It resembles a compact disc in a 2.5-inch square housing. MD recorders use a data compression scheme called ATRAC.

MIX—1. To combine two or more different signals into a common signal. 2. A control on an effects unit that varies the ratio between the dry signal and the effected signal.

MIXDOWN—The process of playing recorded tracks through a mixer and mixing them to (usually) two stereo channels for recording on a two-track recorder.

MIXER—A device that mixes or combines audio signals and controls the relative levels of the signals.

MIXING CONSOLE—A large mixer with additional functions such as equalization or tone control, pan pots, monitoring controls, solo functions, channel assigns, and control of signals sent to external signal processors.

MODULAR DIGITAL MULTITRACK (MDM)—A multitrack tape recorder that records 8 tracks digitally on a videocassette. Several 8-track modules can be connected together to add more tracks in sync. Two examples of MDMs are the Alesis ADAT-XT and TASCAM DA-38.

MONAURAL—Referring to listening with one ear. Often incorrectly used to mean monophonic.

MONITOR—A loudspeaker in a control room, or headphones, used for judging sound quality. Also, a video display screen used with a computer.

MONITORING—Listening to an audio signal with a monitor.

MONO, MONOPHONIC—1. Referring to a single channel of audio. A monophonic program can be played over one or more loudspeakers, or one or more headphones. 2. Describing a synthesizer that plays only one note at a time (not chords).

419

MONO-COMPATIBLE—A characteristic of a stereo program, in which the program channels can be combined to a mono program without altering the frequency response or balance. A mono-compatible stereo program has the same frequency response in stereo or mono because there is no delay or phase shift between channels to cause phase interference.

MOVING-COIL MICROPHONE—A dynamic microphone in which the conductor is a coil of wire moving in a fixed magnetic field. The coil is attached to a diaphragm which vibrates when struck with sound waves. Usually called a dynamic microphone.

M-S RECORDING—*See* Mid-Side.

MUDDY—Unclear sounding; having excessive leakage, reverberation, or overhang.

MULTIEFFECTS PROCESSOR—*See* Multiprocessor.

MULTIPLE-D MICROPHONE—A directional microphone which has multiple sound-path lengths between its front and rear sound entries. This type of microphone has minimal proximity effect.

MULTIPROCESSOR—A signal processor that can perform several different signal-processing functions.

MULTITIMBRAL—In a synthesizer, the ability to produce two or more different patches or timbres at the same time.

MULTITRACK—Referring to a recorder or tape-recorder head that has more than two tracks.

MUTE—To turn off an input signal on a mixing console by disconnecting the input-module output from channel assign and direct out. During mixdown, the mute function is used to reduce tape noise during silent portions of tracks, or to turn off unused performances. During recording, mute is used to turn off mic signals.

NEAR COINCIDENT—A stereo microphone technique in which two directional microphones are angled apart symmetrically on either side of center and spaced a few inches apart horizontally.

NEAR-FIELD MONITORING—A monitor-speaker arrangement in which the speakers are placed very near the listener (usually just behind the mixing console) to reduce the audibility of control-room acoustics.

NOISE—Unwanted sound, such as hiss from electronics or tape. An audio signal with an irregular, non-periodic waveform.

NOISE GATE—A gate used to reduce or eliminate noise between notes.

NOISE-REDUCTION SYSTEM—A signal processor (Dolby or dbx) used to reduce tape hiss (and sometimes print-through) caused by the recording process. Some of these systems compress the signal during recording and expand it in a complementary fashion during playback.

NON-DESTRUCTIVE EDITING—In a digital audio workstation, editing done by changing pointers (location markers) to information on the hard disk. A non-destructive edit can be undone.

NONLINEAR—1. Referring to a storage medium in which any data point can be accessed or read almost instantly. Examples are a hard disk, compact disc, and MiniDisc. *See* Random Access. 2. Referring to an audio device that is distorting the signal.

OCTAVE—The interval between any two frequencies where the upper frequency is twice the lower frequency.

OFF-AXIS—Not directly in front of a microphone or loudspeaker.

OFF-AXIS COLORATION—In a microphone, the deviation from the on-axis frequency response that sometimes occurs at angles off the axis of the microphone. The coloration of sound (alteration of tone quality) for sounds arriving off-axis to the microphone.

OMNIDIRECTIONAL MICROPHONE—A microphone that is equally sensitive to sounds arriving from all directions.

ON-LOCATION RECORDING—A recording made outside the studio, in a room or hall where the music usually is performed or practiced.

OPEN TRACKS—On a multitrack tape recorder, tracks that have not yet been used, or have already been bounced and are available for use.

ORTF—Named after the French broadcasting network (Office de Radiodiffusion Television Française), a near-coincident stereo mic technique which uses two cardioid mics angled 110 degrees apart and spaced 17 cm horizontally.

OUTBOARD EQUIPMENT—Signal processors that are external to the mixing console.

OUTPUT—A connector in an audio device from which the signal comes, and feeds successive devices.

OUT-TAKE—A take, or section of a take, that is to be removed or not used.

OVERDUB—To record a new musical part on an unused track in synchronization with previously recorded tracks.

OVERHANG—The continuation of a signal at the output of a device after the input signal has ceased. Sometimes called ringing.

OVERLOAD—The distortion that occurs when an applied signal exceeds a system's maximum input level.

OVERTONE—In a complex wave, a frequency component that is higher than the fundamental frequency.

PAD—*See* Attenuator.

PAN POT—Abbreviation for panoramic potentiometer. In each input module in a mixing console, a control that divides a signal between two channels in an adjustable ratio. By doing so, a pan pot controls the location of a sonic image between a stereo pair of loudspeakers.

PARAMETRIC EQUALIZER—An equalizer with continuously variable parameters, such as frequency, bandwidth, and amount of boost or cut.

PATCH—1. To connect one piece of audio equipment to another with a cable. 2. A setting of synthesizer parameters to achieve a sound with a certain timbre.

PATCH BAY (PATCH PANEL)—An array of connectors, usually in a rack, to which equipment inputs and outputs are wired. A patch bay makes it easy to interconnect various pieces of equipment in a central, accessible location.

PATCH CORD—A short cable with a phone plug on each end, used for signal routing in a patch bay. Stereo (tip-ring-sleeve) phone plugs are used with balanced signals.

PEAK—On a graph of a sound wave or signal, the highest point in the waveform where voltage or sound pressure is highest. On a frequency-response curve, a high point in the curve.

PEAK AMPLITUDE—*See* Amplitude, Peak.

PEAK PROGRAM METER (PPM)—A meter that responds fast enough to closely follow the peak levels in a program.

PEAKING EQUALIZER—An equalizer that provides maximum cut or boost at one frequency, so that the resulting frequency response of a boost resembles a mountain peak.

PERIOD—The time between the peak of one wave and the peak of the next. The time between corresponding points on successive waves. Period is the inverse of frequency.

PERSONAL STUDIO—A minimal group of recording equipment set up for one's personal use, usually using a 4-track recorder-mixer. Also, a portable 4-track recorder-mixer for one's personal use.

PERSPECTIVE—In the reproduction of a recording, the audible sense of distance to the musical ensemble, the point of view. A close perspective has a high ratio of direct sound to reverberant sound; a distant perspective has a low ratio of direct sound to reverberant sound.

PFL—Abbreviation for Pre-Fader Listen. *See also* Solo.

PHANTOM POWER—A DC voltage (usually 12 to 48 volts) applied to microphone signal conductors to power condenser microphones.

PHASE—The degree of progression in the cycle of a wave, where one complete cycle is 360 degrees.

PHASE CANCELLATION, PHASE INTERFERENCE—The cancellation of certain frequency components of a signal that occurs when the signal is combined with its delayed replica. At certain frequencies, the direct and delayed signals are of equal level and opposite polarity (180 degrees out of phase), and when combined, they cancel out. The result is a comb-filter frequency response having a periodic series of peaks and dips. Phase interference can occur between the signals of two microphones picking up the same source at different distances, or can occur at a microphone picking up both a direct sound and its reflection from a nearby surface.

PHASE SHIFT—The difference in degrees of phase angle between corresponding points on two waves. If one wave is delayed with respect to another, there is a phase shift between them of $2\pi FT$, where $\pi = 3.14$, F = frequency in Hz, and T = delay in seconds.

PHASING—A special effect in which a signal is combined with its phase-shifted replica to produce a variable comb-filter effect. *See also* Flanging.

PHONE PLUG—A cylindrical, co-axial plug (usually 1/4-inch diameter). An unbalanced phone plug has a tip for the hot signal and a sleeve for the shield or ground. A balanced phone plug has a tip for the hot signal, a ring for the return signal, and a sleeve for the shield or ground.

PHONO PLUG—A coaxial plug with a central pin for the hot signal and a ring of pressure-fit tabs for the shield or ground. Also called RCA plug.

PICKUP—A piezoelectric transducer that converts mechanical vibrations to an electrical signal. Used in acoustic guitars, acoustic basses, and fiddles. Also, a magnetic transducer in an electric guitar that converts string vibration to a corresponding electrical signal.

PINCH ROLLER—In a tape-recorder transport, the rubber wheel that pinches or traps the tape between itself and the capstan, so that the capstan can move the tape.

PING-PONGING—*See* Bouncing Tracks.

PINK NOISE—A noise signal containing all frequencies (unless band-limited), with equal energy per octave. Pink noise is a test signal used for equalizing a sound system to the desired frequency response, and for testing loudspeakers.

PITCH—The subjective lowness or highness of a tone. The pitch of a tone usually correlates with the fundamental frequency.

PITCH CONTROL—A control on a tape recorder that varies the tape speed (or buffer readout speed), thereby varying the pitch of the playback signal. The pitch control can be used to match the pitch of prerecorded instruments with that of an instrument to be overdubbed. It is also used for special effects, such as "chipmunk voices," and to play prerecorded tracks slowly so that fast musical passages can be overdubbed more easily.

PITCH SHIFTER—A signal processor that changes the pitch of a signal without changing its duration.

PLAYBACK EQUALIZATION—In tape-recorder electronics, fixed equalization applied to the signal during playback to compensate for certain losses.

PLAYBACK HEAD—The head in a tape recorder that picks up a recorded magnetic signal from the moving tape and converts it to a corresponding electrical signal. The playback head is not the same as the sel-sync or sync head.

PLAYLIST—*See* Edit Decision List.

PLUG—A male connector that inserts into a jack.

PLUG-IN—Software effects that you install in your computer. The plug-in software becomes part of another program you are using, such as a digital editing program.

POLAR PATTERN—The directional pickup pattern of a microphone. A plot of microphone sensitivity vs. angle of sound incidence. Examples of polar patterns are omnidirectional, bidirectional, and unidirectional. Subsets of unidirectional are cardioid, supercardioid, and hypercardioid.

POLARITY—Referring to the positive or negative direction of an electrical, acoustical, or magnetic force. Two identical signals in opposite polarity are 180 degrees out-of-phase with each other at all frequencies.

424

POLYPHONIC—Describing a synthesizer that can play more than one note at a time (chords).

POP—1. A thump or little explosion sound heard in a vocalist's microphone signal. Pop occurs when the user says words with "p," "t," or "b" so that a turbulent puff of air is forced from the mouth and strikes the microphone diaphragm. 2. A noise heard when a mic is plugged into a monitored channel, or when a switch is flipped.

POP FILTER—A screen placed on a microphone grille that attenuates or filters out pop disturbances before they strike the microphone diaphragm. Usually made of open-cell plastic foam or silk, a pop filter reduces pop and wind noise.

PORTABLE STUDIO—A combination recorder and mixer in one portable chassis.

POST-ECHO—A repetition of a sound, following the original sound, caused by print-through.

POWER AMPLIFIER—An electronic device that amplifies or increases the power level fed into it to a level sufficient to drive a loudspeaker.

POWER GROUND (SAFETY GROUND)—A connection to the power company's earth ground through the U-shaped hole in a power outlet. In the power cable of an electronic component with a 3-prong plug, the U-shaped prong is wired to the component's chassis. This wire conducts electricity to power ground if the chassis becomes electrically hot, preventing shocks.

PREAMPLIFIER (PREAMP)—In an audio system, the first stage of amplification that boosts a mic-level signal to line level. A preamp is a stand-alone device or a circuit in a mixer.

PREDELAY—Short for pre-reverberation delay. The delay (about 30 to 150 milliseconds) between the arrival of the direct sound and the onset of reverberation. Usually, the longer the predelay, the greater the perceived room size.

PRE-ECHO—A repetition of a sound that occurs before the sound itself, caused by print-through.

PREFADER/POSTFADER SWITCH—A switch that selects a signal either ahead of the fader (prefader) or following the fader (postfader). The level of a prefader signal is independent of the fader position; the level of a postfader signal follows the fader position.

PREPRODUCTION—Planning in advance what you're going to do at a recording session, in terms of track assignments, overdubbing, studio layout, and microphone selection.

PRESENCE—The audible sense that a reproduced instrument is present in the listening room. Some synonyms are closeness, definition, and punch. Presence is often created by an equalization boost in the midrange or upper midrange.

PRESSURE ZONE MICROPHONE—A boundary microphone constructed with the microphone diaphragm parallel with, and facing, a reflective surface.

PREVERB—A special effect in which the reverberation of a note precedes it, rather than follows it. Preverb is achieved by playing an instrument's track backwards while adding reverberation to it, and recording the reverberation on an unused track. When the tape is reversed so that the instrument's track plays forward, preverb is heard as the reverberation plays backwards.

PRINT—To record on tape or disc.

PRINT-THROUGH—The transfer of a magnetic signal from one layer of tape to the next on a reel, causing an echo preceding or following the program. Not a problem with digital recording.

PRODUCTION—1. A recording that is enhanced by special effects. 2. The supervision of a recording session to create a satisfactory recording. This involves getting musicians together for the session, making musical suggestions to the musicians to enhance their performance, and making suggestions to the engineer for sound balance and effects.

PROGRAM BUS—A bus or output that feeds an audio program to a recorder track.

PROGRAM MIXER—In a mixing console, a mixer formed of input module outputs, combining amplifiers and program buses.

PROXIMITY EFFECT—The bass boost that occurs with a single-D directional microphone when it is placed a few inches from a sound source. The closer the microphone, the greater the low-frequency boost due to proximity effect.

PUNCH IN/OUT—A feature in a multitrack recorder that lets you insert a recording of a corrected musical part into a previously recorded track by going into and out of record mode as the tape is rolling.

PURE WAVEFORM—A waveform of a single frequency; a sine wave. A pure tone is the perceived sound of such a wave.

QUARTER-TRACK—A tape track recorded across one-quarter of the width of the tape. A quarter-track recorder usually records two stereo programs (one in each direction).

RACK—A 19-inch-wide wooden or metal cabinet used to hold audio equipment.

RADIO-FREQUENCY INTERFERENCE (RFI)—Radio-frequency electromagnetic waves induced in audio cables or equipment, causing various noises in the audio signal.

RANDOM ACCESS—Referring to a storage medium in which any data point can be accessed or read almost instantly. Examples are a hard disk, compact disc, and MiniDisc.

RAREFACTION—The portion of a sound wave in which molecules are spread apart, forming a region with lower-than-normal atmospheric pressure. The opposite of compression.

R-DAT—*See* DAT.

REAL-TIME RECORDING—1. Recording notes into a sequencer in the correct tempo, for later playback at the same tempo as recorded. 2. A recording made direct to lacquer disc or direct to 2-track without any overdubs.

RECIRCULATION (REGENERATION)—Feeding the output of a delay device back into its input to create multiple echoes. Also, the control on a delay device that affects how much delayed signal is recycled to the input.

RECORD—To store an event in permanent form. Usually, to store an audio signal in magnetic form on magnetic tape. Recording is also possible on hard disk, on compact disc (CD-R), on magneto-optical disk, MiniDisc, and in RAM.

RECORD EQUALIZATION—In tape-recorder electronics, equalization applied to the signal during recording to compensate for certain losses.

RECORD HEAD—The head in a tape recorder that puts the audio signal on tape by magnetizing the tape particles in a pattern corresponding to the audio signal.

RECORDER-MIXER—A combination multitrack recorder and mixer in one chassis.

RECORDING/REPRODUCTION CHAIN—The series of events and equipment that are involved in sound recording and playback.

REFLECTED SOUND—Sound waves that reach the listener after being reflected from one or more surfaces.

REGENERATION—*See* Recirculation.

REGION—In a digital audio editing program, a defined segment of the audio program.

RELEASE—The final portion of a note's envelope in which the note falls from its sustain level back to silence.

RELEASE TIME—In a compressor, the time it takes for the gain to return to normal after the end of a loud passage.

REMIX—To mix again; to do another mixdown with different console settings or different editing.

REMOTE RECORDING—*See* On-Location Recording.

REMOVABLE HARD DRIVE—A hard disk drive that can be removed and replaced with another, used in a digital audio workstation to store a long program temporarily.

RESISTANCE—The opposition of a circuit to a flow of current. Resistance is measured in ohms, abbreviated Ω, and may be calculated by dividing voltage by current.

RESISTOR—An electronic component that opposes current flow.

RETURN-TO-ZERO—*See* Memory Rewind.

REVERBERATION—Natural reverberation in a room is a series of multiple sound reflections which makes the original sound persist and gradually die away or decay. These reflections tell the ear that you're listening in a large or hard-surfaced room. For example, reverberation is the sound you hear just after you shout in an empty gymnasium. A reverb effect simulates the sound of a room—a club, auditorium, or concert hall—by generating random multiple echoes that are too numerous and rapid for the ear to resolve. The timing of the echoes is random, and the echoes increase in number with time as they decay. An echo is a discrete repetition of a sound; reverberation is a continuous fade-out of sound after each note.

REVERBERATION TIME (RT60)—The time it takes for reverberation to decay to 60 dB below the original steady-state level.

REVERSE ECHO—A multiple echo that precedes the sound that caused it, building up from silence into the original sound. This special effect is created in a manner similar to preverb.

RFI—*See* Radio Frequency Interference.

RHYTHM TRACKS—The recorded tracks of the rhythm instruments (guitar, bass, drums, and sometimes keyboards).

RIBBON MICROPHONE—A dynamic microphone in which the conductor is a long metallic diaphragm (ribbon) suspended in a magnetic field.

RIDE GAIN—To turn down the volume of a microphone when the source gets louder, and turn up the volume when the source gets quieter, in an attempt to reduce dynamic range.

RINGING—*See* Overhang.

ROOM MODES—*See* Standing Wave.

RT60—*See* Reverberation Time.

SAFETY COPY—A copy of the master tape, to be used if the master tape is lost or damaged.

SAFETY GROUND—*See* Power Ground.

SAMPLE—1. To digitally record a short sound event, such as a single note or a musical phrase, into computer memory. 2. A recording of such an event.

SAMPLING—Recording a short sound event into computer memory. The audio signal is converted into digital data representing the signal waveform, and the data is stored in memory chips, tape, or disc for later playback.

SATURATION—Overload of magnetic tape. The point at which a further increase in magnetizing force does not cause an increase in magnetization of the tape oxide particles. Distortion is the result.

SCRATCH VOCAL—A vocal performance that is done simultaneously with the rhythm instruments so that the musicians can keep their place in the song and get a feel for the song. Because it contains leakage, the scratch-vocal recording is usually erased. Then the singer overdubs the vocal part that is to be used in the final recording.

SCRUB—To manually move an open-reel tape slowly back and forth across a recorder playback head in order to locate an edit point. Some digital editing software has an equivalent scrubbing function.

SENSITIVITY—1. The output of a microphone in volts for a given input in sound pressure level. 2. The sound pressure level a loudspeaker produces at one meter when driven with one watt of pink noise. *See also* Sound Pressure Level.

SEQUENCE—A MIDI data file of musical-performance note parameters, recorded by a sequencer.

SEQUENCER—A device that records a musical performance done on a MIDI controller (in the form of note numbers, note on, note off, etc.) into computer memory or hard disk for later playback. A computer can act as a sequencer when it runs a sequencer program. During playback, the sequencer plays synthesizer sound generators or samples.

SESSION—1. A time period set aside for recording musical instruments, voices, or sound effects. 2. On a CD-R, a lead-in, program area, and lead-out.

SHELVING EQUALIZER—An equalizer that applies a constant boost or cut above or below a certain frequency, so that the shape of the frequency response resembles a shelf.

SHIELD—A conductive enclosure (usually metallic) around one or more signal conductors, used to keep out electrostatic fields that cause hum or buzz.

SHOCK MOUNT—A suspension system which mechanically isolates a microphone from its stand or boom, preventing the transfer of mechanical vibrations.

SIBILANCE—In a speech or vocal recording, excessive frequency components in the 5 to 10 kHz range, which are heard as an overemphasis of "s" and "sh" sounds.

SIDE-ADDRESSED—Referring to a microphone whose main axis of pickup is perpendicular to the side of the microphone. You aim the side of the mic at the sound source. *See also* End-Addressed.

SIGNAL—A varying electrical voltage that represents information, such as a sound.

SIGNAL PATH—The path a signal travels from input to output in a piece of audio equipment.

SIGNAL PROCESSOR—A device used to alter a signal usually to create an effect.

SIGNAL-TO-NOISE RATIO (S/N)—The ratio in decibels between signal voltage and noise voltage. An audio component with a high S/N has little background noise accompanying the signal; a component with a low S/N is noisy.

SINE WAVE—A wave following the equation $y = \sin x$, where x is degrees and y is voltage or sound pressure level. The waveform of a single frequency. The waveform of a pure tone without harmonics.

SINGLE-D MICROPHONE—A directional microphone having a single distance between its front and rear sound entries. Such a microphone has proximity effect.

SINGLE-ENDED—1. An unbalanced line. 2. A single-ended noise reduction system is one that works only during tape playback (unlike Dolby or dbx, which work both during recording and playback).

SLAP, SLAPBACK—An echo following the original sound by about 50 to 200 milliseconds, sometimes with multiple repetitions.

SLATE—At the beginning of a recording, a recorded announcement of the name of the tune and its take number. The term is derived from the slate used in the motion-picture industry to identify the production and take number being filmed.

SMPTE TIME CODE—A modulated 1200 Hz square-wave signal used to synchronize two or more tape transports or other multitrack recorders. SMPTE is an abbreviation for the Society of Motion Picture and Television Engineers, who developed the time code.

SNAKE—A multipair or multichannel mic cable. Also, a multipair mic cable attached to a connector junction box.

SOLO—On an input module in a mixing console, a switch that lets you monitor that particular input signal by itself. The switch routes only that input signal to the monitor system.

SOUND—Longitudinal vibrations in a medium (such as air) in the frequency range 20 Hz to 20,000 Hz.

SOUND CARD—A circuit card that plugs into a computer, and converts an audio signal into computer data for storage in memory or on hard disk. The sound card also converts computer data into an audio signal.

SOUND MODULE (SOUND GENERATOR)—1. A synthesizer without a keyboard, containing several different timbres or voices. These sounds are triggered or played by MIDI signals from a sequencer program, or by a MIDI controller. 2. An oscillator.

SOUND PRESSURE LEVEL (SPL)—The acoustic pressure of a sound wave, measured in decibels above the threshold of hearing. The higher the SPL of a sound, the louder it is. dB SPL = $20 \log (P/P_{ref})$, where P = the measured acoustic pressure and P_{ref} = 0.0002 dyne/cm^2.

SOUND WAVE—The periodic variations in sound pressure radiating from a sound source.

SPACED-PAIR—A stereo microphone technique using two identical microphones spaced several feet apart horizontally, usually aiming straight ahead toward the sound source.

SPATIAL PROCESSOR—A signal processor that allows images to be placed beyond the limits of a stereo pair of speakers—even behind the listener or toward the sides.

S/PDIF—Sony Philips Digital Interface (IEC 958 Type II). A digital signal interface format which uses a 75 ohm coaxial cable terminated with RCA or BNC connectors. *See also* AES/EBU.

SPEAKER—*See* Loudspeaker.

SPECIAL EFFECTS—*See* Effects.

SPECTRUM—The output vs. frequency of a sound source, including the fundamental frequency and overtones.

SPL—*See* Sound Pressure Level.

SPLICE—Two lengths of magnetic tape joined end-to-end by a piece of sticky "splicing" tape. Or a length of magnetic tape joined end-to-end to a length of leader tape.

SPLICING BLOCK—*See* Editing Block.

SPLIT CONSOLE—A console with a separate monitor-mixer section.

SPLITTER—A transformer or circuit used to divide a microphone signal into two or more identical signals to feed different sound systems.

SPOT MICROPHONE—In classical music recording, a close-placed microphone that is mixed with more-distant microphones to add presence or to improve the balance.

STANDING WAVE—An apparently stationary waveform, created by multiple reflections between opposite room surfaces. At certain points along the standing wave, the direct and reflected waves cancel, and at other points the waves add together or reinforce each other.

STEP-TIME RECORDING—Recording notes into a sequencer one at a time without regard to tempo, for later playback at a normal tempo.

STEREO, STEREOPHONIC—An audio recording and reproduction system with correlated information between two channels (usually discrete channels), and meant to be heard over two or more loudspeakers to give the illusion of sound-source localization and depth.

STEREO BAR, STEREO MICROPHONE ADAPTER—A microphone stand adapter that mounts two microphones on a single stand for convenient stereo miking.

STEREO IMAGING—The ability of a stereo recording or reproduction system to form clearly defined audio images at various locations between a stereo pair of loudspeakers.

STEREO MICROPHONE—A microphone containing two mic capsules in a single housing for convenient stereo recording. The capsules usually are coincident.

STUDIO—A room used or designed for sound recording.

SUBMASTER—1. A master volume control for an output bus. 2. An unedited recorded tape that is used to form an edited master tape.

SUBMIX—A small preset mix within a larger mix, such as a drum mix, keyboard mix, vocal mix, etc. Also a cue mix, monitor mix, or effects mix.

SUBMIXER—A smaller mixer within a mixing console (or stand-alone) that is used to set up a submix, a cue mix, an effects mix, or a monitor mix.

SUPERCARDIOID MICROPHONE—A unidirectional microphone that attenuates side-arriving sounds by 8.7 dB, attenuates rear-arriving sounds by 11.4 dB, and has two nulls of maximum sound rejection at 125 degrees off-axis.

SUPPLY REEL—*See* Feed Reel.

SURROUND SOUND—A multichannel recording and reproduction system that plays sound all around the listener. The 5.1 surround system uses the following speakers—front-left, center, front-right, left-surround, right-surround, and subwoofer.

SUSTAIN—The portion of the envelope of a note in which the level is constant. Also, the ability of a note to continue without noticeably decaying, often aided by compression.

SWEETENING—The addition of strings, brass, harmony singers, etc. to a previously recorded tape of the basic rhythm tracks.

SYNC, SYNCHRONIZATION—Aligning two separate audio programs in time, and maintaining that alignment as the programs play.

SYNC, SYNCHRONOUS RECORDING—Using a record head temporarily as a playback head during an overdub session, to keep the overdubbed parts in synchronization with the recorded tracks.

SYNC TONE—*See* Tape Sync.

SYNC TRACK—A track of a multitrack recorder that is reserved for recording an FSK sync tone or SMPTE time code. This allows audio tracks to synchronize with virtual tracks recorded with a sequencer. A sync track also can synchronize two audio tape machines or an audio recorder and a video recorder, and can be used for console automation.

SYNTHESIZER—A musical instrument (usually with a piano-style keyboard) that creates sounds electronically, and allows control of the sound parameters to simulate a variety of conventional or unique instruments.

TAIL-OUT—Referring to a reel of tape wound with the end of the program toward the outside of the reel. Tape stored tail out is less likely to have audible print-through.

TAKE—A recorded performance of a song. Usually, several takes are done of the same song, and the best one—or the best parts of several—become the final product.

TAKE SHEET—A list of take numbers for each song, plus comments on each take.

TAKE-UP REEL—The right-side reel on a tape recorder that winds up the tape as it is playing or recording.

TALKBACK—An intercom in the mixing console for the engineer and producer to talk to the musicians in the studio.

TAPE—*See* Magnetic Recording Tape.

TAPE LOOP—An endless loop formed from a length of recording tape spliced end-to-end, used for continuous repetition of several seconds of recorded signal.

TAPE RECORDER—A device that converts an electrical audio signal into a magnetic audio signal on magnetic tape, and vice versa. A tape recorder includes electronics, heads, and a transport to move the tape.

TAPE SYNC—A frequency-modulated signal recorded on a tape track, used to synchronize a tape recorder to a sequencer. Tape sync also permits the synchronized transfer of sequences to tape. *See also* Sync Track.

3-PIN CONNECTOR—A 3-pin professional audio connector used for balanced signals. Pin 1 is soldered to the cable shield, pin 2 is soldered to the signal hot lead, and pin 3 is soldered to the signal return lead. *See also* XLR-Type Connector.

THREE-TO-ONE RULE—A rule in microphone applications. When multiple mics are mixed to the same channel, the distance between mics should be at least three times the distance from each mic to its sound source. This prevents audible phase interference.

THRESHOLD—In a compressor or limiter, the input level above which compression or limiting takes place. In an expander, the input level below which expansion takes place.

TIE—To connect electrically, for example, by soldering a wire between two points in a circuit.

TIGHT—1. Having very little leakage or room reflections in the sound pickup. 2. Referring to well-synchronized playing of musical instruments. 3. Having a well-damped, rapid decay.

TIMBRE—The subjective impression of spectrum and envelope. The quality of a sound that allows us to differentiate it from other sounds. For example, if you hear a trumpet, piano, and a drum, each has a different timbre or tone quality that identifies it as a particular instrument.

TIME CODE—A modulated 1200-Hz square-wave signal used to synchronize two or more tape or disc transports. *See also* Tape Sync, Sync Track, SMPTE.

TONAL BALANCE—The balance or volume relationships among different regions of the frequency spectrum, such as bass, midbass, midrange, upper midrange, and highs.

TRACK—A path on magnetic tape containing a single channel of audio. A group of bytes in a digital signal (on tape, on hard disk, on compact disc, or in a data stream) that represents a single channel of audio or MIDI. Usually one track contains a performance of one musical instrument.

TRANSDUCER—A device that converts energy from one form to another, such as a microphone or loudspeaker.

TRANSFORMER—An electronic component made of two magnetically coupled coils of wire. The input signal is transferred magnetically to the output, without a direct connection between input and output.

TRANSIENT—A short signal with a rapid attack and decay, such as a drum stroke, cymbal hit, or acoustic-guitar pluck.

TRANSIENT RESPONSE—The ability of an audio component (usually a microphone or loudspeaker) to follow a transient accurately.

TRANSPORT—The mechanical system in a reproduction device that moves the media past the read/write heads. In a tape recorder, the transport controls tape motion during recording, playback, fast forward, and rewind.

TRIM—1. In a mixing console, a control for fine adjustment of level, as in a Bus Trim control. 2. In a mixing console, a control that adjusts the gain of a mic preamp to accommodate various signal levels.

TUBE—A vacuum tube, an amplifying component made of electrodes in an evacuated glass tube. Tube sound is characterized as being "warmer" than solid-state or transistor sound.

TWEETER—A high-frequency loudspeaker.

UNBALANCED LINE—An audio cable having one conductor surrounded by a shield that carries the return signal. The shield is at ground potential.

UNIDIRECTIONAL MICROPHONE—A microphone that is most sensitive to sounds arriving from one direction—in front of the microphone. Examples are cardioid, supercardioid, and hypercardioid.

VALVE—British term for vacuum tube. *See* Tube.

VIRTUAL CONTROLS—Audio-equipment controls that are simulated on a computer monitor screen. You adjust them with a mouse.

VIRTUAL TRACK—1. A sequencer recording of a single musical line, recorded as data in computer memory. A virtual track is the computer's equivalent of a tape track on a multitrack tape recorder. 2. In a random-access recorder, a recording of one take of a performance of a single musical instrument.

VU METER—A voltmeter with a specified transient response, calibrated in VU or volume units, used to show the relative volume of various audio signals, and to set recording level.

WAVEFORM—A graph of a signal's sound pressure or voltage vs. time. The waveform of a pure tone is a sine wave.

WAVELENGTH—The physical length between corresponding points of successive waves. Low frequencies have long wavelengths; high frequencies have short wavelengths.

WEBER—A unit of magnetic flux.

WEIGHTED—Referring to a measurement made through a filter with a specified frequency response. An A-weighted measurement is taken through a filter that simulates the frequency response of the human ear.

WINDSCREEN—*See* Pop Filter.

WOOFER—A low-frequency loudspeaker.

WORKSTATION—A system of MIDI- or computer-related equipment that works together to help you compose and record music. Usually, this system is small enough to fit on a desktop or equipment stand. *See also* Keyboard Workstation and Digital Audio Workstation.

WOW—A slow periodic variation in tape speed.

XLR-TYPE CONNECTOR—An ITT Cannon part number which has become the popular definition for a 3-pin professional audio connector. *See also* 3-Pin Connector.

XMIDI—Extended MIDI, a proposed compatible upgrade to MIDI with enhanced features and capabilities, and more functions.

X-Y—*See* Coincident-Pair.

Y-ADAPTER—A cable that divides into two cables in parallel to feed one signal to two destinations.

Z—Abbreviation for impedance.

ZONE—*See* Region.

INDEX

458

Focal Press

Related Title

Modern Recording Techniques
Fourth Edition

*by David Miles Huber and
Robert E. Runstein*

As the most complete, up-to-date, accurate and authoritative recording guide, this book reflects the latest developments in digital audio and hard-disk recording techniques. It provides in-depth insights into hands-on operation of studio recording equipment, such as project studio, and explores the latest in digital technology, multitrack systems, MIDI, and the electronic musical instrument industry. It is the perfect book for anyone wanting to learn professional recording – producers, musicians, multimedia developers, audio for video professionals, universities, schools, and audio enthusiasts.

David Miles Huber is widely acclaimed in the recording industry as an author, musician, digital audio consultant, engineer, and guest lecturer.

Robert E. Runstein has been associated with all aspects of the recording industry, working as a performer, sound mixer, electronics technician, A & R specialist, and record producer.

1995 • 496pp • Paperback • 0-240-80308-6

Visit the Focal Press Web Site at: http://www.bh.com/focalpress

**Available from all better book stores or in case of difficulty call:
1-800-366-2665 in the U.S. or +44 1865 310366 in Europe.**

Focal Press

Related Title

The MIDI Manual

by David Miles Huber

This comprehensive guide to Musical Instrument Digital Interfacing puts MIDI to work for any beginning-and-intermediate-level user. It studies the multiple uses of MIDI and provides introductory coverage of electronic music technology. Also included is a reference and equipment guide with advice on which system to purchase.

Selected Contents:
- Sequencing
- Editor and Librarian
- Music Printing Programs
- Synchronization
- Digital Audio
- Mixing
- Animation

1990 • 250pp • Paperback • 0-240-80320-5

Visit the Focal Press Web Site at: http://www.bh.com/focalpress

Available from all better book stores or in case of difficulty call:
1-800-366-2665 in the U.S. or +44 1865 310366 in Europe.

 **Focal Press**

Related Titles

Sound Synthesis and Sampling
Edited by Francis Rumsey
Martin Russ
January 1997   224pp   Paperback   0 240 51429 7

Sound Synthesis and Sampling – The CD-ROM
Martin Russ
September 1997   Compact Disk   0 240 51497 1

Sound and Recording: An Introduction, Third Edition
Francis Rumsey and Tim McCormick
April 1997   384pp   Paperback   0 240 51487 4

Acoustics and Psychoacoustics
Edited by Francis Rumsey
David Howard and James Angus
1996   224pp   Paperback   0 240 51428 9

The Audio Workstation Handbook
Francis Rumsey
1996   286pp   Paperback   0 240 51450 5

Sound for Film and Television
Tomlinson Holman
April 1997   253pp   Paperback   0 240 80291 8

Visit the Focal Press Web Site at: http://www.bh.com/focalpress

Available from all better book stores or in case of difficulty call:
1-800-366-2665 in the U.S. or +44 1865 310366 in Europe.

WITHDRAWN

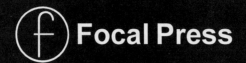

Focal Press

Focal Press is a leading international publisher of books and CD-ROM products in all areas of media and communications, including: photography and imaging, broadcasting, video, film and multimedia production, media writing, communication technology, theatre and live performance technology. To find out more about media titles available from Focal Press, request your catalog today.

Free Catalog

For a free copy of the Focal Press Catalog, call 1-800-366-2665 in the U.S. (ask for Item Code #560); and in Europe call +44 1865 310366.

Mailing List

An e-mail mailing list giving information on latest releases, special promotions/offers and other news relating to Focal Press titles is available.

To subscribe, send an e-mail message to: majordomo@world.std.com Include in message body (not in subject line) subscribe focal-press.

Web Site

Visit the Focal Press Web Site at:
http://www.bh.com/focalpress